ASP.NET网站设计教程

（微课视频版）

陶永鹏 郭鹏 刘建鑫 张立杰 主编

清華大学出版社
北京

内 容 简 介

本书以实用为原则，弱化了 ASP.NET 框架的基础知识，以 Visual Studio 2019 为开发平台，以 C#为程序设计语言，使用 SQL Server 2012 为后台数据库，以大量的实例介绍动态控件的属性和相关应用，以工程实践环节巩固这些方法和技术。本书将控件按功能进行分类，并细化每个控件的属性、事件及基本功能，使读者能够清晰、熟练地掌握每个动态控件。书中实例侧重实用性和启发性，趣味性强，分布合理，通俗易懂，使读者能够快速掌握 ASP.NET 网站设计的基础知识与编程技能，为实战应用打下坚实的基础。

本书可作为计算机相关专业高职、本科生 ASP.NET 网站设计课程的教材，也可作为 ASP.NET 网页开发爱好者的自学参考书。

图书在版编目(CIP)数据

ASP.NET 网站设计教程：微课视频版/陶永鹏等主编. —北京：清华大学出版社，2023.5 (2025.1重印)
(清华开发者学堂)
ISBN 978-7-302-62801-9

Ⅰ.①A… Ⅱ.①陶… Ⅲ.①网页制作工具-程序设计-教材 Ⅳ.①TP393.092.2

中国国家版本馆 CIP 数据核字(2023)第 032174 号

责任编辑：张 玥 常建丽
封面设计：刘艳芝
责任校对：韩天竹
责任印制：刘海龙

出版发行：清华大学出版社
网 址：https://www.tup.com.cn，https://www.wqxuetang.com
地 址：北京清华大学学研大厦 A 座 **邮 编**：100084
社 总 机：010-83470000 **邮 购**：010-62786544
投稿与读者服务：010-62776969，c-service@tup.tsinghua.edu.cn
质量反馈：010-62772015，zhiliang@tup.tsinghua.edu.cn
课件下载：https://www.tup.com.cn，010-83470236
印 装 者：三河市铭诚印务有限公司
经 销：全国新华书店
开 本：185mm×260mm **印 张**：25 **字 数**：593 千字
版 次：2023 年 5 月第 1 版 **印 次**：2025 年 1 月第 2 次印刷
定 价：79.50 元

产品编号：097621-01

前言

ASP.NET 是 Microsoft 公司力推的 Web 开发编程技术，也是当今最热门的 Web 开发编程之一。为了方便广大读者学习，作者通过多年一线教学的积累，以实用为原则，将教学中的案例加以整理提升编写了本书。本书以 Visual Studio 2019 为开发平台，以 C# 为程序设计语言，使用 SQL Server 2012 为后台数据库。

本书独特地将控件按功能进行分类，细化每个控件的属性、事件及基本功能，使读者能够清晰、熟练地掌握每一个基本控件；书中实例侧重实用性和启发性，趣味性强，分布合理，通俗易懂，使读者能够快速掌握 ASP.NET 网站设计的基础知识与编程技能，为实战应用打下坚实的基础；本书中的三层架构、综合实例章节从开发环境构建、基本流程、基本配置以及开发步骤、数据绑定和表单标签、文件上传和下载、输入校验等详细设计展开讲解，使读者对 ASP.NET 网站设计有全面的理解。通过学习本书，读者能够在较短时间内对 ASP.NET 编程有基本的认识，掌握 Web 开发的主要技能。

本书共 13 章内容。

第 1 章主要介绍 ASP.NET 基础和.NET 平台的历史以及发展，讲解开发环境的使用及如何高效地开发 Web 应用程序。

第 2 章详细介绍 ASP.NET 4.5 应用程序中提供的基本控件，分类讲解内容显示控件、按钮控件和选择控件，以类比的形式讲解每种控件的共有属性、方法和事件，加深读者对控件的理解。

第 3 章详细介绍 ASP.NET 4.5 应用程序中提供的高级控件，着重讲解视图区域控件、文件上传控件、日历控件、广告控件、向导控件等 ASP.NET 4.5 高级控件的使用方法和技巧。

第 4 章主要介绍客户端验证和服务器端验证的概念和具体应用，以及 ASP.NET 中的各种服务器验证控件。

第 5 章主要介绍 ASP.NET 4.5 中内置对象的概念和具体应用，以及全局应用程序类 Global.asax 中的各种事件。

第 6 章主要介绍主题、母版页和用户控件，重点介绍用 ASP.NET 4.5 应用程序进行样式控制的方法和技巧。

第 7 章主要介绍导航控件的使用，详细讲解了三种导航控件及站点地图的应用。

第 8 章主要介绍 AJAX 技术，详细介绍如何在 ASP.NET 4.5 中进行 AJAX 应用程序的开发。

第 9 章主要介绍 ADO.NET 的基础，对 ADO.NET 中的类进行了详细讲解，通过示例实现了对数据库数据的增、删、改、查操作。

第 10 章主要介绍 ASP.NET 中的数据绑定，对列表控件和数据控件的绑定进行了详细讲解。

第 11 章介绍 Web 系统中的三层架构，讲解如何在 ASP.NET 中创建三层架构的项目。

第 12 章从需求分析、数据库设计、项目模块设计、三层架构等具体步骤、模块着手，详细讲解"美妆网"实例开发，使读者有实际项目的体会，从而能够深刻了解本书讲解的知识，并达到实战的效果。

第 13 章首先对档案管理系统业务逻辑进行分析，然后详细讲解系统数据库的设计，对系统中的项目层次划分进行讲解，最后对系统进行页面设计和后台代码实现。

在编写过程中，编者得到家人和同仁的大力支持，在此一并表示感谢。尽管编者在编写过程中尽了最大努力，但由于水平有限，本书的疏漏之处在所难免，恳请读者批评指正。

编者

2022 年 8 月

目录

第 3 章 Web 高级控件 /51

第 4 章 服务器验证控件 /75

第11章　Web系统中的三层架构　/253

第12章　美妆网的设计与实现　/264

第1章 .NET框架与ASP.NET

本章学习目标

- 了解.NET 框架
- 了解 ASP.NET 的主要特点和工作原理
- 熟练掌握 ASP.NET 中的 Web 开发方法

本章首先介绍.NET 的基本框架，然后讲解 ASP.NET 的主要特点和工作原理，最后介绍 Visual Studio 2019 中创建 Web 页面的方法，讲解各窗口的使用。

1.1 .NET 框架结构

2000 年 6 月，微软推出 Microsoft.NET 战略，创建.NET 框架的目的是便于开发人员更容易地建立 Web 应用程序和 Web 服务，可使 Internet 上的各种应用程序之间通过 Web 服务进行沟通，从而达到“任何”时刻，通过“任何”设备，访问“任何”数据库的目的。

.NET 框架是一个基于多语言组件的开发和执行环境，提供了一个跨语言的统一编程环境。开发平台允许创建各种应用程序，如 XML Web 服务、Web 窗体、Win32 GUI 程序、Win32 CUI 应用程序、Windows 服务、实用程序以及独立的组件模块等。相比于以前的开发平台，.NET 平台可以提供更多的技术，如代码重用、代码专业化、资源管理、多语言开发、安全、部署与管理等。.NET 框架结构如图 1.1 所示。

.NET 框架结构主要有以下几部分。

- 公共语言运行时(Common Language Runtime，CLR)

公共语言运行时和 Java 虚拟机类似，是一个运行时环境，负责内存分配和垃圾收集等资源管理操作，同时作为一种多语言执行环境，支持众多的数据类型和语言特性，可以简化代码管理，提高平台可靠性，达到面向事务电子商务应用的稳定级别。同时，CLR 还负责监

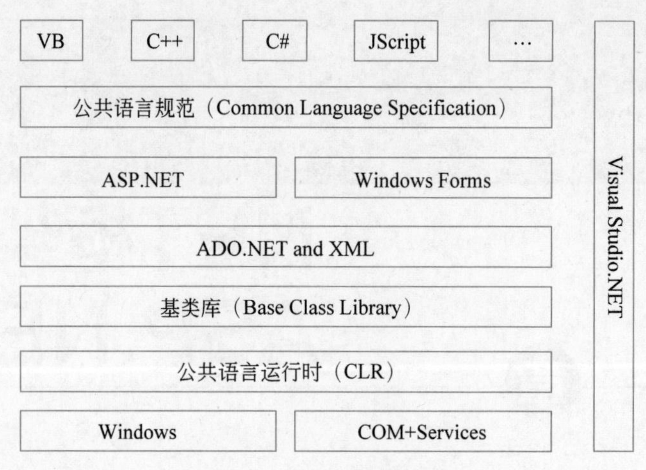

图 1.1 .NET 框架结构

视程序运行等其他任务。

• 框架类库(Framework Class Library,FCL)

.NET 框架类库包含了数以千计的类,并按照其功能用命名空间(Namespace)进行组织。在.NET 平台中使用的语言只定义了一些规则,在实际运用中调用 FCL 中的类型,可以使用比较少的语言知识来创建丰富的程序。

• 公共语言规范(Common Language Specification,CLS)

公共语言规范是所有应用程序必须遵守的一套基本语言功能,以.NET 平台为目标语言所必须支持的最小特征。定义了在多种语言中都可使用的功能,从而增强和确保语言的互用性。同时,它还建立了 CLS 遵从性要求,可帮助用户确定托管代码是否符合 CLS。

1.2 ASP.NET 简介

ASP.NET 是 Microsoft 公司推出的基于.NET 框架的动态 Web 应用程序开发平台,其作为.NET 框架的一部分,可以使用任何在.NET 中兼容的语言,如使用 VB.NET、C#、F# 等编写 ASP.NET 应用程序。本书采用 C#语言进行网站程序的开发。

ASP.NET 又称为 ASP+,但不仅是 ASP 的简单升级,它吸收了 ASP 以前版本的优点并参照 Java、VB 语言的开发优势加入了许多新的特色,同时修正了 ASP 版本的运行错误。ASP.NET 具备开发网站应用程序的一切解决方案,包括验证、缓存、状态管理、调试和部署等全部功能。在代码撰写方面具有页面逻辑和业务逻辑分离的特色,使网页的撰写更容易,同时兼容 CSS、JavaScript、DIV、AJAX 等技术和方法,能方便地进行 Web 开发。

扫一扫

1.2.1 ASP.NET 技术的发展

1996 年,ASP 1.0(Active Server Pages)版本问世,降低了动态网页开发的难度,引起 Web 开发的新革命。在此之前开发动态网页需要编写大量的代码,编程效率比较低,并且开发者需要掌握一定的编程技巧,而 ASP 使用简单的脚本语言,能够将代码直接嵌入

HTML,使设计 Web 页面变得更简单。

1998 年,微软公司发布了 ASP 2.0,ASP 2.0 与 ASP 1.0 的主要区别是其外部组件可以初始化,组件都有了独立的内存空间,并且可以进行事务处理。2000 年 6 月,微软公司宣布了.NET 框架,ASP.NET 1.0 正式发布,2003 年 ASP.NET 升级为 1.1 版本。ASP.NET 1.1 对网络技术有巨大的推动作用。随后,微软公司提出“减少 70%代码”的目标,在 2005 年 11 月又发布了 ASP.NET 2.0。ASP.NET 2.0 的发布是.NET 技术走向成熟的标志,增加了方便、实用的新特性,使 Web 开发人员能够更加快捷方便地开发 Web 应用程序。ASP.NET 2.0 以高安全性、易管理性和高扩展性等特点著称。

伴着强劲的发展势头,2008 年微软推出 ASP.NET 3.5。ASP.NET 3.5 是建立在 ASP.NET 2.0 CLR 基础上的框架,底层类库依然调用.NET 2.0 封装好的所有类,但在此基础上增加了很多新特性,如 LINQ 数据库访问技术等,使网络程序开发更倾向于智能开发。

2010 年发布了 ASP.NET 4.0,其集中化支持 ASP.NET 路由、增强支持 Web 标准,以及支持更新的浏览器,增加了 Chart 控件。ASP.NET 4.0 加入了 MVC 框架,并在 Microsoft Ajax 库中增加了基于客户端 Ajax 应用程序的附加支持。2012 年发布了 ASP.NET 4.5,其在 ASP.NET Web Forms4.5 中新增了强类型数据控件,针对 HTML5 进行了更新,增加了部分控件的枚举属性值及功能,并增加了对 HTML5 表单的验证,极大地降低了验证的代码量。

2015 年发布了 ASP.NET 5,通常也称之为 Core50 或.NET Core。.NET Core 可实现开源跨平台 WebApp 组件化开发,并支持云部署。作为.NET 的重构版本,ASP.NET 5 只保留了原有框架中的最小化功能集,对其他的特性和功能(如 Session、MVC 等)可以通过 Nuget 下载,之后以插件的形式运行。

2017 年.NET Framework 升级到 4.7 版本,解决了 D3DCompiler 的依赖问题,同时发布了 Core 2.0。2019 年发布了.NET Core 3.0,微软官方推出.NET Framework 4.7.2 版本,支持 Azure Active Directory 的通用和多重身份验证,同时兼容.NET 4.6、4.5 及以下多个版本,兼容性好,并支持 Page、Custom Handler 和 User Control 的依赖注入,可以控制 Cookies 中的 SameSite 属性。

1.2.2 ASP.NET 的主要特点

扫一扫

ASP.NET 在 Web 编程方面主要具有如下特点。

1. 强大性和适应性

ASP.NET 基于公共语言的编译运行程序,具有强大的适应性,几乎可以运行在 Web 应用软件开发的全部平台上。

2. 简单性和易学性

基于事件驱动编程,.aspx 页面与.cs 文件分离,即显示逻辑与处理逻辑分离,便于分工、美工和编程,支持丰富的服务器控件,减少了大量代码的编写。

3. 高效可管理性

ASP.NET 使用一种以字符为基础的分级的配置系统,使服务器环境和应用程序的设置更加简单。配置信息都保存在简单文本中,新的设置不需要启动本地的管理员工具就可

以实现。

1.2.3 ASP.NET 的工作原理

ASP.NET 使用 HTTP 和 HTML 技术，首先，客户端向服务器请求一个文件(default.aspx)，ASP.NET 的运行库和 ASP.NET 辅助进程开始工作，针对文件 default.aspx 的第一次请求，会启动 ASP.NET 分析器，编译器将该文件和与.aspx 文件相关的 C# 文件一起编译，创建程序集，然后.NET 运行库的 JIT 编译器把程序集编译为机器代码。ASP.NET 的基本工作流程如图 1.2 所示。

程序集中包含一个 Page 类，通过调用该类，将 HTML 代码返回到客户端，同时这个 Page 对象会被删除，但是仍会保留在程序集，用于以后的请求。当第二次请求时，就不需要再编译程序集，直接返回 HTML 代码给客户端即可。

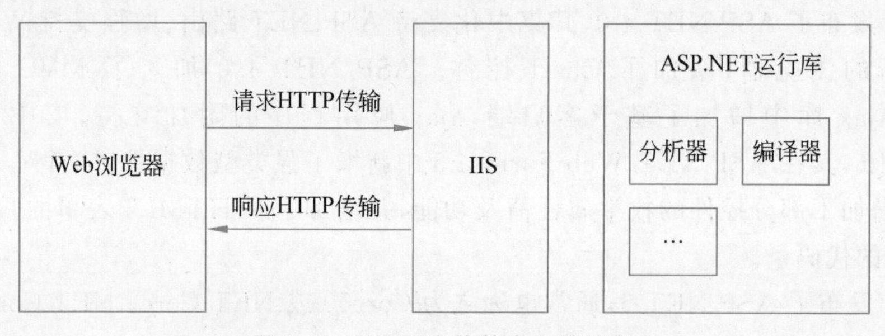

图 1.2 ASP.NET 的基本工作流程

1.3 ASP.NET 开发页面简介

1.3.1 第一个 ASP.NET 网站

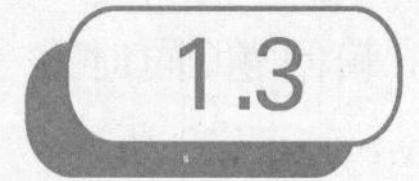

扫一扫

【示例 1-1】 在 E 盘的 ASP.NET 项目代码目录中创建 chapter1 子目录，将其作为网站根目录，创建名为 example1-1 的网页，页面内包含一个按钮，单击该按钮会显示"第一个 ASP.NET 页面"。创建步骤详细如下。

(1) 通过"程序"或桌面快捷方式打开 Visual Studio 2019。Visual Studio 2019 应用程序图标如图 1.3 所示。

图 1.3 Visual Studio 2019 应用程序图标

(2) 在菜单中选择"创建新项目"→"ASP.NET Web 应用程序"选项，建立网站，如图 1.4 所示。

(3) 在"配置新项目"窗口设置项目名称为 example1，位置为 E:\ASP.NET 项目代码\chapter1，将解决方案和项目放在同一目录中，框架选择".NET Framework4.7.2"，具体设置如图 1.5 所示。

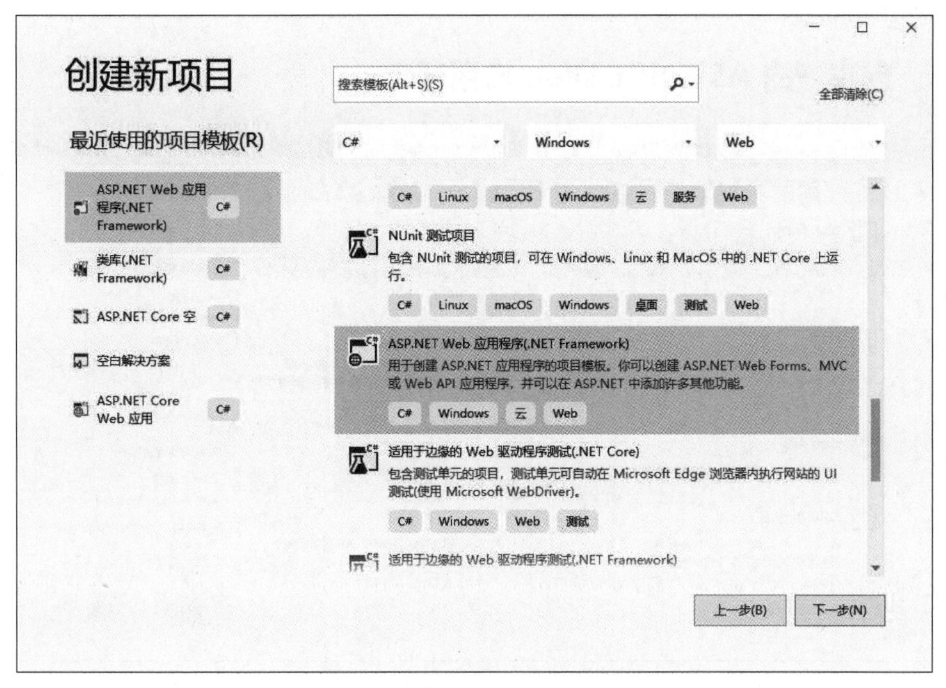

图 1.4　建立网站操作图

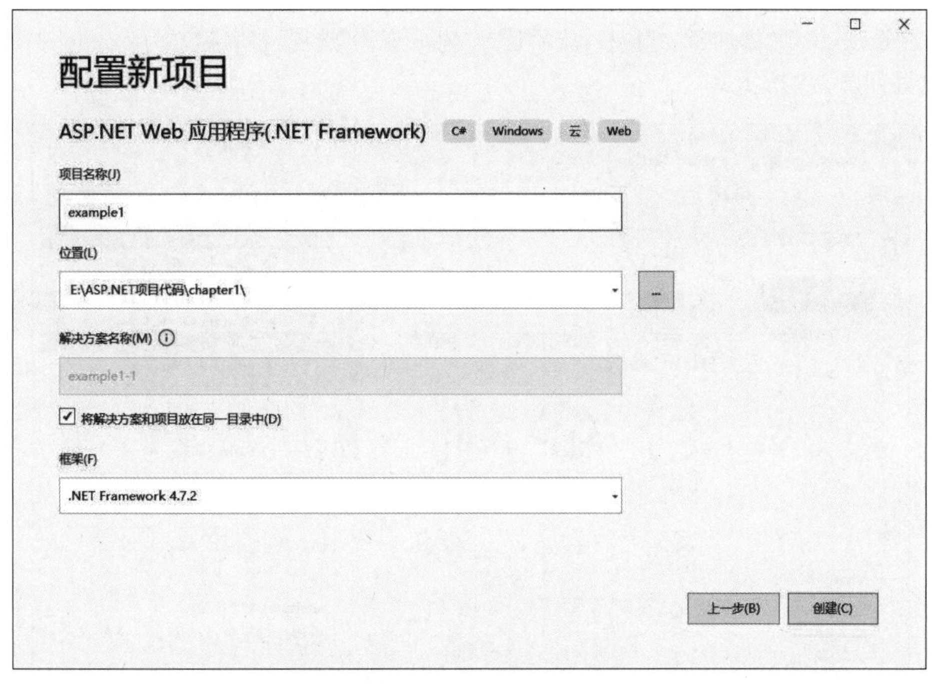

图 1.5　新项目设置图

(4) 在“创建新的 ASP.NET Web 应用程序”窗口选择“空”模板，如图 1.6 所示。

(5) 在新窗口右侧的“解决方案资源管理器”目录中，初始只有 Web.config 配置文件等

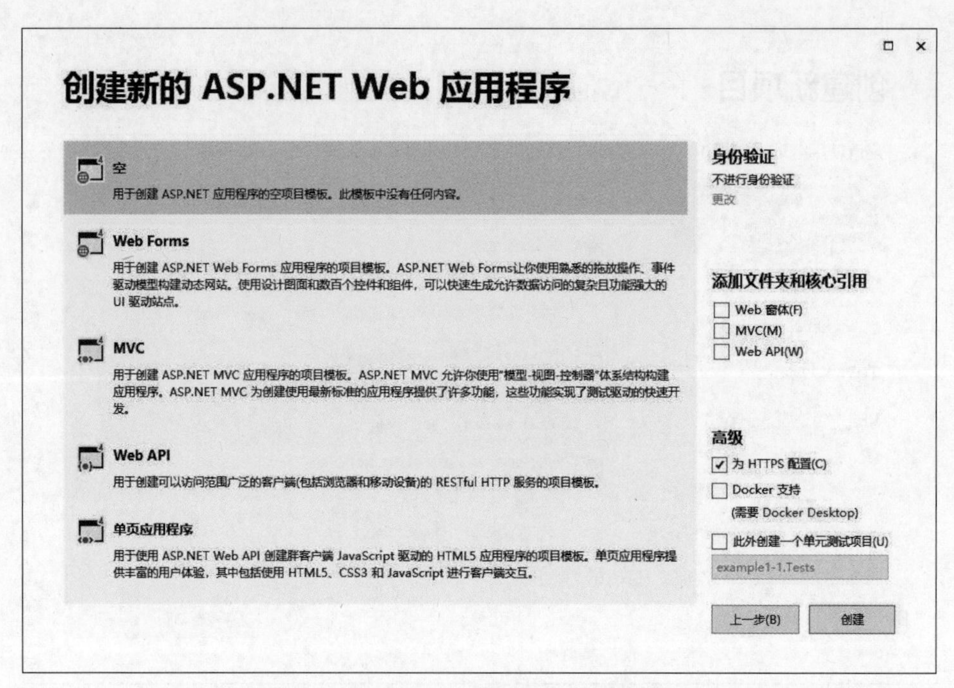

图 1.6 ASP.NET Web 应用程序模板选择图

系统文件，如图 1.7 所示。右击网站根目录 chapter1，在弹出的快捷菜单中选择"添加"→"新建项"选项，出现"添加新项"窗口，之后选择"Web 窗体"选项，将名称改为 example1-1.aspx，具体设置如图 1.8 所示。

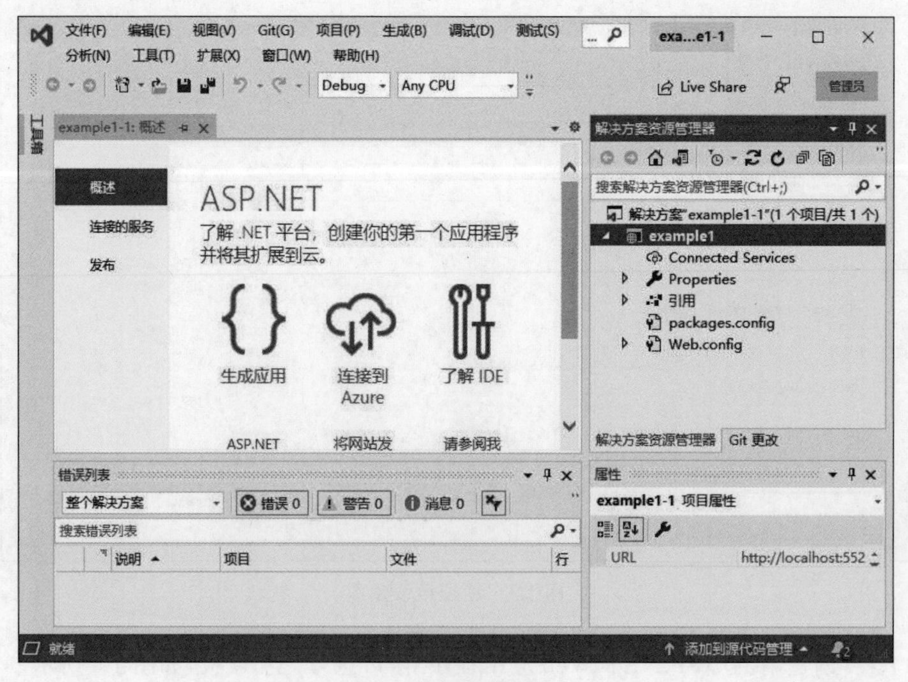

图 1.7 初始空网站

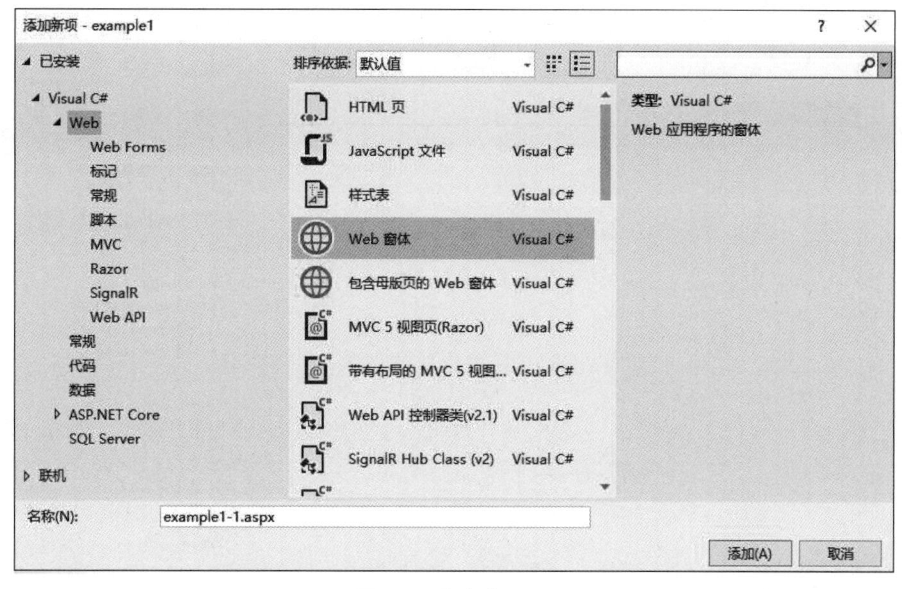

图 1.8　添加新页面

(6) 创建的第一个页面默认显示“源”视图，显示页面的 HTML 代码，如图 1.9 所示。可通过底部选项卡切换视图显示，“设计”视图显示界面设计，如图 1.10 所示。“拆分”视图显示 HTML 代码和界面设计，如图 1.11 所示。三种视图可根据不同需要灵活切换，编辑时两部分内容联动。

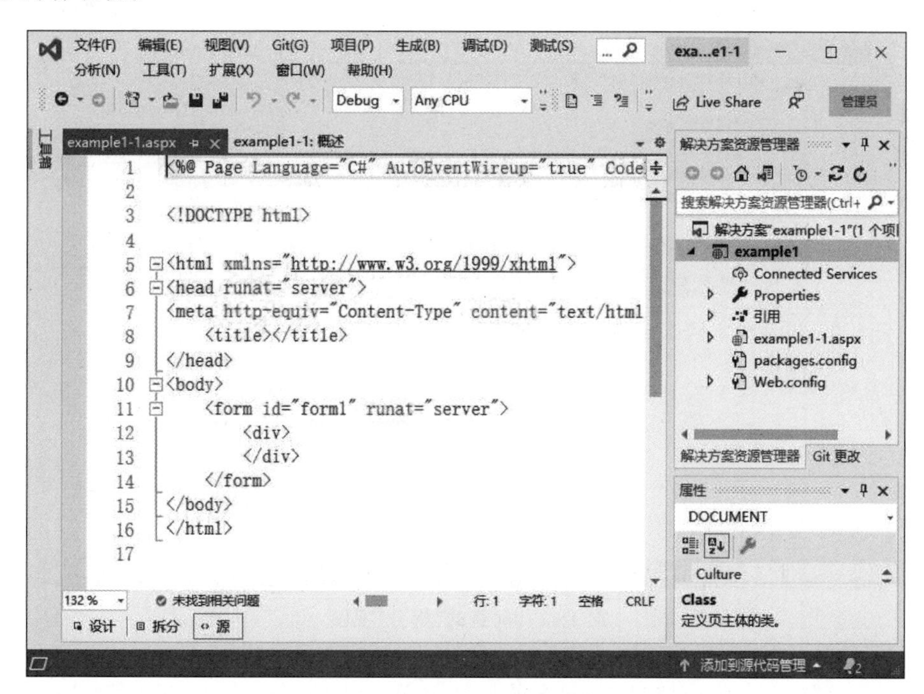

图 1.9　网页的“源”视图

(7) 按住“工具箱”中 Button 按钮控件将其拖曳到右侧“设计”视图的任一空白区域，如

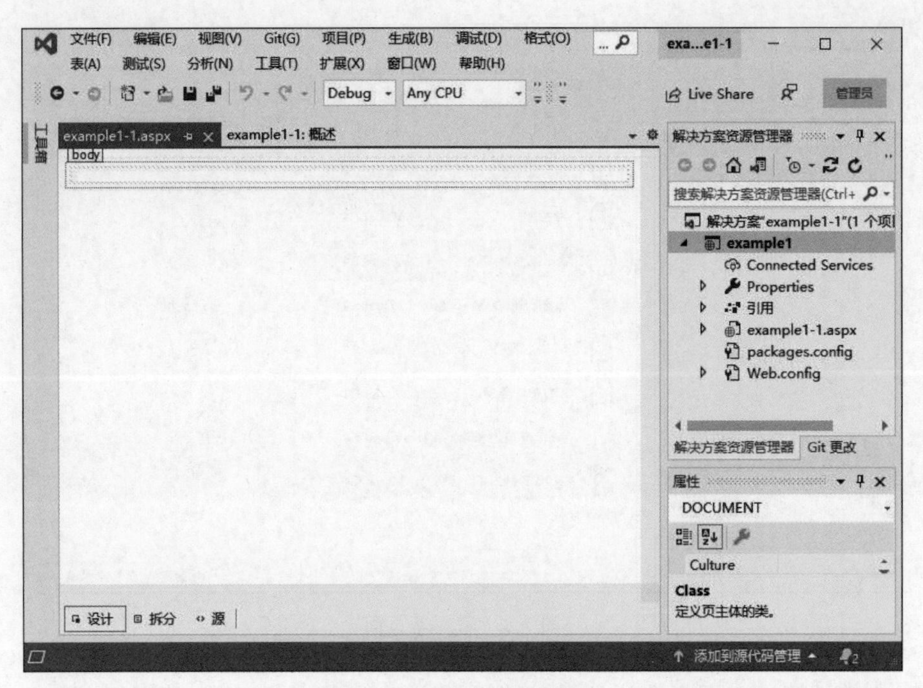

图 1.10 网页的"设计"视图

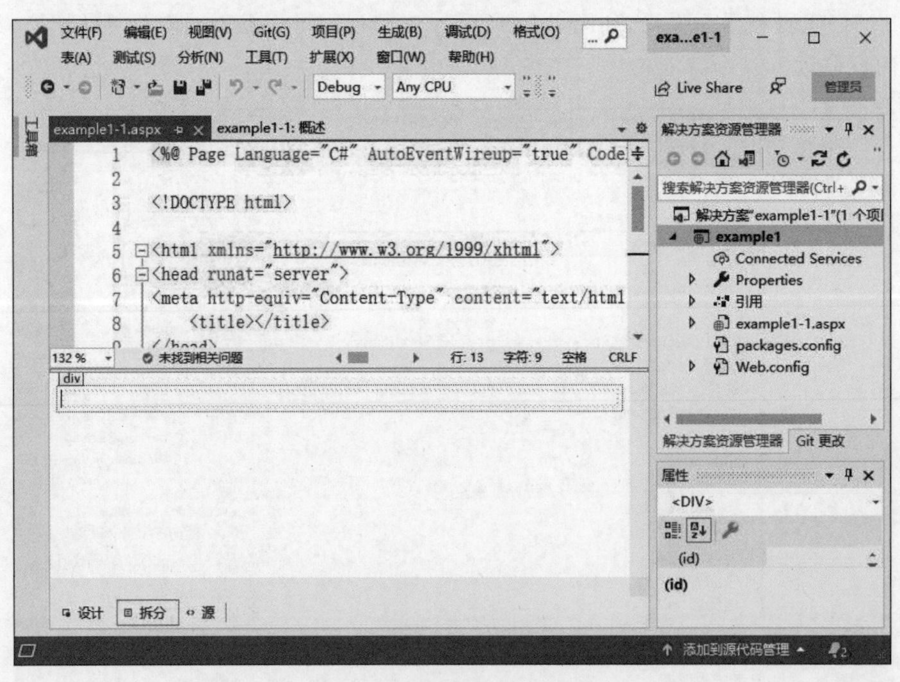

图 1.11 网页的"拆分"视图

图 1.12 所示。右击 Button 按钮，在弹出的快捷菜单中选择"属性"，出现"属性"窗口，修改 ID 属性如图 1.13 所示。

(8) 在"属性"窗口选择"⚡"闪电图标，显示按钮控件可添加的部分事件，如图 1.14 所

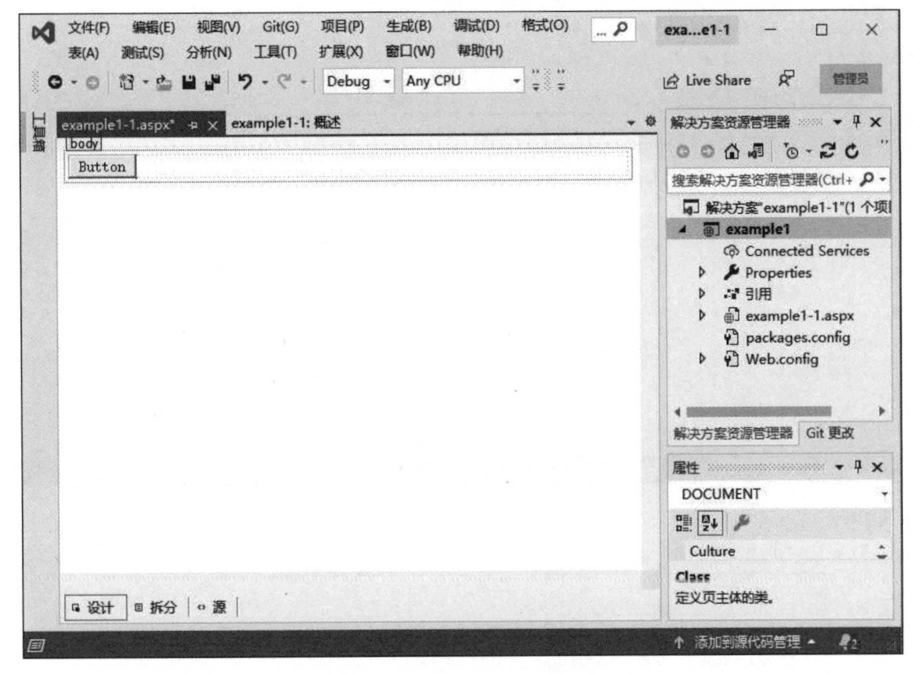

图 1.12　添加 Button 控件

示。在 Click 事件操作后面的空白处双击，进入事件的后台 C# 代码编辑区域。

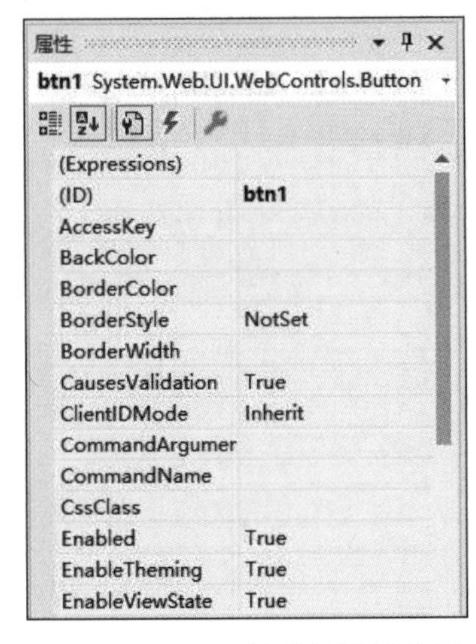

图 1.13　Button 控件的属性设置

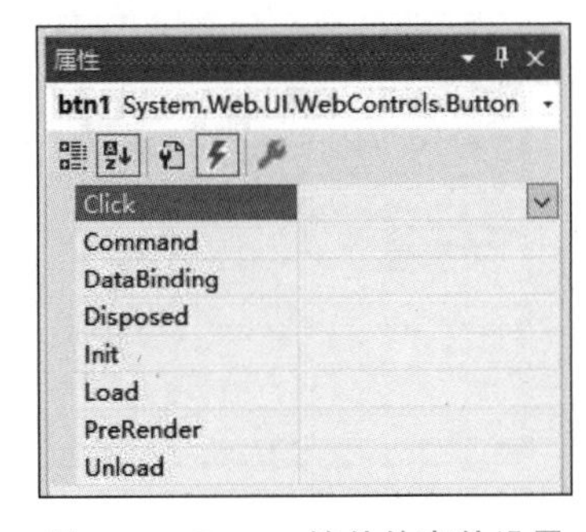

图 1.14　Button 控件的事件设置

(9) 编辑 example1-1.aspx.cs 代码，在 btn1_Click()方法中添加代码，如图 1.15 所示。

(10) 在“解决方案”中右击 example1-1.aspx，在弹出的快捷菜单中选择“在浏览器中查看”，网页运行如图 1.16 所示。

(11) 单击“按钮”控件，按钮上方会出现文字“第一个 ASP.NET 页面”，如图 1.17 所示。

1.3.2　ASP.NET 中的特殊文件夹

Visual Studio 2019 开发 ASP.NET 网站程序时，会将 C# 类以及 Web Services 等文件存放

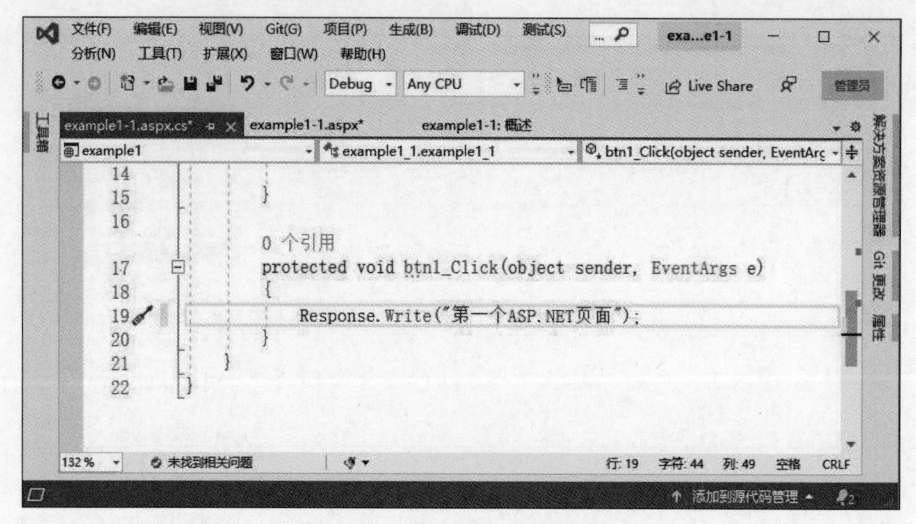

图 1.15　Button 控件事件代码

图 1.16　初始运行页面

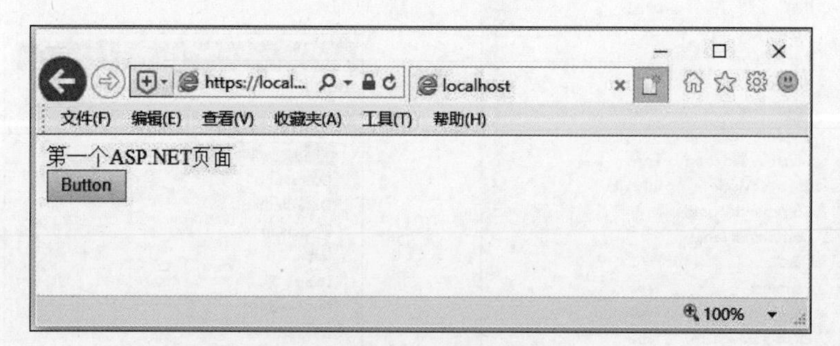

图 1.17　触发按钮单击事件的浏览页面

在一些特殊的文件夹中。与普通文件夹不同，特殊的文件夹中的程序和文件只允许应用程序访问，对网页的请求不予响应，无法读取文件内容。部分特殊文件夹如表 1.1 所示。

表 1.1　部分特殊文件夹

文　件　夹	说　　明
App_Browsers	存储浏览器定义(.browse)文件，通过文件识别并判断浏览器
App_Code	存储公用程序的源代码(如.cs、.vb 和.js 等)文件，将会编译为应用程序的一部分

续表

文　件　夹	说　　明
App_Data	存储应用程序的数据文件(如.md 和.xm 等)
App_GlobalResources	存储资源(.resx 和.resources)文件,将会编译成具有全局范围的组件
App_LocalResources	存储资源(.resx 和.resources)文件,将会与特定的页面、用户控件或应用程序的主页面(.MasterPage)进行关联
主题	存储主题文件(如.skin 和.css 等),用于定义网页和控件的外观

1.3.3 ASP.NET 中的文件类型

在网站文件夹中会看到各种类型的文件,尤其是大型工程项目通常包含眼花缭乱的.NET Framework 文件。本节将详细讲解 ASP.NET 中的文件类型及其扩展名。

1. Visual Studio 的文件类型

项目开发时 C# 中的通用文件如表 1.2 所示。

表 1.2 项目开发时 C# 中的通用文件

文　件　名	后　缀　名	说　　明
解决方案文件	.sln	一个或多个项目的集合
用户选项文件	.suo	特定用户、存储 Web 项目的转换表、项目的离线状态,以及其他项目构建的设置信息
C# 项目文件	.csproj	参考内容、名称、版本等项目细节
C# 项目的用户文件	.csproj.user	用户的相关信息

2. 普通 Web 开发文件

打开网站,在网站项目上右击,在弹出的快捷菜单中选择“添加”→“新建项”,弹出“新建项”窗口,该窗口呈现出网站项目中可以使用的所有文件类型,其中 Windows 服务和 Web 开发通用的文件如表 1.3 所示。

表 1.3 Windows 服务和 Web 开发通用的文件

文　件　名	后　缀　名	说　　明
C# 文件	.cs	C# 源代码文件。
XML 文件	.xml	XML 文件与数据标准文件
数据库文件	.mdf	SQL Server 数据库文件
类图文件	.cd	类图表文件
脚本文件	.js	JavaScript 代码文件
配置文件	.config	存储程序设置的程序配置文件
图标文件	.ico	图标样式的图像文件
文本文件	.txt	普通文本文件

3. ASP.NET 的文件类型

ASP.NET Web 开发有时还使用一些特定的文件类型，如表 1.4 所示。

表 1.4　Web 文件类型说明

文件名	后缀名	说明
Web 窗体文件	.aspx	代码分离的 Web 窗体
全局程序文件	.asax	以代码形式处理程序全局事件的应用文件，一个项目最多只包括一个 global.asax 文件
静态页面文件	.htm/.HTML	标准的 HTML 页
样式文件	.css	设置外观的层叠样式表
站点地图文件	.sitemap	表示页面间层次关系的站点地图
皮肤文件	.skin	用于指定服务器控件的主题
用户控件文件	.ascx	用户自主创建的 Web 控件
浏览器文件	.browser	定义浏览器相关信息的文件

1.4　Visual Studio 2019 开发环境介绍

1.4.1　菜单栏和工具栏

Visual Studio 2019 的菜单栏继承了 Visual Studio 早期版本的所有命令功能，如“文件”“编辑”“视图”“窗口”“帮助”的核心功能，还有“生成”“调试”“测试”等程序设计专用的功能菜单。菜单栏下方为标准工具栏，可以快速访问菜单栏中的常用功能，如图 1.18 所示。

图 1.18　菜单栏和标准工具栏

1. 显示工具箱及属性等窗口

单击“视图”菜单，显示“解决方案资源管理器”“属性窗口”“工具栏”“错误列表”等窗口，除“帮助”窗口，其他窗口及内容的显示都可以在“视图”菜单设置，如图 1.19 所示。

2. 程序执行及断点调试

单击“调试”菜单可以进行程序调试、执行等编译，在代码内部新建以及取消断点，对程序进行逐语句、逐过程(直接调用函数、属性的模块，不逐条执行模块内的语句)调试，如图 1.20 所示。

3. 代码文本编辑

选择“工具”→“选项”菜单，在“环境”选项卡中的“字体和颜色”子选项卡可以设置代码

图 1.19 “视图”菜单

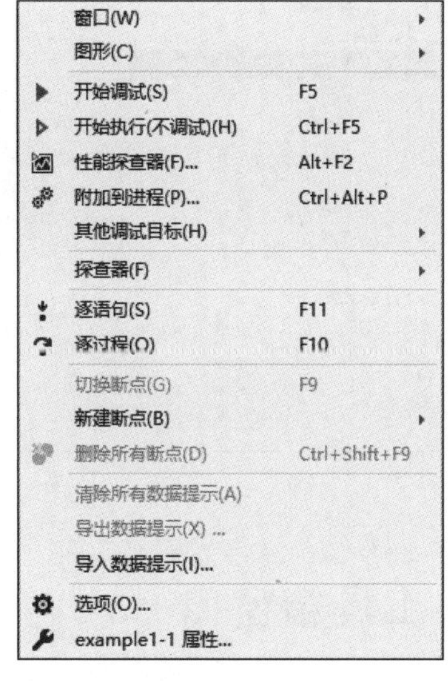

图 1.20 调试菜单

编辑区域文本的字号、颜色、背景色等属性，如图 1.21 所示。

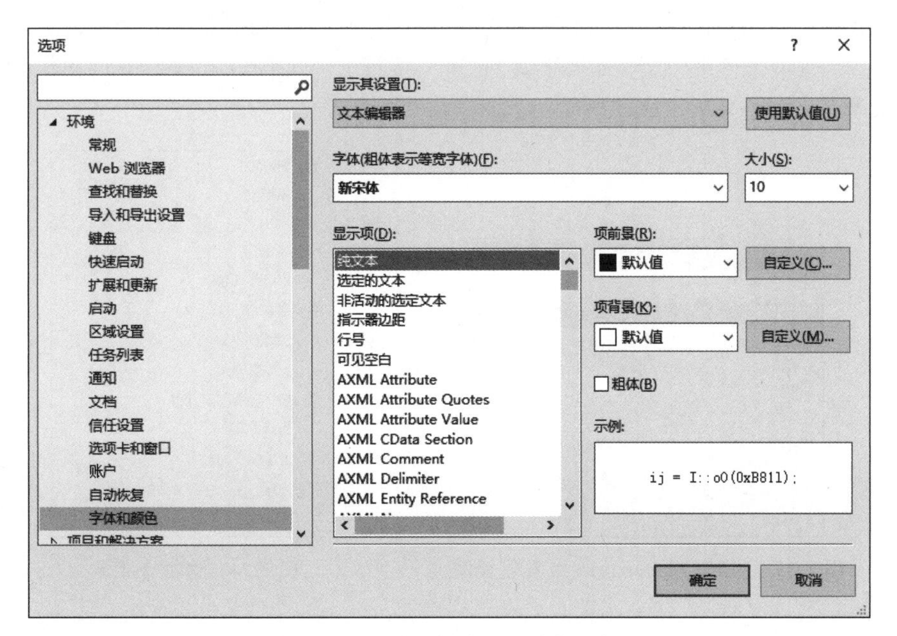

图 1.21 “字体和颜色”子选项卡

“文本编辑器”选项卡中的 C# 子选项卡可以设置自动换行、显示行号等属性，如图 1.22 所示。

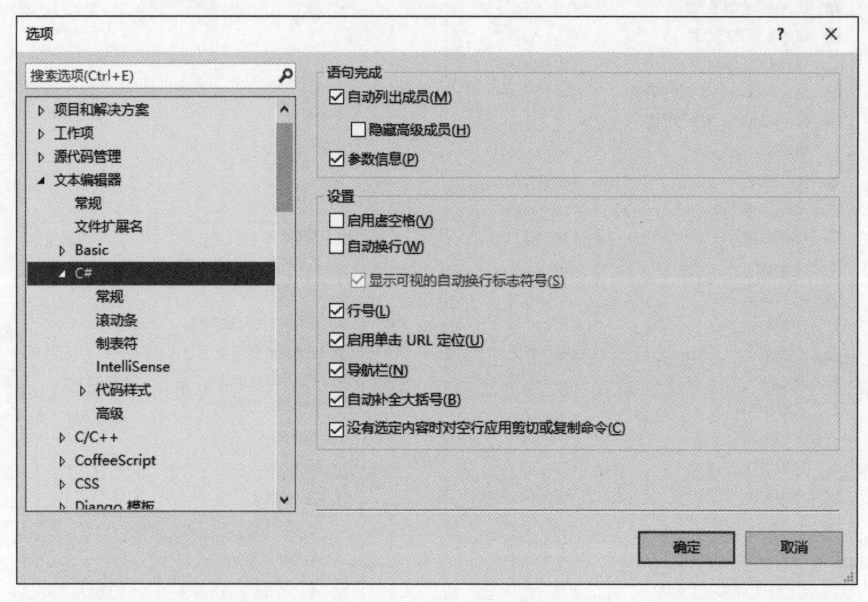

图 1.22　C# 子选项卡

1.4.2　工具箱窗口

Visual Studio 2019 集成开发环境的左侧是控件工具箱，Web 开发使用的控件均在此列出，如图 1.23 所示。使用控件时从工具箱中将其拖曳到界面上即可，这样可极大地节省编写代码的时间。

在工具箱中右击，显示“工具箱内容菜单”，可对“选项卡”进行添加、删除、重命名等操作，如图 1.24 所示。单击“选择项”，弹出“选择工具箱项”对话框，可以为工具箱添加其他可选控件及第三方组件，如图 1.25 所示。

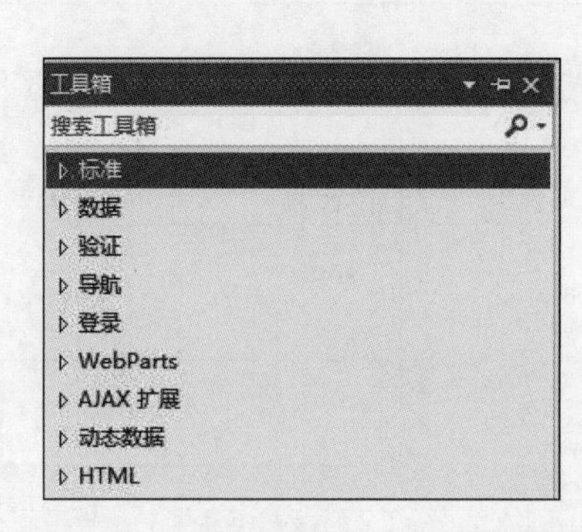

图 1.23　Visual Studio 2019 工具箱视图

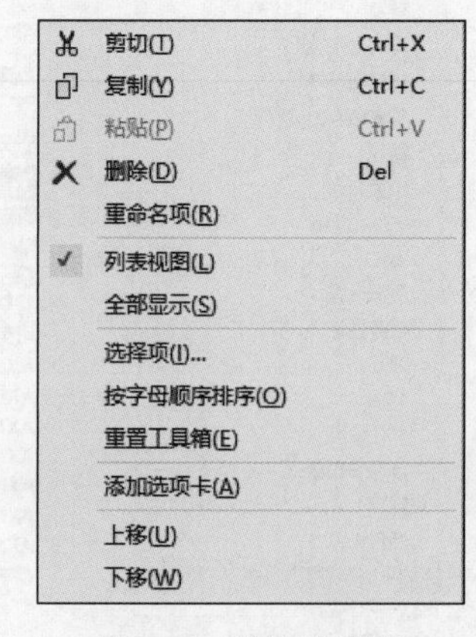

图 1.24　选项卡菜单

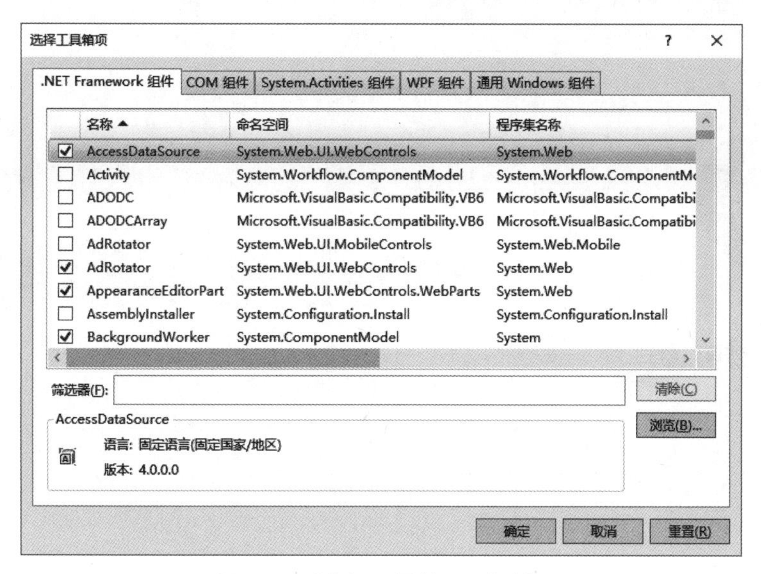

图 1.25 “选择工具箱项”对话框

1.4.3 解决方案资源管理器

Visual Studio 2019 集成开发环境右侧的“解决方案资源管理器”窗口，提供了网站项目及文件的组织结构视图，如各个类库、数据库文件以及系统配置文件等。在“解决方案资源管理器”窗口可以添加或者删除文件，也可以添加系统或用户文件夹来实现文件的管理，当解决方案资源管理器中显示内容与网站实际结构不符时，可单击“”进行同步刷新。“解决方案资源管理器”窗口如图 1.26 所示。

1.4.4 属性窗口

Visual Studio 2019 集成开发环境右下角的“属性”窗口，可以查看对象属性，也可以对页面及页面中的控件进行量值化的属性值设置。“属性”窗口最顶部的下拉列表，可以选择待设置属性的对象，“”图标表示信息列表按字母排序，“”图标表示信息列表按分类排序。修改对象属性值后，属性值自动到 HTML 源代码中添加，反之亦然。“属性”窗口如图 1.27 所示。

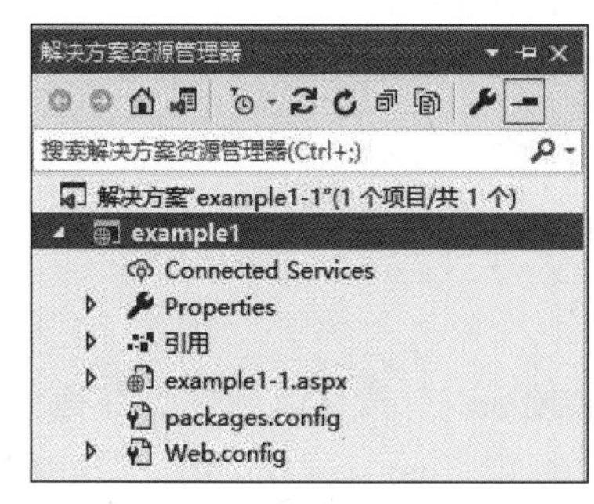

图 1.26 “解决方案资源管理器”窗口

图 1.27 “属性”窗口

第2章 Web基本控件

本章学习目标

- 掌握 Web 基本控件的使用方法
- 了解控件属性、事件源文件与.cs 代码的对应关系
- 熟练掌握 ASP.NET 中内容显示控件的使用方法
- 熟练掌握 ASP.NET 中按钮控件的使用方法
- 熟练掌握 ASP.NET 中选择控件的使用方法

本章首先介绍 ASP.NET 服务器控件的基本概念、属性和事件，然后通过实例介绍内容显示控件、按钮控件、选择控件特有的属性、事件，最后进行相关应用。

2.1 控件简介

扫一扫

ASP.NET 服务端控件是对 HTML 封装的一种特殊对象，当客户端请求服务器的网页时，控件在服务器上运行并向客户端呈现 HTML 解析。Web 标准控件也称为服务器控件，可以通过写 HTML 标记创建，也可以直接从工具箱中拖曳添加，服务端控件使 ASP.NET 使用方便、容易上手，下面分别采用 HTML 标签和拖曳的两种方法添加标准控件。

- HTML 标签法

切换到“源”视图，填写标签<asp:Button ID="btn1" runat="server" Text="按钮1"/>，便在页面上添加了一个按钮。

- 拖曳法

在“工具箱”的“标准”选项卡中选中“ab Button”，按住鼠标左键并拖曳到“设计”视图或“源”视图，添加按钮控件，自动生成 HTML 标记<asp:Button ID="Button1" runat="server" Text="Button"/>，可在“源”视图或“属性”窗口中修改属性。控件添加效果图

如图 2.1 所示。

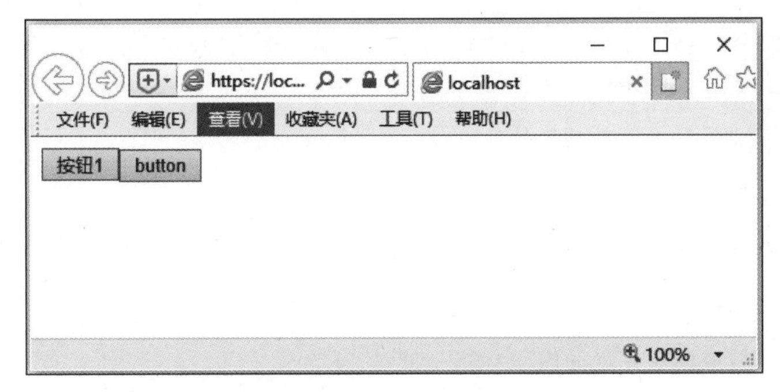

图 2.1　控件添加效果图

服务器控件的页面源文件如图 2.2 所示，可见服务器控件在客户端也是以 HTML 标签进行解析显示。

```
HtmlPage1.html
1  <!DOCTYPE html>
2  <html>
3  <head>
4      <meta charset="utf-8" />
5      <title></title>
6  </head>
7  <body>
8      <input id="Button1" type="button" value="按钮1" /><input id="Button1" type="button" value="button" />
9  </body>
10 </html>
```

图 2.2　服务器控件的页面源文件

ASP.NET 标准控件都继承自 System.Web.UI.Control.WebControl 类，具有部分相同的属性、方法和事件，如表 2.1～表 2.3 所示。相关系统资源都包含在 System.Web.UI.WebControls 命名空间中，需要添加对该命名空间的引用，页面才可以使用标准控件。

表 2.1　标准控件的公共属性

扫一扫

属　性　名	说　　明
AccessKey	获取或设置访问控件的快捷键，按下 Alt 键加上指定键可选择该控件
BackColor	获取或设置控件的背景色，可设置颜色名称或者 #RRGGBB 的十六进制颜色格式
BorderColor	获取或设置控件的边框颜色，颜色设置同背景色
BorderStyle	获取或设置控件的边框样式，可选择系统提供的边框样式
BorderWidth	获取或设置控件的边框宽度，以像素为单位的整数值
CssClass	获取或设置应用到控件的 CSS 类，可设置控件的样式
Enabled	获取或设置指示是否启用控件，True 代表控件可正常使用
Font	获取或设置控件的字体属性，内容、格式与 Office 中基本一致

续表

属 性 名	说 明
EnableTheming	获取或设置是否为控件启用主题
ForeColor	获取或设置控件的前景色
Height	获取或设置控件的高度,以像素为单位的整数值
IsEnabled	获取或设置一个值,该值指示是否启用控件
SkinID	获取或设置控件的皮肤,设置控件应用的皮肤 ID
Style	获取或设置控件的内联 CSS 样式
TabIndex	获取或设置控件的 Tab 键控制次序,从 0 开始的整数值,默认值为 0
ToolTip	获取或设置鼠标指针移动到控件上时显示的文本
Width	获取或设置控件的宽度,以像素为单位的整数值
Visible	获取或设置指示该控件是否可见并被呈现出来

表 2.2 标准控件的公共方法

方 法 名	说 明
DataBind()	绑定数据源,将数据源绑定到该服务器控件或其子控件
Dispose()	清理操作,从内存中释放该控件所占用的资源
Focus()	获得焦点,为该控件设置输入焦点
GetType()	获得类型,获取当前使用控件的类型

表 2.3 标准控件的公共事件

事 件 名	说 明
DataBinding	当服务器控件绑定到数据源时触发
Disposted	从内存中释放该控件时触发
Init	初始化该控件时触发
Load	该控件所在页面加载时触发

2.2 内容显示控件

不论静态内容还是动态内容,都必须依托于某一控件呈现在页面中,这种以内容显示为主要功能的控件统称为内容显示控件,主要包含标签控件(Label)、文本控件(TextBox)、特殊文本控件(Literal)和图片控件(Image),本节将对这四种控件进行详细讲解。

2.2.1 标签控件 Label

扫一扫

Label 控件又称标签控件,在工具箱中图标为“A Label”,封装在 System.Web.UI.

Control.WebControl 命名空间的 Label 类中，用于显示文本信息，其最主要的属性是 Text，可以获取和设置该控件显示的文本。

【示例 2-1】 在 E 盘 ASP.NET 项目代码目录中创建 chapter2 子目录，将其作为网站根目录，创建名为 example2-1 的网页，页面内包含一个“今天”按钮和一个空白标签，单击“今天”按钮可显示当前日期。

(1) 按图 2.3 所示添加相应的 Button 控件和 Label 控件。

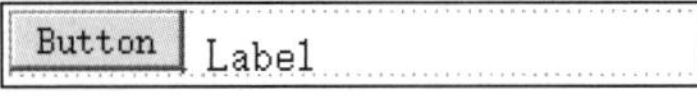

图 2.3 Label 控件应用页面设计

(2) 在“属性”窗口或“源”视图中设置控件属性如下。

```
<form id="form1" runat="server">
    <div>
        <asp:Button ID="btnToday" runat="server" Text="今天"/>
         <asp:Label ID="lblToday" runat="server" Text=""></asp:Label>
    </div>
</form>
```

(3) 为 btnToday 按钮添加 Click 事件，并编辑代码如下。

```
protected void btnToday_Click(object sender, EventArgs e)
{
    lblToday.Text=DateTime.Now.ToLongDateString();
}
```

(4) 运行网站，单击“今天”按钮，执行效果如图 2.4 所示。

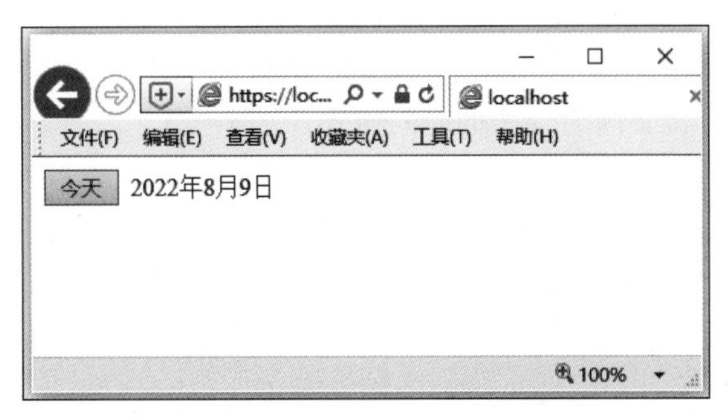

图 2.4 Label 控件应用网页演示

2.2.2 文本控件 TextBox

扫一扫

TextBox 控件又称文本控件，在工具箱中图标为“ TextBox ”，封装在 System.Web.UI.Control.WebControl 命名空间的 TextBox 类中，用于在网页中显示或输入文本信息。TextBox 的常用属性如表 2.4 所示。TextBox 控件的常用事件如表 2.5 所示。

表 2.4 TextBox 控件的常用属性

属 性 名	说 明
(ID)	获取或设置该控件的编程标识符
AutoPostBack	设置文本变化后是否自动回发

续表

属性名	说明
CausesValidation	设置文本控件是否触发验证
MaxLength	获取或设置 TextBox 中文本的最大长度
ReadOnly	获取或设置是否控件内容只读
Text	获取或设置文本内容
TextMode	获取或设置文本框的行为模式

说明：

(1) 属性 AutoPostBack 表示自动回发。“回发”是指浏览器/服务器(B/S)结构中，在浏览器(Browser)上通过单击控件或通过某些行为触发事件，引发网页从浏览器向服务器(Server)发送网页。之所以称作“回发”，是为了区分从浏览器向服务器第一次发送消息请求服务器网页。

(2) TextMode 属性用于控制 TextBox 控件的文本显示方式，最常用的几种枚举值如下。

- SingleLine：单行模式，只能在一行中输入信息，可以选择限制控件接收的字符数。
- MultiLine：多行模式，允许用户输入多行文本并执行换行。
- Password：密码模式，将用户输入的字符用黑点(●)屏蔽，以隐藏这些信息。
- Date：日期模式，必须为合法的日期格式。
- DateTime：日期时间模式，必须为合法的日期加时间格式。
- Time：时间模式，必须为合法的时间格式。

表 2.5　TextBox 控件的常用事件

事件名	说明
TextChanged	在前后两次回发服务器时 TextBox 的 Text 属性发生改变时触发

【示例 2-2】　在 chapter2 网站根目录下创建名为 example2-2 的网页，页面内包含若干文本框，练习使用 TextBox 控件的相关属性及事件。

(1) 添加相应的 Button 控件和 TextBox 控件，如图 2.5 所示。

(2) 在“属性”窗口或“源”视图中修改控件属性如下。

```
<form id="form1" runat="server">
    <asp:TextBox ID="txtID" runat="server" Width="140px"></asp:TextBox>
    <asp:TextBox ID="txtPwd" runat="server" TextMode="Password" Width="140px"></asp:TextBox>
    <asp:TextBox ID="txtRePwd" runat="server" TextMode="Password" Width="140px"></asp:TextBox>
    <asp:TextBox ID="txtBirthDay" runat="server" TextMode="Date" AutoPostBack="True" OnTextChanged="txtBirthDay_TextChanged" Width="140px"></asp:TextBox>
    <asp:TextBox ID="txtAge" runat="server" ReadOnly="True" Width="140px" >自动计算年龄</asp:TextBox>
```

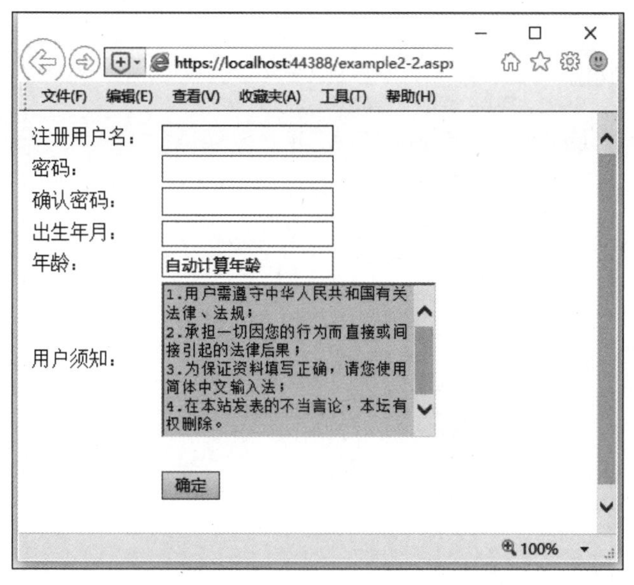

图 2.5 TextBox 控件应用页面设计

```
        <asp:TextBox ID="txtInfo" runat="server" Height="121px"  TextMode="
MultiLine" Width="224px" BackColor="#00FFCC" ReadOnly="True">
    1. 用户需遵守中华人民共和国有关法律、法规;
    2. 承担一切因您的行为而直接或间接引起的法律后果;
    3. 为保证资料填写正确,请您使用简体中文输入法;
    4. 在本站发表的不当言论,本网站有权删除。</asp:TextBox>
      <asp:Button ID="btnOK" runat="server" Text="确定"/>
</form>
```

(3) 为 txtBirthDay 控件添加事件,并编辑代码如下。

```
protected void Page_Load(object sender, EventArgs e)
{
    //页面加载,为 txtID 控件设置焦点
    txtID.Focus();
}
protected void txtBirthDay_TextChanged(object sender, EventArgs e)
{
    //当前年份减生日年份计算出年龄
    int age=DateTime.Now.Year-DateTime.Parse(txtBirthDay.Text).Year;
    //为 txtAge 控件的 Text 属性赋值
    txtAge.Text=age.ToString();
}
```

(4) 运行网站,密码框和确认密码框为加密显示,出生年月为日期格式,自动"回发"服务器,触发事件计算年龄,年龄为只读模式,用户须知为多行、只读模式,执行效果如图 2.6 所示。

2.2.3 特殊文本控件 Literal

Literal 控件又称特殊文本控件,在工具箱中图标为" Literal ",封装在 System.

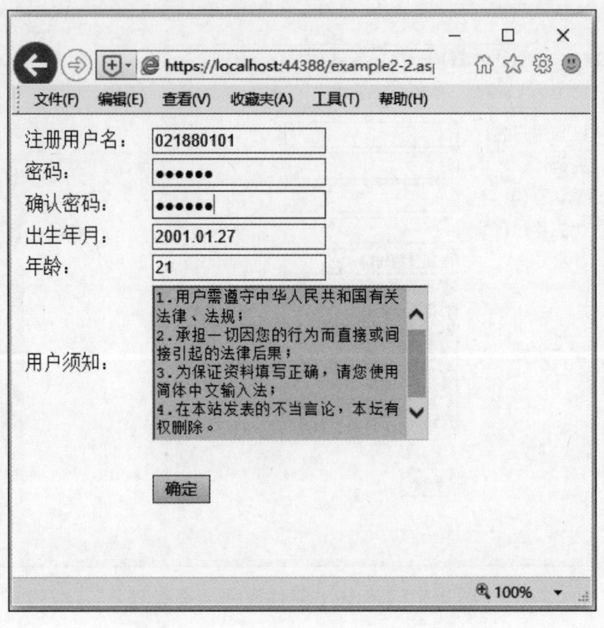

图 2.6　TextBox 控件应用网页演示

Web.UI.Control.WebControl 命名空间的 Literal 类中。与标签控件 Label 类似，Literal 控件可用于显示文本信息，同时支持 Mode 属性，用于指定控件对文本内容的处理方式。Mode 属性的枚举值如下。

- Transform：将对添加到控件中的任何标记进行转换，以适应请求浏览器的协议。
- PassThrough：添加到控件中的任何标记都将按原样呈现在浏览器中。
- Encode：将使用 HtmlEncode()方法对添加到控件中的任何标记进行编码，会将 HTML 编码转换为其文本表示形式，如<b>标记将呈现为 <b>，这样有助于防止在浏览器中执行恶意标记，显示来自不受信任的源的字符串时推荐使用此设置。

【示例 2-3】　在 chapter2 网站根目录下创建名为 example2-3 的网页，页面内包含一个 Literal 控件，练习使用 Literal 控件的 Mode 属性。

(1) 按如下源文件添加控件并设置相关属性值。

```
<form id="form1" runat="server">
    <div>
         PassThrough 模式 :<br>
       <asp:Literal ID="Literal1" runat="server" Text="<h1>1 级标题</h1>"
Mode="PassThrough"></asp:Literal>
    Encode 模式:<br>
       <asp:Literal ID="Literal2" runat="server" Text="<h1>1 级标题</h1>"
Mode="Encode"></asp:Literal>
    </div>
</form>
```

(2) 运行网站，执行效果如图 2.7 所示。

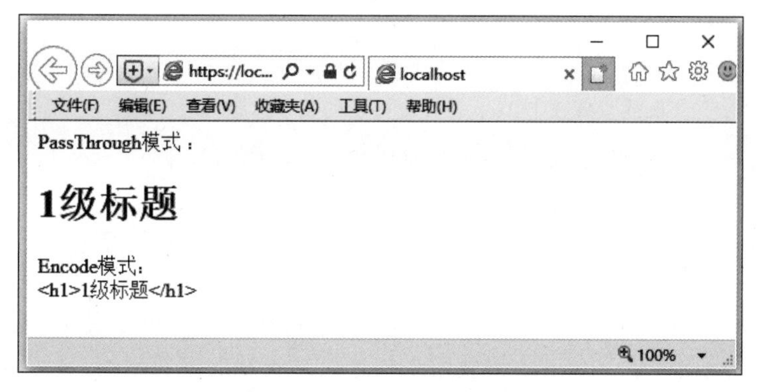

图 2.7 Literal 控件应用网页演示

2.2.4 图片控件 Image

Image 控件又称图片控件，在工具箱中图标为“ Image ”，封装在 System.Web.UI.Control.WebControl 命名空间的 Image 类中。其工作原理与 ImageButton 控件的图像显示功能类似，只是不支持鼠标单击事件，具体使用方法可以参照 2.3.3 节的 ImageButton 控件。

2.3 按钮控件

按钮也是页面中常见的元素，可以显示文本，也可以显示超链接或者图像，允许用户通过单击来执行操作。当按钮被单击时，可以触发“单击”事件，将页面“回发”给服务器，执行服务器端 C# 代码。主要有普通按钮控件(Button)、超链接按钮控件(LinkButton)、图片按钮控件(ImageButton)和特殊的热点图控件(ImageMap)，本节将对这四种控件进行详细讲解。

2.3.1 按钮控件 Button

扫一扫

Button 控件又称普通按钮控件，在工具箱中图标为“ Button ”，封装在 System.Web.UI.Control.WebControl 命名空间的 Button 类中，允许用户通过单击来执行操作。Button 控件的常用属性如表 2.6 所示。Button 控件的常用事件如表 2.7 所示。

表 2.6 Button 控件的常用属性

属 性 名	说 明
(ID)	获取或设置该控件的编程标识符
CommandArgument	获取或设置可选参数，该参数与关联的 CommandName 一起被传递到 Command 事件
CommandName	获取或设置命令名，该命令名与传递给 Command 事件的 Button 控件相关联
OnClientClick	获取或设置在引发某个 Button 控件的 Click 事件时所执行的浏览器端脚本

续表

属 性 名	说 明
Text	获取或设置在 Button 控件上显示的文本标题
CausesValidation	设置该按钮控件提交表单前是否触发客户端验证
PostBackUrl	获取和设置单击该 LinkButton 控件从当前页面发送到的 URL

说明：

URL(Uniform Resource Locator)又称为统一资源定位器，可以是某一文件资源，也可以是 WWW 页的地址，是 Internet 领域用来描述信息资源的字符串。

表 2.7 Button 控件的常用事件

事件名称	说 明
Click	在单击 Button 控件时触发此事件
Command	在单击 Button 控件并定义关联的命令时触发此事件

默认的 Button 按钮类似于 HTML 中提交类型的按钮 Submit，单击之触发 Click 事件将表单提交给服务器处理。可以通过设置 CommandName 和 CommandArgument 属性，使按钮成为 Command"命令"按钮，此时单击按钮时既可以触发 Click 事件，又可以触发 Command 事件，并可以通过 CommandArgument 设置参数值，后续的视图与多视图、数据控件章节将对此部分内容进行进一步讲解。

Button 按钮控件的 Click 事件前面章节已讲解，下面通过两个示例介绍 OnClientClick 属性和 Command 事件。

扫一扫

【示例 2-4】 在 chapter2 网站根目录下创建名为 example2-4 的网页，页面内包含一个 Button 控件，练习使用 Button 控件的 OnClientClick 属性。

(1) 按如下源文件添加控件并设置相关属性值。

```
<form id="form1" runat="server">
    <div>
        <asp:TextBox ID="txtDoc" runat="server"></asp:TextBox>
        <asp:Button ID="btnClear" runat="server" OnClick="btnClear_Click" Text=
"清空文本框内容" OnClientClick="return confirm('确认删除?');"/>
    </div>
</form>
```

(2) 为 btnClear 控件添加事件，并编辑代码如下。

```
protected void btnClear_Click(object sender, EventArgs e)
{
    //设置 txtDoc 控件的 Text 属性为空字符串
    txtDoc.Text=string.Empty;
}
```

(3) 运行网站，单击" 确认"框中的"确定"按钮，清空 txtDoc 文本框的内容；单击"取消"按钮，不执行清空操作。执行效果如图 2.8 所示。

图 2.8 Button 控件应用网页演示

说明：

“return confirm('确认删除?');”为 js 脚本命令，其功能为弹出“确认”框并接受其返回值，“确定”按钮返回值为 True，继续触发按钮控件的单击事件，“回发”服务器执行相关 C# 代码；“取消”按钮返回值为 False，不触发按钮控件的单击事件。

【示例 2-5】 在 chapter2 网站根目录下创建名为 example2-5 的网页，页面内包含若干个 Button 控件和 TextBox 控件，练习使用 Button 控件的 Command 事件。

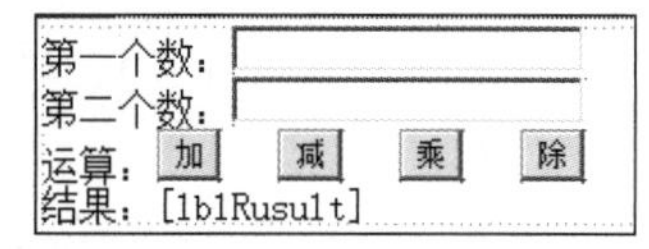

图 2.9 Button 控件属性应用网页设计

(1) 添加相应控件，如图 2.9 所示。

(2) 按如下源文件设置控件的相关属性值。

```
<form id="form1" runat="server">
    <div>
        第一个数:<asp:TextBox ID="txtNum1" runat="server"></asp:TextBox>
        <br/>
        第二个数:<asp:TextBox ID="txtNum2" runat="server"></asp:TextBox>
        <br/>
        运算:<asp:Button ID="btnAdd" runat="server" Text="加" CommandName=
"Add" OnCommand="btn_Command"/>
        <asp:Button ID="btnSub" runat="server" Text="减" CommandName="Sub"
OnCommand="btn_Command"/>
        <asp:Button ID="btnMul" runat="server" Text="乘" CommandName="Mul"
OnCommand="btn_Command"/>
        <asp:Button ID="btnDiv" runat="server" Text="除" CommandName="Div"
OnCommand="btn_Command"/>
        <br/>
        结果:<asp:Label ID="lblResult" runat="server"></asp:Label>
    </div>
</form>
```

(3) 为控件添加事件，并编辑代码如下。

```
protected void btn_Command(object sender, CommandEventArgs e)
{
    //将 txtNum1 和 txtNum2 中的 Text 属性值转换为 Double 类型赋值给 Num1 和 Num2 变量
```

```
        double num1=double.Parse(txtNum1.Text);
        double num2=double.Parse(txtNum2.Text);
        double reselut=0.0;
        //根据 CommandName 属性判断触发事件的控件要执行的操作
        switch (e.CommandName)
        {
            case "Add": reselut=num1+num2; break;
            case "Sub": reselut=num1-num2; break;
            case "Mul": reselut=num1 * num2; break;
            case "Div": reselut=num1/num2; break;
        }
        lblResult.Text=reselut.ToString();
    }
```

(4) 运行网站，正确填写两个数值，单击按钮触发 Command 事件，完成加、减、乘、除的基本运算，执行效果如图 2.10 所示。

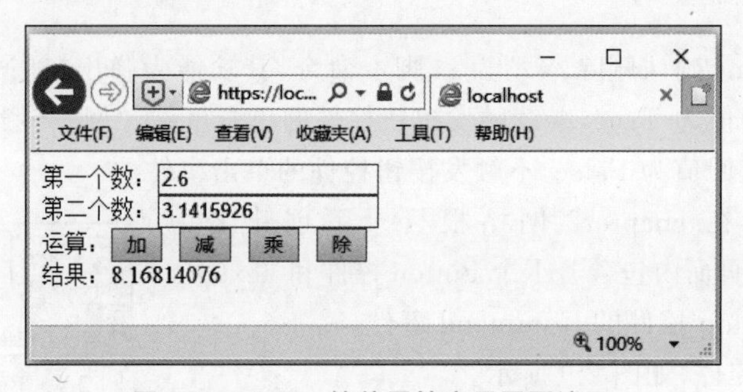

图 2.10 Button 控件属性应用网页演示

扫一扫

2.3.2 超链接按钮控件 LinkButton

LinkButton 控件又称超链接按钮控件，在工具箱中图标为“LinkButton”，封装在 System.Web.UI.Control.WebControl 命名空间的 LinkButton 类中。其工作原理与 Button 控件类似，只是呈现的外观为超链接样式，具体使用方法可以参照 Button 控件。

扫一扫

2.3.3 图片按钮控件 ImageButton

ImageButton 控件又称图片按钮控件，在工具箱中图标为“ImageButton”，封装在 System.Web.UI.Control.WebControl 命名空间的 ImageButton 类中。其工作原理与 Button 控件类似，只是呈现的外观为图像，外形更加美观，使用方法与 Button 控件类似，特殊属性为 ImageUrl，可以获取和设置该控件上显示的图像资源。

对于 ImageButton 控件的单击事件，通过服务器端编写 C# 程序可以获取到鼠标单击的像素点坐标，从而在单击同一按钮不同位置时实现不同的功能。

【示例 2-6】 在 chapter2 网站根目录下创建名为 example2-6 的网页，页面内包含 ImageButton 控件，练习 ImageButton 控件鼠标单击像素点位置的应用。

(1) 在 chapter2 网站根目录下添加 image 文件夹，右击 image 文件夹，在弹出的快捷菜单中选择“添加现有项”，添加一张图片，本书以 bird.jpg 图片为例，按图 2.11 所示添加相应控件。

鼠标单击位置的坐标：([lblX]，[lblY])

图 2.11　ImageButton 控件应用页面设计

（2）按如下源文件设置控件的相关属性值。

```
<form id="form1" runat="server">
    <div>
        <asp:ImageButton ID="imgBtn1" runat="server" BorderColor="#333300"
BorderWidth="1px" ImageUrl="~/image/bird.jpg" Width="200px" OnClick="imgBtn1_
Click"/>
        <br/>
        鼠标单击位置的坐标:(<asp:Label ID="lblX" runat="server" Text=""></asp:
Label> ,<asp:Label ID="lblY" runat="server" Text=""></asp:Label>)
        </div>
  </form>
```

（3）为控件添加事件，并编辑代码如下。

```
protected void imgBtn1_Click(object sender, ImageClickEventArgs e)
{
    //将鼠标单击坐标点的横坐标 X 和纵坐标 Y 赋值给标签控件的 Text 属性
    lblX.Text=e.X.ToString();
    lblY.Text=e.Y.ToString();
}
```

（4）运行网站，单击图片按钮的任意位置，可获得该点的坐标值。其图像的左上角点为坐标原点(0,0)，水平向右为横坐标 X 正值，竖直向下为纵坐标 Y 正值，执行效果如图 2.12 所示。

图 2.12　ImageButton 控件应用页面演示 1

（5）通过获取若干坐标点，可知道两只鸟的大致坐标值。添加如下代码后，单击同一按钮不同区域可实现不同功能，如图 2.13 所示。

```
protected void imgBtn1_Click(object sender, ImageClickEventArgs e)
{
    //将鼠标单击坐标点的横坐标 X 和纵坐标 Y 赋值给标签控件的 Text 属性
    lblX.Text=e.X.ToString();
    lblY.Text=e.Y.ToString();
    if (e.Y<120)
    {
        Response.Write("<script>alert('点击的是第一只鸟，正在展翅飞翔！');</
script>");
    }
    else
    {
        Response.Write("<script>alert('点击的是第二只鸟，正在养精蓄锐！');</
script>");
    }
}
```

图 2.13　ImageButton 控件应用页面演示 2

说明：

Response.Write("<script>alert('点击的是第一只鸟，正在展翅飞翔！');</script>");实现的功能是向客户端浏览器输出字符串“"<script>alert('点击的是第一只鸟，正在展翅飞翔！');</script>"”，该字符串 js 脚本会被浏览器解析执行，弹出一个警示框，包含文字“点击的是第一只鸟，正在展翅飞翔！”。关于 Response 内置对象的详细讲解见本书第 5 章。

扫一扫

2.3.4　热点图控件 ImageMap

ImageMap 控件又称热点图控件，在工具箱中图标为“ImageMap”，封装在 System.Web.UI.Control.WebControl 命名空间的 ImageMap 类中。在网页上显示图像同时可以指定若干热区，单击热区时触发相应事件“回发”到服务器或者导航到指定的 URL。

ImageMap 控件的常用属性如表 2.8 所示。ImageMap 控件的常用事件如表 2.9 所示。

表 2.8　ImageMap 控件的常用属性

属 性 名	说　　明
(ID)	获取或设置该控件的编程标识符
ImageUrl	获取和设置该控件上显示的图像资源
HotSpotMode	获取或设置单击热点区域后的默认行为方式
HotSpots	获取或设置 HotSpot 对象集合

说明：

(1) ImageMap 控件的 HotSpotMode 属性有几个枚举值，可以为控件整体设置属性值，也可以为每个热点区域设置单独的 HotSpot. HotSpotMode 属性值，并且 HotSpot. HotSpotMode 属性的优先级更高。

- Inactive：无任何操作，此时就如同没有热点区域的普通图片。
- NotSet：未设置项，同时也是默认项，将执行定向操作，链接到指定的 URL 地址。如果未指定 URL 地址，则默认链接到应用程序根目录下。
- Navigate：定向操作项，链接到指定的 URL 地址，若未指定 URL 地址，则默认链接到应用程序根目录下。
- PostBack：回传操作值，可设置回传值。单击热点区域后，将触发控件的 Click 事件。

(2) HotSpots 属性是 HotSpot 对象集合，每个 HotSpot 对象指定一个热点区域，包含圆形热区(CircleHotSpot)、矩形热区(RectangleHotSpot)和多边形热区(PolygonHotSpot)三种枚举值。

- CircleHotSpot：用于在图像映射中定义一个圆形区域，区域设置包含三个属性 X、Y 和 Radius，X 和 Y 表示圆心的坐标为(X，Y)，Radius 表示圆的半径。
- RectangleHotSpot：用于在图像映射中定义一个矩形区域，区域设置包含 Top、Left、Bottom 和 Right 四个属性，其中 Top 和 Left 表示左上角点(Left，Top)，Bottom 和 Right 表示右下角点(Right，Bottom)，过这两个点分别做一条水平线和竖直线，四条线中间围成的区域形成一个矩形热点区域。
- PolygonHotSpot：用于在图像映射中定义一个不规则形状区域。区域设置只包含一个属性 Coordinates，其值为一个“数值列表”字符串，格式如“x1，x2，x3，x4，x5，x6，…”形式，表示点(x1，x2)、点(x3，x4)、点(x5，x6)……依次首尾相连构建的封闭热点区域。

表 2.9　ImageMap 控件的常用事件

事件名称	说　　明
Click	单击 ImageMap 控件的任一 HotSpot 区域触发此事件

【示例 2-7】　在 chapter2 网站根目录下创建名为 example2-7 的网页，页面内包含 ImageMap 控件，练习 ImageMap 控件的热区设置及热区模式的应用。

（1）在 chapter2 网站根目录下的 image 文件夹中添加图片 house.jpg。

（2）添加一个 ImageMap 控件，并设置 ImageUrl 属性，使其显示 house.jpg 图片，对应源文件如下。

```
<asp:ImageMap ID="imgMap1" runat="server" ImageUrl="~/image/house.jpg" Width
="600px">
```

（3）从 ImageMap 控件的属性窗口中选择 HotSpots 属性，单击后面的"..."图标，弹出"HotSpot 集合编辑器"窗口，如图 2.14 所示。

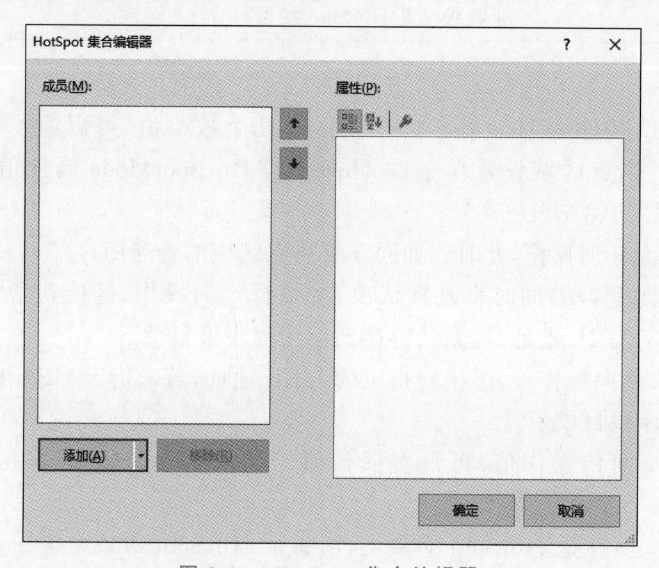

图 2.14　HotSpots 集合编辑器

（4）单击"添加(A)"右侧的下拉列表，选择添加一个 CircleHotSpot 热区，按图 2.15 所示设置相关属性。

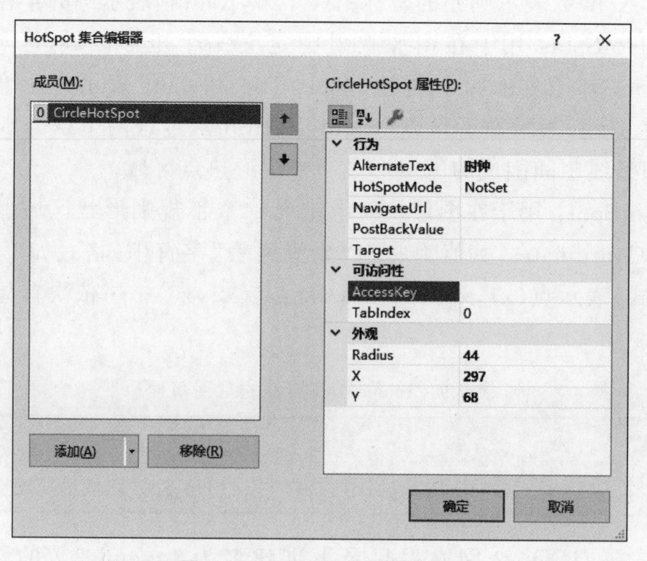

图 2.15　HotSpots 属性设置

(5) 执行后鼠标指针移动到时钟区域(以(297,68)为圆心,44 为半径的热区)时,出现"时钟"提示,单击鼠标左键可"回发"服务器,具体效果如图 2.16 所示。

图 2.16　ImageMap 控件应用页面设计

(6) 按如下源文件添加其他热区,并设置相关属性。

```
<form id="form1" runat="server">
    <div>
        <asp:ImageMap ID="imgMap1" runat="server" ImageUrl="~/image/house.
jpg" OnClick="imgMap1_Click" Width="600px">
            <asp:CircleHotSpot AlternateText="时钟" Radius="44" X="297" Y="68"/>
            <asp:RectangleHotSpot Bottom="123" HotSpotMode="PostBack" Left=
"35" PostBackValue="table lamp" Right="98" Top="81"/>
            <asp:PolygonHotSpot Coordinates="83,303,78,216,106,202,146,202,
147,191,151,190,151,178,244,173,338,173,423,173,472,174,523,171,525,202,551,
200,590,220,585,307,85,306" HotSpotMode="PostBack" PostBackValue="sofa"/>
        </asp:ImageMap>
    </div>
    </form>
```

(7) 为控件添加事件,并编辑代码如下。

```
protected void imgMap1_Click(object sender, ImageMapEventArgs e)
{
    if (e.PostBackValue=="table lamp")
    {
        Response.Write("<script>alert('您选择的是台灯!');</script>");
    }
    if (e.PostBackValue=="sofa")
    {
        Response.Write("<script>alert('您选择的是沙发!');</script>");
    }
}
```

（8）单击沙发所在多边形热区，可以“回发”服务器，传回 PostBackValue 属性值 sofa，执行效果如图 2.17 所示。

图 2.17　ImageMap 控件应用页面演示

通过上述示例可以感受 ImageMap 控件功能的强大，但如何设置热区的位置点一直是该控件使用的难点，下面结合 ImageButton 控件的属性编写一个小程序，可视化地获取图片热区的位置点。

【示例 2-8】　在 chapter2 网站根目录下创建名为 example2-8 的网页，页面内包含 ImageButton 控件，练习使用 ImageButton 控件可视化地获取图片热区的位置点。

（1）添加一个 ImageButton 控件，并按示例 2-7 中的 ImageMap 控件设置 ImageUrl 和 Width 属性。

（2）按如下源文件添加其他控件并设置相关属性值。

```
<form id="form1" runat="server">
    <div>
        <asp:HiddenField ID="HiddenField1" runat="server"/>
    </div>
        您构建的区域字符串:<asp:Label ID="lblPoints" runat="server"></asp:
Label>
        <br/>
        <asp:ImageButton ID="ImageButton1" runat="server" ImageUrl="~/image/
    house.jpg" OnClick="ImageButton1_Click" Width="600px"/>
</form>
```

（3）为控件添加事件，并编辑代码如下。

```
protected void ImageButton1_Click(object sender, ImageClickEventArgs e)
{
    //使用 HiddenField1 控件的 Value 属性保存鼠标单击点的横、纵坐标
    HiddenField1.Value+=e.X.ToString()+","+e.Y.ToString()+",";
```

```
        //移除 HiddenField1 控件的 Value 属性中的最后一个字符','
        lblPoints.Text=HiddenField1.Value.Remove(HiddenField1.Value.Length-1, 1);
    }
```

(4) 运行网页,依次顺序围绕多边形热区边缘单击一圈,即可得到多边形区域的"数值列表",执行效果如图 2.18 所示。

图 2.18 ImageMap 控件"数值列表"获取

2.4 选择控件

选择控件可以为用户提供选择项,按指定形式显示文本或图像,允许执行单选或者复选操作,当选中项发生改变时,可以触发相关事件,将页面"回发"给服务器,执行服务器端 C# 代码。选择控件主要包含单选按钮控件 RadioButton、单选按钮列表控件 RadioButtonList、复选框控件 CheckBox、复选框列表控件 CheckBoxList、下拉列表控件 DropDownList、列表框控件 ListBox 和子弹列表控件 BulletedList 等,本节将对这些控件进行详细讲解。

2.4.1 单选按钮控件 RadioButton

扫一扫

RadioButton 控件又称单选按钮控件,在工具箱中图标为" RadioButton ",封装在 System.Web.UI.Control.WebControl 命名空间的 RadioButton 类中,以文字形式呈现选择项,允许用户互斥地从选择项中选择一个选项,是实现单选功能最常使用的一种方式。RadioButton 控件的常用属性如表 2.10 所示。RadioButton 控件的常用事件如表 2.11 所示。

表 2.10 RadioButton 控件的常用属性

属 性 名	说 明
(ID)	获取或设置该控件的编程标识符
AutoPostBack	获取或设置单选按钮选的选中状态变化时是否自动回发到服务器
GroupName	获取或设置单选按钮所属的组名
Checked	获取或设置单选按钮的选中状态
Text	获取或设置在单选按钮控件上显示的文本

说明：

(1) GroupName 属性相同的单选按钮为逻辑上的同一组，同组内实现单选。

(2) 当单击某一“未选中”状态的 RadioButton 控件时，状态变为“选中”状态，并自动清除该单选组中的其他选中项。

表 2.11 RadioButton 控件的常用事件

事件名称	说 明
CheckedChanged	在 RadioButton 控件的 Checked 属性值改变时触发此事件

【示例 2-9】 在 chapter2 网站根目录下创建名为 example2-9 的网页，页面内包含若干个 RadioButton 控件，练习使用 RadioButton 控件的属性和事件。

(1) 按如下源文件添加控件并设置相关属性值。

```
<form id="form1" runat="server">
      <p>请选择正确答案:</p>
      <asp:RadioButton ID="rbtnA" runat="server" Text="A.地球是圆的" GroupName
="question1"/>
      <br/>
      <asp:RadioButton ID="rbtnB" runat="server" Text="B.地球是方的" GroupName
="question1"/>
      <br/>
       <asp:RadioButton ID="rbtnC" runat="server" Text="C.地球是扁的"
GroupName="question1"/>
      <br/>
       <asp:RadioButton ID="rbtnD" runat="server" Text="D.地球是椭球的"
GroupName="question1"/>
      <br/>
 <br/>
  
       <asp:Button ID="btnOK" runat="server" Text="确定" OnClick="btnOK_
Click"/>
      <br/>
  </form>
```

(2) 为 btnOK 控件添加事件，并编辑代码如下。

```
protected void btnOK_Click(object sender, EventArgs e)
{
    if (rbtnD.Checked)
```

```
        {
            Response.Write("<script>alert('恭喜你答对了!');</script>");
        }
        else
        {
            Response.Write("<script>alert('很遗憾答错了,正确答案 D。');</script>");
        }
    }
```

（3）运行网站，执行效果如图 2.19 所示。

图 2.19　RadioButton 控件应用网页演示

2.4.2　单选按钮列表控件 RadioButtonList

扫一扫

RadioButtonList 控件又称单选按钮列表控件，也可归结为列表控件类型，在工具箱中图标为“RadioButtonList”，封装在 System.Web.UI.Control.WebControl 命名空间的 RadioButtonList 类中。在一个控件内以文字形式呈现多个选择项，构建单选按钮列表，允许用户互斥地从选择项中选择一个，是实现多选项中单选功能的一种常用方式。RadioButtonList 控件的常用属性如表 2.12 所示。RadioButtonList 控件的常用事件如表 2.13 所示。

表 2.12　RadioButtonList 控件的常用属性

属 性 名	说　明
(ID)	获取或设置该控件的编程标识符
AutoPostBack	获取或设置单选按钮列表中的项变化时是否自动回发到服务器
Items	控件中所有项的集合
SelectedIndex	获取或设置选定项索引

续表

属 性 名	说 明
SelectedItem	获取或设置列表中的选中项
SelectedValue	获取控件中选定项的值
Text	获取或设置单选按钮控件上 SelectedValue 的值
RepeatDirection	获取或设置列表中单选按钮的排列方向
RepeatColumns	获取或设置列表中单选按钮显示的列数
TextAlign	获取或设置列表中单选按钮的文本对齐方式

说明：

(1) RepeatDirection 属性包含 Horizontal 和 Vertical 两个枚举值，Horizontal 表示列表中的项以行的形式水平显示，从左到右，从上至下；Vertical 表示列表中的项以列的形式水平显示，从上到下，从左至右。

(2) TextAlign 属性包含 Left 和 Right 两个枚举值，Left 表示关联文本显示在单选框的左侧，Right 表示关联文本显示在单选框的右侧。

表 2.13　RadioButtonList 控件的常用事件

事 件 名 称	说 明
CheckedIndexChanged	在 RadioButtonList 控件的选中项的索引改变时触发
TextChanged	当 Text 和 SelectValue 属性改变时触发

说明：

控件中的每一个选择项称为 Items 中的一个元素，具有 Item.Index 和 Item.Text 等属性，其索引值为从 0 开始的连续正整数，Text 值为任意字符串，CheckedIndex 值一定是不同的，而 SelectValue 值却可以相同，故只要触发 TextChanged 事件，就一定会触发 CheckedIndexChanged 事件；反之，触发 CheckedIndexChanged 事件，却不一定会触发 TextChanged 事件。

【示例 2-10】　在 chapter2 网站根目录下创建名为 example2-10 的网页，页面内包含一个 RadioButtonList 控件，练习 RadioButtonList 控件中 ListItem 集合编辑器的应用。

(1) 添加一个 RadioButtonList 控件，在其属性窗口中选择 Items 属性，单击后面的“...”图标(或者选中该控件，单击控件右上角的“>”图标)，弹出“ListItem 集合编辑器”窗口，如图 2.20 所示。

(2) 添加其他控件，按如下源文件添加控件并设置相关属性值。

```
<form id="form1" runat="server">
    <div>
        <br/>
        请选择正确答案:<asp:RadioButtonList ID="rbtnSelect" runat="server">
            <asp:ListItem Value="A">A.地球是圆的</asp:ListItem>
            <asp:ListItem Value="B">B.地球是方的</asp:ListItem>
            <asp:ListItem Value="C">C.地球是扁的</asp:ListItem>
```

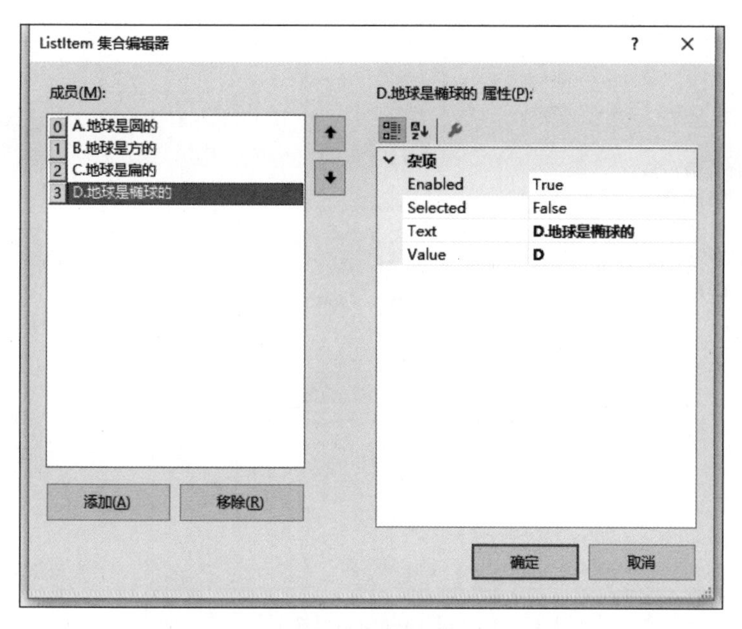

图 2.20 ListItem 属性设计

```
            <asp:ListItem Value="D">D.地球是椭球的</asp:ListItem>
        </asp:RadioButtonList>
        <br/>
        <asp:Button ID="btnOK" runat="server" OnClick="btnOK_Click" Text="确定"/>
    </div>
</form>
```

(3) 为 btnOK 控件添加事件,并编辑代码如下。

```
protected void btnOK_Click(object sender, EventArgs e)
{
    if (rbtnSelect.SelectedValue=="D")
    {
        Response.Write("<script>alert('恭喜你答对了!');</script>");
    }
    else
    {
        Response.Write("<script>alert('很遗憾答错了,正确答案 D。');</script>");
    }
}
```

(4) 运行网站,执行效果如图 2.21 所示。

上述示例使用 RadioButtonList 控件实现了与示例 2-9 完全相同的功能,在控件添加、属性设置以及事件后台代码等方面有明显改进。此外,RadioButtonList 控件更适合选择项动态变化的应用。

【示例 2-11】 在 chapter2 网站根目录下创建名为 example2-11 的网页,页面内包含一个 RadioButtonList 控件,练习在 RadioButtonList 控件中动态添加 ListItem 项。

(1) 添加 RadioButtonList 控件和其他控件,具体源文件如下所示。

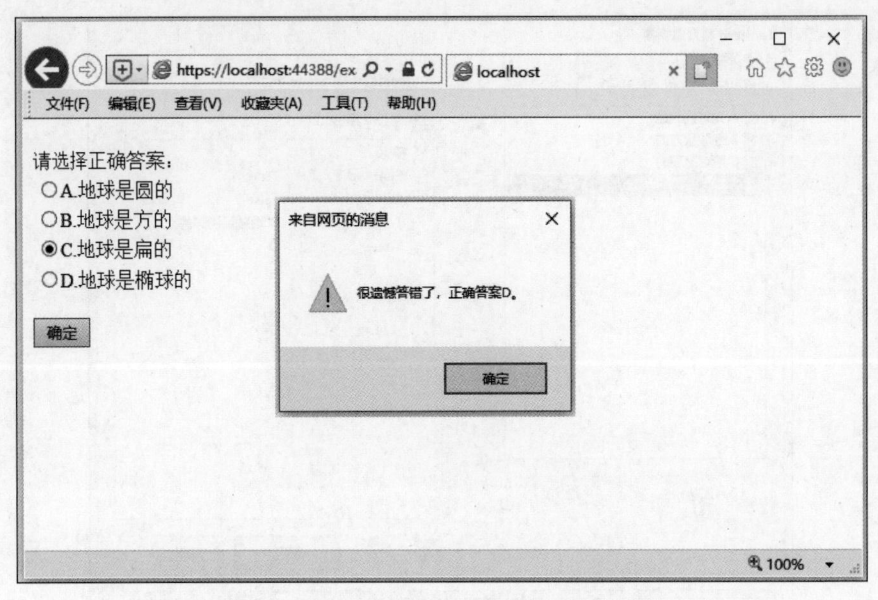

图 2.21 RadioButtonList 控件应用网页演示

```
<form id="form1" runat="server">
    请选择你的幸运数字：
    <asp:RadioButtonList ID="rbtnNums" runat="server" AutoPostBack="True"
OnSelectedIndexChanged="rbtnNums_SelectedIndexChanged" RepeatColumns="10"
RepeatDirection="Horizontal">
    </asp:RadioButtonList>
</form>
```

(2) 添加相关事件,并编辑代码如下。

```
protected void Page_Load(object sender, EventArgs e)
{
    if (!IsPostBack)
    {
        //定义 ListItem 类型对象 li
        ListItem li;
        for (int i=1; i<=100; i++)
        {
            li=new ListItem(i.ToString(), i.ToString());
            //将对象 li 添加到 rbtnNums 控件的 Items 集合
            rbtnNums.Items.Add(li);
        }
    }
}
protected void rbtnNums_SelectedIndexChanged(object sender, EventArgs e)
{
    Response.Write("<script>alert('您选择的幸运数字是:"+rbtnNums.
SelectedValue+"');</script>");
}
```

(3) 运行网站,执行效果如图 2.22 所示。

IsPostBack 属性表示页面是否为“回发”,只有页面第一次加载时! IsPostBack 属性才

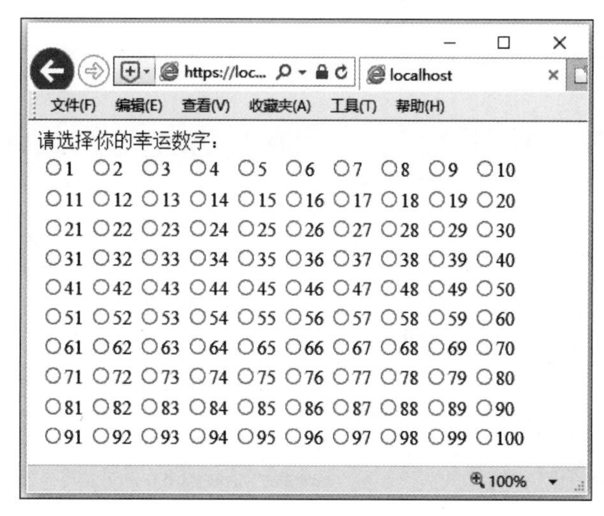

图 2.22　动态 RadioButtonList 控件页面演示

为 True，详细讲解见本书第 5 章。

2.4.3　复选框控件 CheckBox

扫一扫

CheckBox 控件又称复选框控件，在工具箱中图标为“☑ CheckBox”，封装在 System.Web.UI.Control.WebControl 命名空间的 CheckBox 类中，以文字形式呈现选择项，允许从选择项中选择多个，是实现多选功能最常使用的一种方式。CheckBox 控件常用的属性和事件与 RadioButton 控件基本相同，只是不再需要组的概念，没有 GroupName 属性。

【示例 2-12】　在 chapter2 网站根目录下创建名为 example2-12 的网页，页面内包含若干个 CheckBox 控件，练习使用 CheckBox 控件的属性和事件。

(1) 按如下源文件添加控件并设置相关属性值。

```
<form id="form1" runat="server">
    下列哪些是 C# 中的关键字(多选):<br/>
    <asp:CheckBox ID="chkA" runat="server" Text="A.for"/><br/>
    <asp:CheckBox ID="chkB" runat="server" Text="B.break"/><br/>
    <asp:CheckBox ID="chkC" runat="server" Text="C.so easy"/><br/>
    <asp:CheckBox ID="chkD" runat="server" Text="C.while"/><br/>
    <asp:Button ID="btnOK" runat="server" Text="确定" OnClick="btnOK_
Click"/>
</form>
```

(2) 添加相关事件，并编辑代码如下。

```
protected void btnOK_Click(object sender, EventArgs e)
{
    if (chkA.Checked && chkB.Checked && chkD.Checked)
    {
        Response.Write("<script>alert('恭喜你答对了！');</script>");
    }
    else
```

```
        {
            Response.Write("<script>alert('很遗憾答错了,正确答案 ABD。');</script>");
        }
    }
```

(3) 运行网站,执行效果如图 2.23 所示。

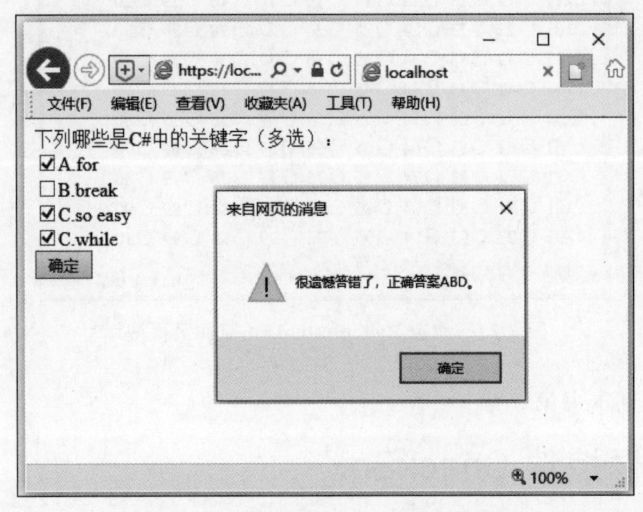

图 2.23 CheckBox 控件应用页面演示

扫一扫

2.4.4 复选框列表控件 CheckBoxList

CheckBoxList 控件又称复选框列表控件,也可归结为列表控件类型,在工具箱中图标为" CheckBoxList ",封装在 System. Web. UI. Control. WebControl 命名空间的 CheckBoxList 类中。可以在一个控件中以文字形式呈现多个选择项,构建多选按钮列表,允许从选择项中选择多个,是实现多选项复选功能的一种常用方式。CheckBoxList 控件常用的属性和事件与 RadioButtonLis 控件基本相同,但 CheckBoxList 控件实现的是复选,因此如下几个属性稍有区别,如表 2.14 所示。

表 2.14 CheckBoxList 控件的常用属性

属 性 名	说 明
SelectedIndex	获取或设置选定项最低序号索引
SelectedItem	获取或设置列表中的索引最小的选中项
SelectedValue	获取控件中索引最小的选定项的值

【示例 2-13】 在 chapter2 网站根目录下创建名为 example2-13 的网页,页面内包含一个 CheckBoxList 控件,练习 CheckBoxList 控件属性与事件的应用。

(1) 添加一个 CheckBoxList 控件,并通过 Items 属性添加相关项,具体源文件如下所示。

```
<form id="form1" runat="server">
    下列哪些是 C# 中的关键字(多选):<br/>
    <asp:CheckBoxList ID="chklSelect" runat="server">
        <asp:ListItem Value="A">for</asp:ListItem>
        <asp:ListItem Value="B">break</asp:ListItem>
        <asp:ListItem Value="C">so easy</asp:ListItem>
        <asp:ListItem Value="D">while</asp:ListItem>
    </asp:CheckBoxList>
    <asp:Button ID="btnOK" runat="server" OnClick="btnOK_Click" Text="确定"/>
  </form>
```

(2) 添加相关事件,并编辑代码如下。

```
protected void btnOK_Click(object sender, EventArgs e)
{
    if (chklSelect.Items[0].Selected&&chklSelect.Items[1].Selected&&chklSelect.
Items[3].Selected)
    {
        Response.Write("<script>alert('恭喜你答对了!');</script>");
    }
    else
    {
        Response.Write("<script>alert('很遗憾答错了,正确答案 ABD。');</script>");
    }
}
```

(3) 运行网站,执行效果如图 2.24 所示。

图 2.24 CheckBoxList 控件应用页面演示

2.4.5 下拉列表控件 DropDownList

DropDownList 控件又称下拉列表控件,也可归结为列表控件类型,在工具箱中图标为"DropDownList",封装在 System. Web. UI. Control. WebControl 命名空间的 DropDownList 类中。下拉列表控件用于把选择项放在一个下拉式选单中,并将每一个下

拉式选单置于主选单的一个选项下，只能实现单选，可替代一组单选按钮，并且比单选按钮列表的占用位置更小，其属性和事件与 RadioButtonList 控件及 CheckBoxList 控件的属性和事件基本相同。

扫一扫

【示例 2-14】 在 chapter2 网站根目录下创建名为 example2-14 的网页，页面内包含多个 DropDownList 控件，设置 DropDownList 控件的属性与事件，实现年、月、日的三级联动。

（1）添加相关控件，设置基本界面如图 2.25 所示。

图 2.25 DropDownList 控件应用页面设计

（2）设置控件的属性值，如下列源文件。

```
<form id="form1" runat="server">
    年<asp:DropDownList ID="ddlYear" runat="server" AutoPostBack="True"
OnSelectedIndexChanged="ddlYearMonth_SelectedIndexChanged"></asp:DropDownList>
    月<asp:DropDownList ID="ddlMonth" runat="server" AutoPostBack="True"
OnSelectedIndexChanged="ddlYearMonth_SelectedIndexChanged"></asp:DropDownList>
    日<asp:DropDownList ID="ddlDay" runat="server"></asp:DropDownList>
    <asp:Button ID="btnOK" runat="server" Text="确定" OnClick="btnOK_Click"/>
    <asp:Button ID="btnToday" runat="server" Text="今天" OnClick="btnToday_
Click"/>
</form>
```

（3）添加相关事件，并编辑代码如下。

```
protected void Page_Load(object sender, EventArgs e)
{
    if (! IsPostBack)
    {
        for (int i=1949; i <=2048; i++)
        {
            ddlYear.Items.Add(i.ToString());
        }
        for (int i=1; i <=12; i++)
        {
            ddlMonth.Items.Add(i.ToString());
        }
        for (int i=1; i <=31; i++)
        {
            ddlDay.Items.Add(i.ToString());
        }
    }
}
/// <summary>
/// 按年、月获得天数
/// </summary>
/// <param name="year">年份</param>
/// <param name="month">月份</param>
/// <returns>天数</returns>
```

```
int GetDaysByYearMonth(int year, int month)
{
    int days=30;
    if (month==1 || month==3 || month==5 || month==7 || month==8 || month==10 || month
==12)
    days=31;
    if (month==2)
    {
        if (year %100 !=0 && year %4==0)
          days=29;
        else if (year% 400==0)
          days=29;
            else days=28;
        }
        return days;
}
protected void ddlYearMonth_SelectedIndexChanged(object sender, EventArgs e)
{
    int year=int.Parse(ddlYear.SelectedValue);
    int month=int.Parse(ddlMonth.SelectedValue);
    int days=GetDaysByYearMonth(year, month);
    ddlDay.Items.Clear();
    for (int i=1; i <=days; i++)
    {
        ddlDay.Items.Add(i.ToString());
    }
}
protected void btnOK_Click(object sender, EventArgs e)
{
    string date=ddlYear.SelectedValue+"年"+ddlMonth.SelectedValue+"月"
            +ddlDay.SelectedValue+"日";
    Response.Write("<script>alert('"+date+"');</script>");
}
protected void btnToday_Click(object sender, EventArgs e)
{
    int year=DateTime.Today.Year;
    int month=DateTime.Today.Month;
    int day=DateTime.Today.Day;
    for (int i=0; i<ddlYear.Items.Count; i++)
    {
        if (ddlYear.Items[i].Text==year.ToString())
        {
            ddlYear.SelectedIndex=i;
            break;
        }
    }
    ddlMonth.SelectedIndex=month-1;
    int days=GetDaysByYearMonth(year, month);
    ddlDay.Items.Clear();
    for (int i=1; i <=days; i++)
    {
        ddlDay.Items.Add(i.ToString());
```

```
        }
        ddlDay.SelectedIndex=day-1;
    }
```

(4) 运行网站,执行效果如图 2.26 所示。

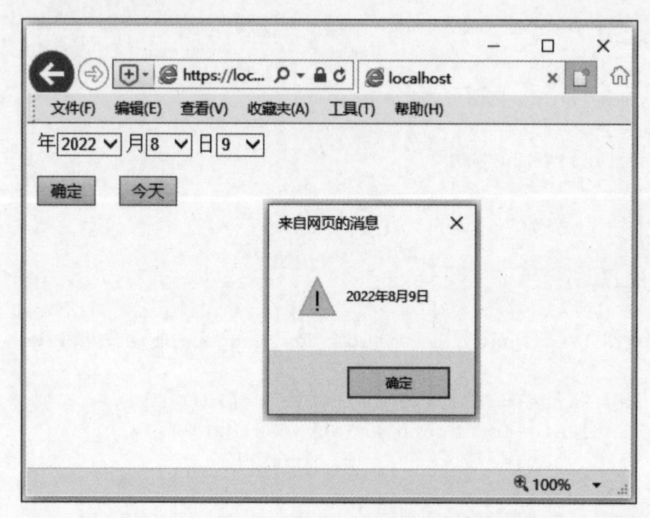

图 2.26 DropDownList 控件应用页面演示

扫一扫

2.4.6 列表框控件 ListBox

ListBox 控件又称列表框控件,也可归结为列表控件类型,在工具箱中图标为"ListBox",封装在 System.Web.UI.Control.WebControl 命名空间的 ListBox 类中。列表框控件可以看作一个扩展的下拉列表,可以将所有选择项放在一个列表框中显示,或者只显示部分选项,其他放置于下拉列表中。列表内的项目可以根据需要设置为多选或单选,分别用来替代一组复选框列表或单选按钮列表。其属性和事件与 DropDownList 控件的属性和事件近似。ListBox 控件的常用属性如表 2.15 所示。

表 2.15 ListBox 控件的常用属性

属 性 名	说 明
Rows	获取或设置列表框中显示的行数
SelectedMode	获取或设置列表框中的选择模式

扫一扫

说明:

SelectedMode 属性可选择 Multiple 和 Single 两个枚举值,Multiple 表示可以通过按住 Ctrl 键后进行多选;Single 表示只能进行单选。

【示例 2-15】 在 chapter2 网站根目录下创建名为 example2-15 的网页,页面内包含多个 ListBox 控件,设置 ListBox 控件属性,实现内容项的添加与删除。

(1) 添加相关控件,并对 table 标签进行必要的页面布局,设置基本界面如图 2.27 所示。

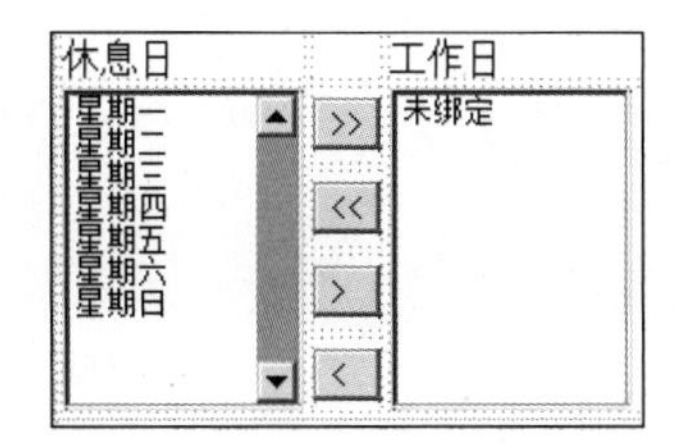

图 2.27 ListBox 控件应用页面设计

(2) 设置控件的属性值,如下列源文件。

```
<form id="form1" runat="server">
    <table>
            <tr>
                <td>休息日</td>
                <td > </td>
                <td>工作日</td>
            </tr>
            <tr>
                <td rowspan="4">
                      <asp:ListBox ID="lbLeft" runat="server" Height="131px"
Width="101px" SelectionMode="Multiple">
                        <asp:ListItem Value="1">星期一</asp:ListItem>
                        <asp:ListItem Value="2">星期二</asp:ListItem>
                        <asp:ListItem Value="3">星期三</asp:ListItem>
                        <asp:ListItem Value="4">星期四</asp:ListItem>
                        <asp:ListItem Value="5">星期五</asp:ListItem>
                        <asp:ListItem Value="6">星期六</asp:ListItem>
                        <asp:ListItem Value="7">星期日</asp:ListItem>
                    </asp:ListBox>
                </td>
                <td >
                    <asp:Button ID="btnLtoRAll" runat="server" Text="&gt;&gt;"
OnClick="btnLtoRAll_Click"/>
                 </td>
                <td rowspan="4">
                     <asp:ListBox ID="lbRight" runat="server"  Height="131px"
Width="101px" SelectionMode="Multiple"></asp:ListBox>
                </td>
            </tr>
            <tr>
                <td>
                    <asp:Button ID="btnRtoLAll" runat="server" Text="&lt;&lt;"
OnClick="btnRtoLAll_Click"/>
                </td>
            </tr>
            <tr>
                <td >
                       <asp:Button ID="btnLtoR" runat="server" Text="&gt; "
OnClick="btnLtoR_Click"/>
                </td>
            </tr>
            <tr>
```

```
                <td >
                        < asp: Button ID = "btnRtoL" runat = "server" Text = "&lt; "
OnClick="btnRtoL_Click"/>
                </td>
            </tr>
        </table>
    </form>
```

（3）添加相关事件,并编辑代码如下。

```
protected void btnLtoRAll_Click(object sender, EventArgs e)
{
    int n=lbLeft.Items.Count;
    for (int i=0; i<n; i++)
    {
        ListItem li=lbLeft.Items[0];
        int index=GetInsertIndex(lbRight.Items, li);
        lbRight.Items.Insert(index, li);
        lbLeft.Items.RemoveAt(0);
    }
    lbRight.SelectedIndex=-1;
}
protected void btnLtoR_Click(object sender, EventArgs e)
{
    for (int i=lbLeft.Items.Count-1; i >=0; i--)
    {
        if (lbLeft.Items[i].Selected)
        {
            ListItem li=lbLeft.Items[i];
            int index=GetInsertIndex(lbRight.Items, li);
            lbRight.Items.Insert(index, li);
            lbLeft.Items.RemoveAt(i);
        }
    }
    lbRight.SelectedIndex=-1;
}
/// <summary>
/// 获取插入项的位置索引
/// </summary>
/// <param name="lic">插入列表的项集合</param>
/// <param name="li">待插入的项</param>
/// <returns></returns>
int GetInsertIndex(ListItemCollection lic, ListItem li)
{
    int i=0;
    for (; i<lic.Count; i++)
    {
        if (int.Parse(li.Value)<=int.Parse(lic[i].Value))
        {
            break;
        }
    }
    return i;
```

```
    }
    protected void btnRtoLAll_Click(object sender, EventArgs e)
    {
        int n =lbRight .Items.Count;
        for (int i=0; i<n; i++)
        {
            ListItem li=lbRight.Items[0];
            int index=GetInsertIndex(lbLeft.Items, li);
            lbLeft.Items.Insert(index, li);
            lbRight.Items.RemoveAt(0);
        }
        lbLeft.SelectedIndex=-1;
    }
    protected void btnRtoL_Click(object sender, EventArgs e)
    {
        for (int i=lbRight.Items.Count-1; i >=0; i--)
        {
            if (lbRight.Items[i].Selected)
            {
                ListItem li=lbRight.Items[i];
                int index=GetInsertIndex(lbLeft.Items, li);
                lbLeft.Items.Insert(index, li);
                lbRight.Items.RemoveAt(i);
            }
        }
        lbLeft.SelectedIndex=-1;
    }
```

（4）运行网站，可以单选或者按住 Ctrl 键进行多选，单击“>”按钮将左侧休息日列表中选中的项移入右侧工作日列表；单击“<”按钮将右侧工作日列表中选中的项移入左侧休息日列表；单击“>>”按钮将左侧休息日列表中的所有项移入右侧工作日列表；单击“<<”按钮将右侧工作日列表中的所有项移入左侧休息日列表。新插入列表中的各项将按星期一至星期日进行排序，执行效果如图 2.28 所示。

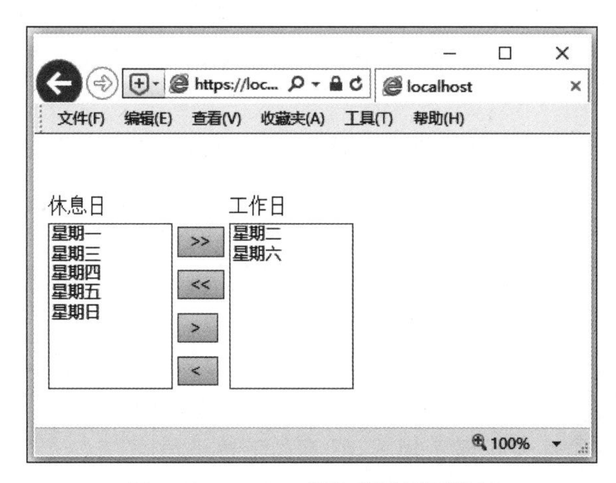

图 2.28　ListBox 控件应用页面演示

扫一扫

2.4.7 子弹列表控件 BulletedList

BulletedList 控件又称子弹列表控件，也可归结为列表控件类型，在工具箱中图标为“BulletedList”，封装在 System.Web.UI.Control.WebControl 命名空间的 BulletedList 类中。控件字面意思为像子弹一样排列的列表，可以在一个控件中以有序列表或者无序列表形式呈现内容项，允许用户对内容项进行单击操作，可以进行页面导航或者触发相应事件。BulletedList 控件的常用属性如表 2.16 所示。

表 2.16 BulletedList 控件的常用属性

属 性 名	说 明
BulletStyle	获取或设置该控件项目符号列表的样式
DisplayMode	该控件显示的列表的类型
FirstBulletNumber	获取或设置有序列表中列表项目的起始数字
BulletImageUrl	获取或设置该控件列表项目图形符号的 URL，在 BulletStyle 属性值为 CustomImage 时使用
CausesValidation	设置该按钮控件提交表单前是否触发客户端验证
PostBackUrl	获取和设置单击 LinkButton 控件从当前页面发送到的 URL

说明：

(1) BulletStyle 属性表示项目符号编号样式值，具有如下枚举值。

- Circle：表示项目符号编号样式设置为“○”空圈圈。
- CustomImage：表示项目符号编号样式设置为自定义图片，其图片由 BulletImageUrl 属性指定。
- Disc：表示项目符号编号样式设置为“●”实圈圈。
- LowerAlpha：表示项目符号编号样式设置为小写字母格式，如 a、b、c、d 等 26 个小写英文字母。
- LowerRoman：表示项目符号编号样式设置为小写罗马数字格式，如ⅰ、ⅱ、ⅲ、ⅳ等小写的罗马数字。
- NotSet：表示不设置项目符号编号样式，此时将以 Disc 样式为默认样式显示。
- Numbered：表示设置项目符号编号样式为数字格式，如 1、2、3、4 等数字格式。
- Square：表示设置项目符号编号样式为“■”实体黑方块。
- UpperAlpha：表示设置项目符号编号样式为大写字母格式，如 A、B、C、D 等 26 个大写英文字母。
- UpperRoman：表示设置项目符号编号样式为大写罗马数字格式，如Ⅰ、Ⅱ、Ⅲ、Ⅳ等大写的罗马数字。

(2) DisplayMode 属性表示显示模式，具有如下枚举值。

- Text：表示以纯文本形式表现项目列表。
- HyperLink：表示以超链接形式表现项目列表。链接文字为某个具体项 ListItem 的 Text 属性，链接目标为 ListItem 的 Value 属性。

- LinkButton：表示以服务器控件 LinkButton 形式表现项目列表。此时每个 ListItem 项都将表现为 LinkButton，同时以 Click 事件回发到服务器端进行相应操作。

BulletedList 控件的常用事件如表 2.17 所示。

表 2.17　BulletedList 控件的常用事件

事件名	说明
Click	单击列表控件中的超链接按钮时触发此事件

说明：

当 BulletedList 控件的 DisplayMode 属性为 LinkButton，BulletedList 控件中的某项被单击时触发 Click 事件。通常将被单击项在所有项目列表中的索引号(从 0 开始)作为传回参数传回服务器端。

【示例 2-16】　在 chapter2 网站根目录下创建名为 example2-16 的网页，页面内包含三个 BulletedList 控件，练习 BulletedList 控件的相关属性和事件。

(1) 在 chapter2 网站根目录下的 image 文件夹中添加图书的图标文件 book.ico。

(2) 添加相关控件，设计页面如图 2.29 所示。

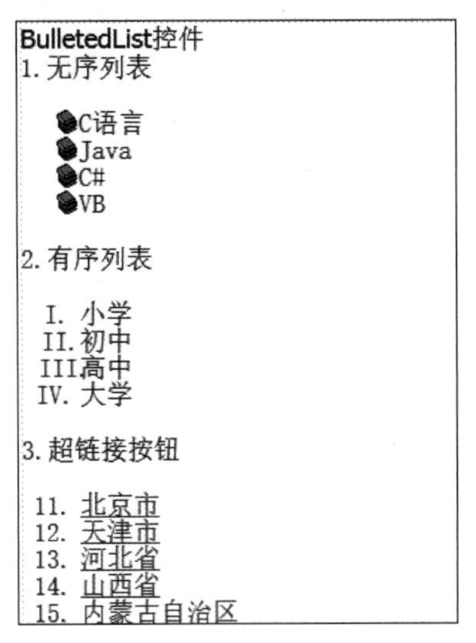

图 2.29　BulletedList 控件应用页面设计

(3) 设置控件属性如下。

```
<form id="form1" runat="server">
    BulletedList 控件<br/>
        1. 无序列表 < asp: BulletedList  ID =" BulletedList1 "  runat =" server "
BulletImageUrl="~/image/book.ico" BulletStyle="CustomImage">
            <asp:ListItem>C 语言</asp:ListItem>
            <asp:ListItem>Java</asp:ListItem>
            <asp:ListItem>C#</asp:ListItem>
            <asp:ListItem>VB</asp:ListItem>
        </asp:BulletedList>
```

```
            2.有序列表<asp:BulletedList ID="BulletedList2" runat="server"
BulletStyle="UpperRoman">
                <asp:ListItem>小学</asp:ListItem>
                <asp:ListItem>初中</asp:ListItem>
                <asp:ListItem>高中</asp:ListItem>
                <asp:ListItem>大学</asp:ListItem>
            </asp:BulletedList>
            3.超链接按钮<asp:BulletedList ID="BulletedList3" runat="server"
BulletStyle="Numbered" DisplayMode="LinkButton" FirstBulletNumber="11"
OnClick="BulletedList3_Click">
                <asp:ListItem>北京市</asp:ListItem>
                <asp:ListItem>天津市</asp:ListItem>
                <asp:ListItem>河北省</asp:ListItem>
                <asp:ListItem>山西省</asp:ListItem>
                <asp:ListItem>内蒙古自治区</asp:ListItem>
            </asp:BulletedList>
        </form>
```

(4) 为 BulletedList3 控件添加如下事件。

```
protected void BulletedList3_Click(object sender, BulletedListEventArgs e)
{
    Response.Write("<script>alert('你选择的是:"+BulletedList3.Items[e.
Index].Text+"');</script>");
}
```

(5) 运行网站，执行效果如图 2.30 所示。

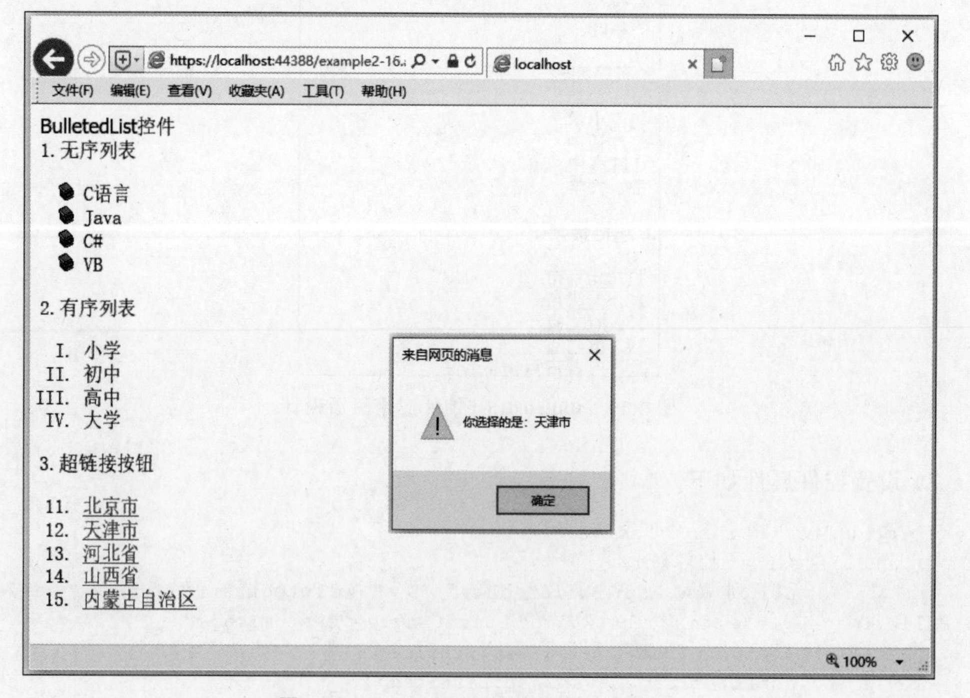

图 2.30 BulletedList 控件应用页面演示

第3章 Web高级控件

本章学习目标

- 熟练掌握 ASP.NET 中视图区域控件的使用方法
- 熟练掌握 ASP.NET 中文件上传控件的使用方法
- 熟练掌握 ASP.NET 中日历控件的使用方法
- 了解 ASP.NET 中广告控件的使用方法
- 了解 ASP.NET 中向导控件的使用方法

本章首先介绍 ASP.NET 服务器部分高级控件应用，对 ASP.NET 中的视图区域控件进行详细讲解，然后通过应用讲解文件上传控件、日历控件和向导控件，最后添加.xml 广告文件，并在广告控件中进行应用。

3.1 简介视图区域控件

视图区域控件是页面分组、布局、外观设置常使用的控件，通过该控件可以将部分控件作为一个单元进行管理，也可以为其他控件提供容器，或者为页面上特定区域设置独特外观。视图区域控件主要包含面板控件(Panel)、占位符控件(PlaceHolder)、视图控件(View)和多视图控件(MultiView)等，本节将对这几种控件进行详细讲解。

3.1.1 面板控件 Panel

扫一扫

Panel 控件又称面板控件，在工具箱中图标为“Panel”，封装在 System.Web.UI.Control.WebControl 命名空间的 Panel 类中。可以在其内部动态地向该控件添加其他子控件和标签，也可以使用该控件将页面拆分为独立排版显示的若干部分。Panel 控件的常用属性如表 3.1 所示。

表 3.1 Panel 控件的常用属性

属 性 名	说 明
BackImageUrl	获取或设置面板控件的背景图像 URL
DefaultButton	获取或设置面板控件的默认按钮
GroupingText	获取或设置面板控件中包含的控件组的标题
Controls	获取面板控件中包含的子控件集合

说明：

(1) 默认按钮属性是指焦点在面板控件内按下 Enter 键时或鼠标“单击”时处理。

(2) Controls 属性中包含 Add()方法，通过该方法可以动态地向面板控件中添加新的子控件。

图 3.1 Panel 控件应用页面布局

【示例 3-1】 在 E 盘的 ASP.NET 项目代码目录中创建 chapter3 子目录，将其作为网站根目录，创建名为 example3-1 的页面，添加面板按钮及其他控件，实现动态添加控件以及隐藏 Panel 功能。

(1) 在页面 example3-1 中添加相应控件，具体如图 3.1 所示。

(2) 设置控件的相关属性，如下列源代码。

```
<form id="form1" runat="server">
      在 Panel 控件中添加元素<br/>
        添加<asp:DropDownList ID="ddlLbl" runat="server">
            <asp:ListItem>0</asp:ListItem>
            <asp:ListItem>1</asp:ListItem>
            <asp:ListItem>2</asp:ListItem>
            <asp:ListItem>3</asp:ListItem>
            <asp:ListItem>4</asp:ListItem>
        </asp:DropDownList>
        个标签控件<br/>
        添加<asp:DropDownList ID="ddlTxt" runat="server">
            <asp:ListItem>0</asp:ListItem>
            <asp:ListItem>1</asp:ListItem>
            <asp:ListItem>2</asp:ListItem>
            <asp:ListItem>3</asp:ListItem>
            <asp:ListItem>4</asp:ListItem>
        </asp:DropDownList>
        个文本框控件<br/>
        <asp:Button ID="btnAdd" runat="server"  Text="动态添加控件" OnClick="
btnAdd_Click"/>
        <asp:CheckBox ID="chkShow" runat="server" AutoPostBack="True" Text="显
示 Panel" OnCheckedChanged="chkShow_CheckedChanged"/>
        <asp:Panel ID="Panel1" runat="server" GroupingText="动态添加的控件">
        </asp:Panel>
    </form>
```

(3) 为控件添加事件代码如下。

```
protected void btnAdd_Click(object sender, EventArgs e)
{
    //设置变量 m 的值为 ddlLbl 控件中选中的数值
    int m=int.Parse(ddlLbl.SelectedItem.Value);
    //循环次数为要添加的 Label 控件的个数
    for (int i=1; i <=m; i++)
    {
        //实例化一个 Label 控件对象 lbl
        Label lbl=new Label();
        //赋值 lbl 对象的 ID 属性为:“lbl”+i
        lbl.ID="lbl"+i;
        //赋值 lbl 对象的 Text 属性为:“标签控件”+i
        lbl.Text="标签控件"+i.ToString()+"<br>";
        //添加 lbl 控件到 Panel 中
        Panel1.Controls.Add(lbl);
    }
    //设置变量 n 的值为 ddlTxt 控件中选中的数值
    int n=int.Parse(ddlTxt.SelectedItem.Value);
    //循环次数为要添加的 TextBox 控件的个数
    for (int i=1; i<=n; i++)
    {
        //实例化一个 TextBox 控件对象 txt
        TextBox txt=new TextBox();
        //赋值 txt 对象的 ID 属性为:“txt”+i
        txt.ID="txt"+i;
        //赋值 txt 对象的 Text 属性为:“文本框控件”+i
        txt.Text="文本框控件"+i.ToString();
        //添加 txt 控件到 Panel 中
        Panel1.Controls.Add(txt);
    }
}
protected void chkShow_CheckedChanged(object sender, EventArgs e)
{
    //如果当前状态是没选中
     if (chkShow.Checked)
     {
        Panel1.Visible=true;             //就把 Panel 选中
        chkShow.Text="显示 Panel";        //CheckBox1 的文本为“显示 Panel”
     }
     else
    {
        Panel1.Visible=false;
        chkShow.Text="隐藏 Panel";
     }
}
```

(4) 网站运行后,根据列表数值,动态向面板控件中加相应数量的标签和文本框控件,单击复选按钮可以显示或隐藏面板控件,部分功能如图 3.2 所示。

3.1.2 占位符控件 PlaceHolder

PlaceHolder 控件又称占位符控件,在工具箱中图标为“⊠ PlaceHolder”,封装在 System.Web.UI.Control.WebControl 命名空间的 PlaceHolder 类中。该控件的功能与面板

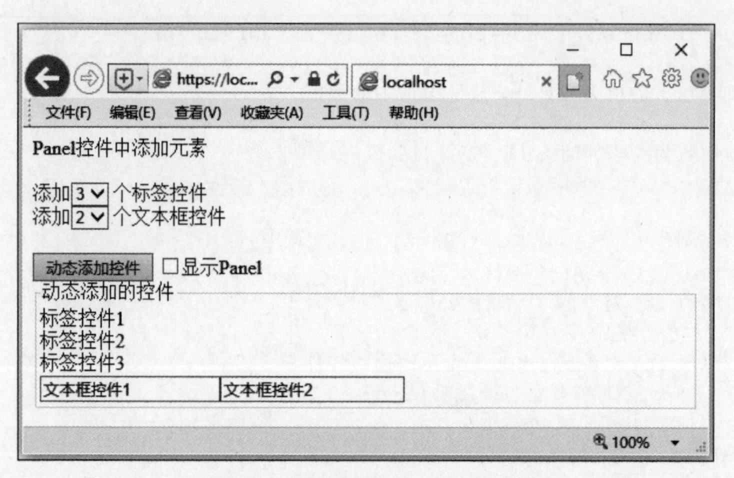

图 3.2 Panel 控件应用页面演示

控件的功能类似，也可以作为页面内的一个容器动态地向其内部添加其他控件，不同的是，PlaceHolder 控件没有基于 HTML 的输出，只是为其他控件标记一个位置。

【示例 3-2】 在 E 盘 ASP.NET 项目代码的 chapter3 目录下创建名为 example3-2 的网页，分别向 Panel 控件和 PlaceHolder 控件内动态添加控件，对比 HTML 源代码的差异。

（1）在页面中添加相应控件，具体如图 3.3 所示。

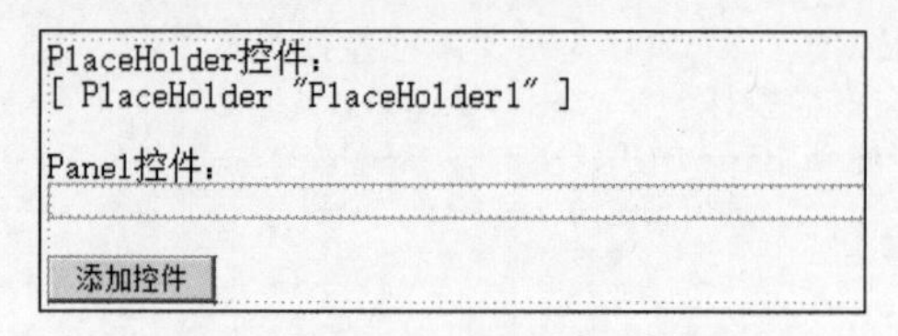

图 3.3 PlaceHolder 控件应用页面布局

（2）设置控件的相关属性，如下列源代码。

```
<form id="form1" runat="server">
    <div>
        PlaceHolder 控件:<br/>
        <asp:PlaceHolder ID="PlaceHolder1" runat="server"></asp:PlaceHolder>
        <br/>
        <br/>
        Panel 控件:<asp:Panel ID="Panel1" runat="server">
        </asp:Panel>
         <br/>
       <asp:Button ID="btnAdd" runat="server" Text="添加控件" OnClick="btnAdd_
Click"/>
    </div>
</form>
```

（3）添加事件代码，如下所示。

```
protected void btnAdd_Click(object sender, EventArgs e)
{
    Button btn1=new Button();
```

```
        btn1.Text="PlaceHolder 内的按纽";
        btn1.ID="btn1";
        PlaceHolder1.Controls.Add(btn1);
        Button btn2=new Button();
        btn2.Text="Panel 内的按纽";
        btn2.ID="btn1";
        Panel1.Controls.Add(btn2);
    }
```

（4）网站运行后单击“添加控件”按钮，如图 3.4 所示。

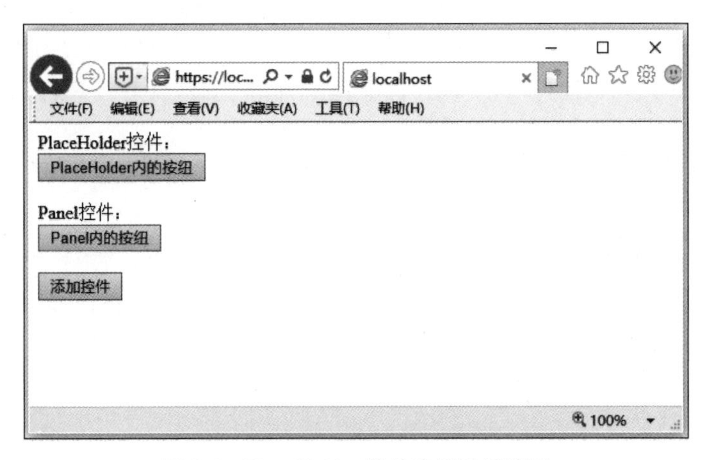

图 3.4　PlaceHolder 控件应用网页演示

（5）对比网页运行后的 HTML 源代码，比较 PlaceHolder 控件和 Panel 控件的客户端脚本，具体如图 3.5 所示。

```
example3-2.aspx
1
2
3 <!DOCTYPE html>
4
5 <html xmlns="http://www.w3.org/1999/xhtml">
6 <head><meta http-equiv="Content-Type" content="text/html; charset=utf-8" /><title>
7
8 </title></head>
9 <body>
10     <form method="post" action="./example3-2.aspx" id="form1">
11 <div class="aspNetHidden">
12 <input type="hidden" name="__VIEWSTATE" id="__VIEWSTATE" value="18AQ+Y5Wnm68Tj/xdPmZpVzcpUrcUTQGiNQptM68KQF2HieNBPFfSvu1X
13 </div>
14
15 <div class="aspNetHidden">
16
17     <input type="hidden" name="__VIEWSTATEGENERATOR" id="__VIEWSTATEGENERATOR" value="71166648" />
18     <input type="hidden" name="__EVENTVALIDATION" id="__EVENTVALIDATION" value="+Nrnow81+/skszsanPBYMS/oqXQ8O4JrYd2c4W6p/
19 </div>
20         <div>
21             PlaceHolder控件: <br />
22         <input type="submit" name="btn1" value="PlaceHolder内的按纽" id="btn1" />
23         <br />
24         <br />
25         Panel控件: <div id="Panel1">
26
27         <input type="submit" name="btn1" value="Panel内的按纽" id="btn1" />
28 </div>
29          <br />
30          <input type="submit" name="btnAdd" value="添加控件" id="btnAdd" />
31     </div>
32     </form>
33 </body>
34 </html>
35
```

图 3.5　PlaceHolder 控件和 Panel 控件的客户端脚本比较

说明：

在网页的 HTML 源文件中，Panel 控件有输出客户端脚本，会产生 DIV 的 HTML 代码，而 PlaceHolder 控件仅在服务器端起分组的作用，不会产生额外的 HTML 代码。在页面中使用控件有进行分组的情况时，如果客户端有对分组进行显示或隐藏，以及改变颜色等操作需求时，应使用 Panel 控件，否则使用 PlaceHolder 控件。

扫一扫

3.1.3 视图控件 View 与多视图控件 MultiView

View 控件又称视图控件，在工具箱中图标为"View"，封装在 System.Web.UI.Control.WebControl 命名空间的 View 类中，用于将页面拆分为独立显示的若干部分。视图必须包含在 MultiView 控件内，可以作为其他控件的容器，没有外观，只有被激活时才会显示。

MultiView 控件又称多视图控件，在工具箱中图标为"MultiView"，封装在 System.Web.UI.Control.WebControl 命名空间的 MultiView 类中。作为使用 View 控件的必选控件，MultiView 控件实际上是作为一个容器使用。在 MultiView 控件中可以包含若干个 View 控件，通过设置 ActiveViewIndex 属性确定在 MultiView 控件内显示的 View 视图，当 ActiveViewIndex=－1 时，所有 View 视图均不显示。MultiView 控件的常用属性如表 3.2 所示。MultiView 控件的常用事件如表 3.3 所示。

表 3.2 MultiView 控件的常用属性

属性名	说明
ActiveViewIndex	获取或设置 MultiView 控件的 ActiveView 控件的索引

表 3.3 MultiView 控件的常用事件

事件名	说明
ActiveViewChanged	当 MultiView 控件的 ActiveView 控件发生变化时触发

说明：

MultiView 控件内各 View 控件的索引不需要设置，按从上至下位置自动赋值，初始索引值为 0。

【示例 3-3】 在 chapter3 网站根目录下创建名为 example3-3 的网页，页面内包含一个 MultiView 控件和 4 个 View 控件以及若干其他控件，练习使用视图的切换，以及按钮的 CommandName 和 CommanArgument 属性。

(1) 在 chapter3 网站根目录上新建"上传文件"文件夹，在"上传文件"文件夹添加卡通数字图片 1.jpg、2.jpg、3.jpg、4.jpg。

(2) 按图 3.6 所示添加相应文本及控件。

(3) 设置控件的相关属性，如源文件所示。

```
<form id="form1" runat="server">
    <div>
      <asp:RadioButtonList ID="RadioButtonList1" runat="server" AutoPostBack=
```

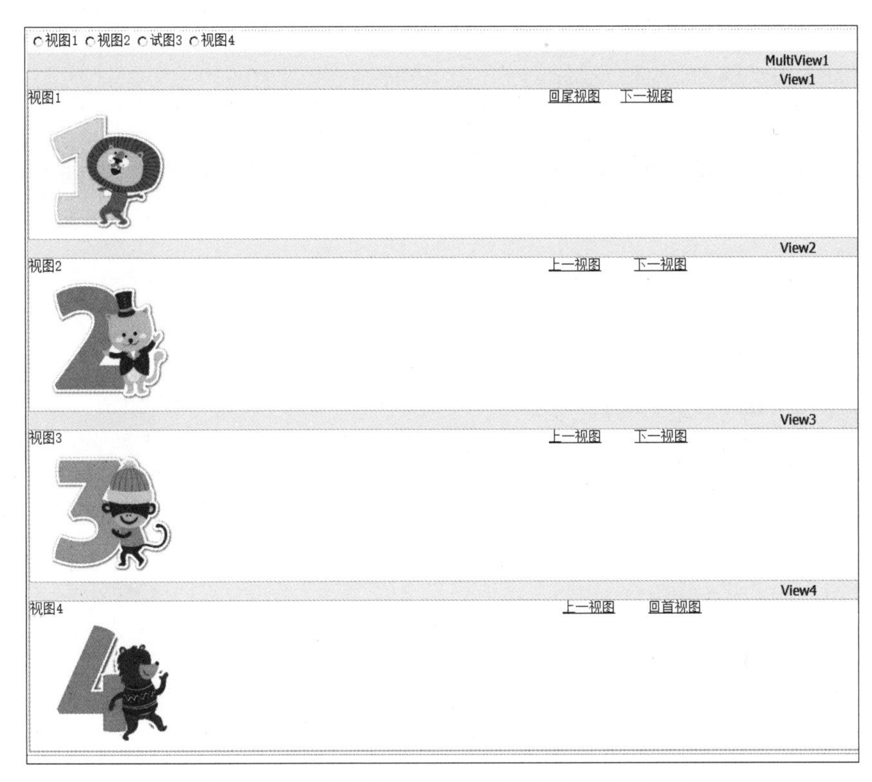

图 3.6 View 控件和 MultiView 控件应用页面布局

```
"True" OnSelectedIndexChanged="RadioButtonList1_SelectedIndexChanged"
RepeatDirection="Horizontal">
            <asp:ListItem>视图 1</asp:ListItem>
            <asp:ListItem>视图 2</asp:ListItem>
            <asp:ListItem>视图 3</asp:ListItem>
            <asp:ListItem>视图 4</asp:ListItem>
        </asp:RadioButtonList>
        <asp:MultiView ID="MultiView1" runat="server" ActiveViewIndex="0"
OnActiveViewChanged="MultiView1_ActiveViewChanged">
            <asp:View ID="View1" runat="server">
                视图 1    
<asp:LinkButton ID="LinkButton7" runat="server" CommandArgument="View4"
CommandName="SwitchViewByID">回尾视图</asp:LinkButton> 
<asp:LinkButton ID="LinkButton1" runat="server" CommandName="NextView">下一视
图</asp:LinkButton><br/>
<asp:Image ID="Image1" runat="server" ImageUrl="~/上传文件/1.jpg" Width="200px"/>
            </asp:View>
            <asp:View ID="View2" runat="server">
            视图 2    
<asp:LinkButton ID="LinkButton4" runat="server" CommandName="PrevView">上一视
图</asp:LinkButton> 
<asp:LinkButton ID="LinkButton2" runat="server" CommandName="NextView">下一视
图</asp:LinkButton><br/>
<asp:Image ID="Image2" runat="server" ImageUrl="~/上传文件/2.jpg" Width="200px"/>
            </asp:View>
```

```
                <asp:View ID="View3" runat="server">
视图 3    
<asp:LinkButton ID="LinkButton9" runat="server" CommandName="PrevView">上一视
图</asp:LinkButton> 
<asp:LinkButton ID="LinkButton10" runat="server" CommandName="NextView">下一
视图</asp:LinkButton>  <br/>
<asp:Image ID="Image3" runat="server" ImageUrl="~/上传文件/3.jpg" Width="
200px"/>
                </asp:View>
                <asp:View ID="View4" runat="server">
视图 4    
<asp:LinkButton ID="LinkButton6" runat="server" CommandName="PrevView">上一视
图</asp:LinkButton> 
<asp:LinkButton ID="LinkButton8" runat="server" CommandArgument="View1"
CommandName="SwitchViewByID">回首视图</asp:LinkButton> <br/>
<asp:Image ID="Image4" runat="server" ImageUrl="~/上传文件/4.jpg" Width="
200px"/>
                </asp:View>
            </asp:MultiView>
        </div>
    </form>
```

(4) 为 RadioButtonList1 控件和 MultiView1 控件的事件添加如下代码。

```
protected void RadioButtonList1_SelectedIndexChanged(object sender, EventArgs e)
{
    MultiView1.ActiveViewIndex=RadioButtonList1.SelectedIndex;
}
protected void MultiView1_ActiveViewChanged(object sender, EventArgs e)
{
    RadioButtonList1.SelectedIndex=MultiView1.ActiveViewIndex;
}
```

(5) 运行网站,可以通过单选按钮列表切换视图,也可以通过每个视图中的链接按钮切换视图,部分运行页面如图 3.7 所示。

图 3.7 View 控件和 MultiView 控件应用网页演示

说明：

通过设置按钮控件的 CommandName 和 CommandArgument 属性可以直接实现视图的切换，CommandName 属性可设置的值如下。

- PrevView：表示切换到上一视图（当前视图索引不可为 0）。
- NextView：表示切换到下一视图（当前视图不可为索引值最大的视图）。
- SwitchViewByID：表示切换到指定 ID 的视图，具体的 ID 属性值需在 CommandArgument 属性中设置，具体见视图 4“回首视图”按钮中属性值的设置。

3.2 文件上传控件 FileUpload

FileUpload 控件又称文件上传控件，在工具箱中图标为“FileUpload”，封装在 System.Web.UI.Control.WebControl 命名空间的 FileUpload 类中，在网页中显示为一个文本框控件和一个“浏览”按钮的组合，用于在网页中通过单击按钮完成待上传文件的选择。FileUpload 控件的常用属性如表 3.4 所示。

扫一扫

表 3.4 FileUpload 控件的常用属性

属性名	说明
FileName	获取客户端使用 FileUpload 控件上传文件的名称
HasFile	获取一个值，该值指示 FileUpload 控件是否包含文件
FileBytes	获取 FileUpload 控件中所包含文件的字节数
PostedFile	获取 FileUpload 控件上传文件的 HttpPostedFile 对象

说明：

(1) HasFile 属性用于验证 FileUpload 控件是否包含文件，若返回值为 True，则表示控件包含文件；若返回值为 False，则表示控件不包含文件。

(2) PostedFile 属性包含若干二级属性，常用的属性如下。

- ContentLength 属性：获取上传文件的大小（以字节为单位）。
- ContentType 属性：获取上传文件的 MIME 类型。
- FileName 属性：获取上传文件的完整路径及文件名。

FileUpload 控件的常用方法如表 3.5 所示。

表 3.5 FileUpload 控件的常用方法

方法名	说明
SaveAs()	将使用 FileUpload 控件上传文件的内容保存到服务器的指定路径

说明：

使用 FileUpload 控件时，可以通过单击“浏览”按钮，在“选择文件”对话框中选择文件，选择文件后使用 HasFile 属性确定是否选择了文件，调用文件上传控件的 SaveAs()方法执行文件的上传。

【示例 3-4】 在 chapter3 网站根目录下创建名为 example3-4 的网页，页面内包含一个文件上传控件和若干其他控件，练习使用文件上传控件的相关属性及事件。

（1）在 example3-4 网页添加相应文件上传控件和其他控件，具体如图 3.8 所示。

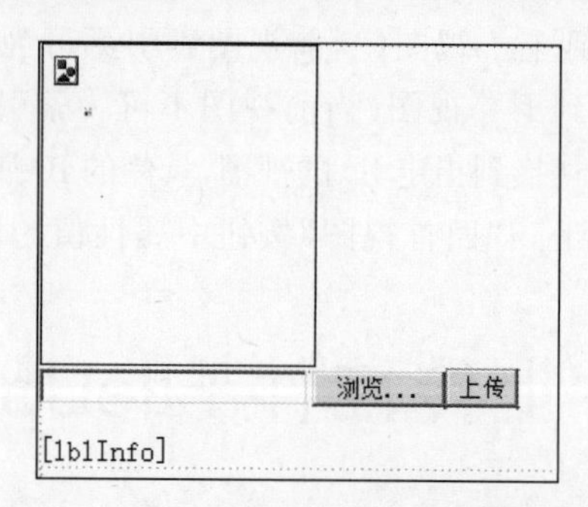

图 3.8 FileUpload 控件应用页面布局

（2）设置控件相关属性，如源文件所示。

```
<form id="form1" runat="server">
    <asp:Image ID="Image1" runat="server" BorderColor="#663300" BorderWidth="
1px" Width="150px"/>
    <br/>
    <asp:FileUpload ID="FileUpload1" runat="server"/>
    <asp:Button ID="btnOK" runat="server" OnClick="btnOK_Click" Text="上传"/>
    <asp:Label ID="lblInfo" runat="server" BorderStyle="None"></asp:Label
</form>
```

（3）为按钮控件的事件添加如下代码。

```
protected void btnOK_Click(object sender, EventArgs e)
{
    //判断文件上传控件中是否包含文件
    if (FileUpload1.HasFile)
    {
        //赋值 fileName 变量为文件上传控件中选择的文件名
        string fileName=FileUpload1.FileName;
        //赋值 lastDotIndex 变量为文件名中最后一个'.'的索引值
        int lastDotIndex=fileName.LastIndexOf('.');
        //赋值 lastName 变量为文件的后缀名
        string lastName=fileName.Remove(0, lastDotIndex);
        //判断选择的文件后缀名是否为 .jpg|.bmp|.jpeg|.gif 中的一种
        if (lastName==".jpg" || lastName==".bmp" || lastName==".jpeg" || lastName==".gif")
        {
            //判断选择的文件是否大于 1MB
            if (FileUpload1.PostedFile.ContentLength<1*1024*1024)
            {
                //"年年年年月月日日时时分分秒秒毫毫毫"格式的时间字符串
                string strTime=DateTime.Now.ToString("yyyyMMddhhmmssfff");
                //赋值 newFileName 变量为原文件名加时间字符串构成的新文件名
                string newFileName=fileName.Insert(lastDotIndex, "("+strTime+")");
                //赋值 path 变量为"上传文件"对应的绝对路径
                string path=Server.MapPath("上传文件");
                //将选中文件保存到指定路径
```

```
                FileUpload1.SaveAs(path+@""+newFileName);
                //赋值 imgUrl 变量为新上传的文件名
                string imgUrl=@"上传文件"+newFileName;
                //为 Image1 控件的 ImageUrl 属性赋值
                Image1.ImageUrl=imgUrl;
                //为 lblInfo 控件的 Text 属性赋值,显示上传文件的文件名、大小、类型属性
                lblInfo.Text="选择的文件名:"+FileUpload1.FileName+"<br>文件大
小:"+ FileUpload1.PostedFile.ContentLength+"b<br>文件类型:"+FileUpload1.
PostedFile.ContentType;
            }
            else
            {
                lblInfo.Text="图像大小必须小于 1MB";
            }
        }
        else
        {
            lblInfo.Text=" 请选择正确图像文件(.jpg|.bmp|.jpeg|.gif)";
        }
    }
    else
    {
        lblInfo.Text="请选择图像";
    }
}
```

(4) 运行网站,如没选择文件时单击“上传”按钮,会提示“请选择图像”;若选择了非图像文件,则提示“请选择正确图像文件(.jpg|.bmp|.jpeg|.gif)”;若选择的文件大于 1MB,提示“图像大小必须小于 1MB”;若选择小于 1MB 的.gif 图像,则可实现图片文件以新文件名上传,如“cat.jpg”,在 image1 控件中显示图像,在 lblInfo 控件中显示相关属性。部分执行效果如图 3.9 所示。

图 3.9 FileUpload 控件应用页面演示

说明:

(1) 在上传文件名中添加时间字符串可有效防止文件同名覆盖问题。

(2) 采用以文件后缀名进行类型验证的方法,不会对文件内容进行检测,对于被恶意修改后缀名的文件无法识别判断。

(3) 路径中使用到的“@”符可以使后续字符串内容不做转义字符处理,如@”//”等价于”////”字符串。

(4) Server.MapPath("上传文件");Server 内置对象中的 MapPath()方法可将相对路径转化为实际物理路径,本书第 5 章将会详细讲解。

(5) 系统默认上传文件大小为 4096KB,如果要上传超过此大小的文件,会出现错误界面。但可在

web.config 文件中进行配置，设置文件大小。在解决方案中找到配置文件 Web.config，添加如下代码。

```
configuration>
    <system.web>
    <httpRuntime maxRequestLength="4096" executionTimeout="120"/>
    </system.web>
</configuration>
```

其中，maxRequestLength 属性限制文件上传的大小，以 KB 为单位，默认值为 4096KB，而最大上限为 2097151KB，大约是 2GB；executionTimeout 属性限制文件上传的时间，以秒(s)为单位，默认值为 90s，可设置该属性值以延长或缩短上传时间。

3.3 日历控件 Calender

Calender 控件又称日历控件，在工具箱中图标为“Calendar”，封装在 System.Web.UI.Control.WebControl 命名空间的 Calender 类中。用户可通过该日历控件导航到某年的某一天，且具有自动套用格式，方便用户选择使用。Calendar 控件是一个比较复杂的控件，具有大量的编程和格式选项，其常用属性如表 3.6 所示。Calender 控件的常用事件如表 3.7 所示。

表 3.6　Calender 控件的常用属性

属 性 名	说　明
Caption	获取或设置与日历控件关联的标题
SelectedDate	获取或设置选定的在控件中突出显示的特定日期
ShowNextPrevMonth	获取或设置是否允许用户进行月份导航
SelectionMode	获取或设置日历控件的选择模式
VisibleDate	获取或设置日期用于确定日历中显示的月份

说明：

(1) 通过 SelectionMode 属性可以设置日历控件的选择模式，具体有如下枚举值。

- None：表示禁用所有日期选择。
- Day：表示用户可以选择一天，每个日期都将包含带有日期编号的链接。
- DayWeek：表示用户可以选择一天或者一周，除日期编号的链接外，日历的左侧会额外添加一个带有周选择链接的列。
- DayWeekMonth：表示用户可以选择一天、一周或一月，除日期编号的链接外，日历的左侧会额外添加一个带有周和月选择链接的列。

(2) 如果 SelectionMode 属性值为 DayWeek 或者 DayWeekMonth，则 SelectedDate 属性值为选中日期中的第一天。

表 3.7 Calender 控件的常用事件

事件名	说明
SelectionChanged	当更改选择的日期时发生

【示例 3-5】 在 E 盘 ASP.NET 项目代码的 chapter3 目录下，创建名为 example3-5 的网页，实现 Calender 控件选择模式的设置以及日期的选择。

(1) 在页面中添加相应控件，具体如图 3.10 所示。

图 3.10 Calender 控件应用页面布局

(2) 单击 Calender 控件右上角的"▣"图标，选择【自动套用格式...】中的"专业型 2"，其他控件的属性设置如下列源文件。

```
<form id="form1" runat="server">
    <div>
        <asp:Calendar ID="Calendar1" runat="server" BackColor="White"
BorderColor="Black" Font-Names="Verdana" Font-Size="9pt" ForeColor="Black"
Height="250px" OnSelectionChanged="Calendar1_SelectionChanged" Width="
330px" BorderStyle="Solid" CellSpacing="1" NextPrevFormat="ShortMonth">
            <SelectedDayStyle BackColor="#333399" ForeColor="White"/>
            <TodayDayStyle BackColor="#999999" ForeColor="White"/>
            <OtherMonthDayStyle ForeColor="#999999"/>
            <DayStyle BackColor="#CCCCCC"/>
            <NextPrevStyle Font-Size="8pt" ForeColor="White" Font-Bold="
True"/>
            <DayHeaderStyle Font-Bold="True" Height="8pt" Font-Size="8pt"
ForeColor="#333333"/>
            <TitleStyle BackColor="#333399" Font-Bold="True" Font-Size="12pt"
ForeColor="White" BorderStyle="Solid" Height="12pt"/>
        </asp:Calendar>
        <br/>
        日历选择模式:<asp:DropDownList ID="ddlMode" runat="server" AutoPostBack=
"True" OnSelectedIndexChanged="ddlMode_SelectedIndexChanged">
            <asp:ListItem Value="None">不选择</asp:ListItem>
            <asp:ListItem Value="Day">天</asp:ListItem>
```

```
            <asp:ListItem Value="DayWeek">天、周</asp:ListItem>
            <asp:ListItem Value="DayWeekMonth">天、周、月</asp:ListItem>
        </asp:DropDownList>
        <br/>
        <br/>
        当前选中的日期是:<asp:Label ID="lblDate" runat="server"></asp:Label>
        <br/>
        选中的天是:<asp:Label ID="lblDay" runat="server"></asp:Label>
        <br/>
        选中的月是:<asp:Label ID="lblMonth" runat="server"></asp:Label>
        <br/>
        选中的年是:<asp:Label ID="lblYear" runat="server"></asp:Label>
        <br/>
    </div>
    </form>
```

(3) 为各控件添加事件代码如下。

```
protected void ddlMode_SelectedIndexChanged(object sender, EventArgs e)
{
    switch (ddlMode.SelectedValue)
    {
        case "None":
            Calendar1.SelectionMode=CalendarSelectionMode.None;
            break;
        case "DayWeekMonth":
            Calendar1.SelectionMode=CalendarSelectionMode.DayWeekMonth;
            break;
        case "DayWeek":
            Calendar1.SelectionMode=CalendarSelectionMode.DayWeek;
            break;
        case "Day":
            Calendar1.SelectionMode=CalendarSelectionMode.Day;
            break;
    }
}
protected void Calendar1_SelectionChanged(object sender, EventArgs e)
{
    lblDate.Text =Calendar1.SelectedDate.ToShortDateString();
    lblDay.Text=Calendar1.SelectedDate.Day.ToString()+"日";
    lblMonth.Text=Calendar1.SelectedDate.Month.ToString()+"月";
    lblYear.Text=Calendar1.SelectedDate.Year.ToString()+"年";
}
```

(4) 网站运行效果如图 3.11 所示。

图 3.11 Calender 控件应用页面演示

3.4 广告控件 AdRotator

AdRotator 控件又称广告控件，在工具箱中图标为“AdRotator”，封装在 System.Web.UI.Control.WebControl 命名空间的 AdRotator 类中，用于在页面上显示广告图像序列，并能通过单击广告图片实现链接跳转。AdRotator 控件的常用属性如表 3.8 所示。

表 3.8 AdRotator 控件的常用属性

属　性	说　明
AdvertisementFile	获取或设置包含 ad 信息的 XML 文件的路径
Target	获取或设置打开 URL 的位置

AdRotator 控件使用 XML 文件存储 ad 信息。XML 文件使用<Advertisements>开始和结束。在<Advertisements>标签内部，有若干个定义每条 ad 的<Ad>标签。<Ad>标签中预定义的元素如表 3.9 所示。

表 3.9 <Ad>标签中预定义的元素

标 签 名	说　明
<ImageUrl>	表示图像文件的路径
<NavigateUrl>	表示该 ad 时所链接的 URL

续表

标签名	说明
<AlternateText>	表示图像的替换文本
<Keyword>	表示 ad 的类别
<Impressions>	表示广告显示权重,所有的 Impressions 之和不能超过 2048,000,000-1

【示例 3-6】 在 chapter3 网站根目录下创建名为 example3-6 的网页,页面内包含一个广告控件,练习使用广告控件和广告文件。

(1) 在 chapter3 网站根目录添加"广告文件"文件夹,其中包含方正电脑.gif、广东证券.gif、雀巢咖啡.gif、招商银行.gif 4 张图片。

(2) 在 chapter3 网站根目录添加 XML 文件,并命名为 AdRotatorFile.xml,编写文件内容如下。

```
<?xml version="1.0" encoding="utf-8" >
<Advertisements>
  <Ad>
    <ImageUrl>广告图片/方正电脑.gif</ImageUrl>
    <NavigateUrl>http://www.founderpc.cn</NavigateUrl>
    <AlternateText>欢迎访问方正电脑</AlternateText>
    <Keyword>方正</Keyword>
    <Impressions>20</Impressions>
  </Ad>
  <Ad>
    <ImageUrl>广告图片/广东证券.gif</ImageUrl>
    <NavigateUrl>http://www.gzs.com.cn</NavigateUrl>
    <AlternateText>欢迎访问广东证券</AlternateText>
    <Keyword>广东证券</Keyword>
    <Impressions>80</Impressions>
  </Ad>
  <Ad>
    <ImageUrl>广告图片/雀巢咖啡.gif</ImageUrl>
    <NavigateUrl>http://www.nescafe.com.cn</NavigateUrl>
    <AlternateText>欢迎访问雀巢咖啡</AlternateText>
    <Keyword>雀巢咖啡</Keyword>
    <Impressions>30</Impressions>
  </Ad>
  <Ad>
    <ImageUrl>广告图片/招商银行.gif</ImageUrl>
    <NavigateUrl>http://www.cmbchina.com</NavigateUrl>
    <AlternateText>欢迎访问招商银行</AlternateText>
    <Keyword>招商银行</Keyword>
    <Impressions>70</Impressions>
  </Ad>
</Advertisements>
```

(3) 在 example3-6 的网页添加 AdRotator 控件,并设置相关属性,如下列源文件。

```
<form id="form1" runat="server">
    <asp:AdRotator ID="AdRotator1" runat="server" AdvertisementFile="~/
AdRotatorFile.xml"/>
</form>
```

(4) 运行页面,将按照权重随机显示一个广告。广告在每次页面载入时更改,每一个广告出现的频率通过 <Impressions>权重</Impressions>属性确定。如本示例中"广东证券"的权重为 80,所有广告的权重和为 200,则每刷新页面 200 次,"广东证券"的广告大约将以 80 次的频数进行显示。部分页面如图 3.12 所示。

图 3.12　AdRotator 控件应用页面演示

3.5 向导控件 Wizard

Wizard 控件又称向导控件,在工具箱中图标为"Wizard",封装在 System.Web.UI.Control.WebControl 命名空间的 Wizard 类中。Wizard 控件提供了一种简单的机制,能够轻松地生成步骤、添加新步骤或重新安排步骤顺序,无须编写代码即可生成线性或非线性的导航,也可实现自定义控件的用户导航。Wizard 控件的常用属性如表 3.10 所示。Wizard 控件的常用事件如表 3.11 所示。

表 3.10　Wizard 控件的常用属性

属 性 名	说　明
ActiveStep	获取 WizardSteps 集合中当前显示给用户的步骤
ActiveStepIndex	获取或设置当前向用户显示的步骤
CancelButtonImageUrl	获取或设置为"取消"按钮显示的图像的 URL
CancelButtonText	获取或设置为"取消"按钮显示的文本标题
CancelButtonType	获取或设置呈现为"取消"按钮的按钮类型
CompleteStep	获取对最终用户账户创建步骤的引用
CreateUserButtonText	获取或设置在"创建用户"按钮上显示的文本标题
CreateUserButtonType	获取或设置呈现为"创建用户"按钮的按钮类型

续表

属 性 名	说 明
Email	获取或设置用户输入的电子邮件地址
FinishDestinationPageUrl	获取或设置当用户单击“完成”按钮时将重定向到的 URL
HeaderText	获取或设置为在控件上的标题区域显示的文本标题
Question	获取或设置用户输入的密码恢复确认问题
QuestionRequiredErrorMessage	获取或设置由于用户未输入密码确认问题而显示的错误信息
StepNextButtonImageUrl	获取或设置为“下一步”按钮显示的图像的 URL
StepNextButtonStyle	获取一个对 Style 对象的引用,该对象定义“下一步”按钮的设置
StepNextButtonText	获取或设置为“下一步”按钮显示的文本标题
StepNextButtonType	获取或设置呈现为“下一步”按钮的按钮类型
StepPreviousButtonImageUrl	获取或设置为“上一步”按钮显示的图像的 URL
StepPreviousButtonStyle	获取一个对 Style 对象的引用,该对象定义“上一步”按钮的设置
StepPreviousButtonText	获取或设置为“上一步”按钮显示的文本标题
StepPreviousButtonType	获取或设置呈现为“上一步”按钮的按钮类型
StepStyle	获取一个对 Style 对象的引用
Style	获取在 Web 服务器控件外部标记上呈现为样式属性的文本属性的集合
TabIndex	获取或设置 Web 服务器控件的选项卡索引
ToolTip	获取或设置当鼠标指针悬停在 Web 服务器控件上时显示的文本
Visible	获取或设置一个值,该值指示服务器控件是否作为 UI 呈现在页面上
WizardSteps	获取一个包含该控件定义的所有 WizardStepsBase 对象集合

说明:

(1) Wizard 控件可用于下列工作。

- 收集多个步骤中的相关信息。
- 将大型输入页面分割成较小的逻辑步骤。
- 允许线性或非线性地导航各个步骤。

(2) Wizard 控件可分成 4 大区域,如图 3.13 所示。

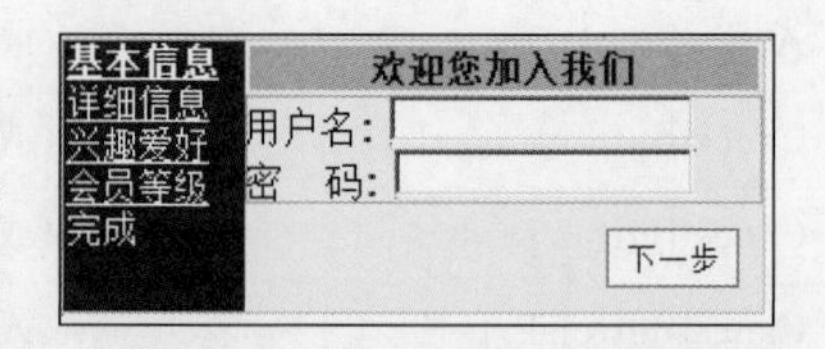

图 3.13 Wizard 控件区域划分

- 向导步骤(WizardStep)区域:Wizard 控件使用多个步骤描绘用户输入的不同部分。每个步骤的内容添加在标记<asp:WizardStep>中,所有的<asp:WizardStep>又都包含在<WizardSteps>标记中。实际应用时,每次只能显示一个<asp:WizardStep>定义的内容。
- 标题(Header)区域:用于在步骤顶部提供一致信息,此项是可选元素。
- 侧栏(Sidebar)区域:此项也是可选元素,通常显示在向导左边,包含所有步骤的列表，并提供在各个步骤间的跳转。

- 导航(Navigation)按钮区域：有 Wizard 内置导航功能，它会根据步骤类型(StepType)设置值的不同，而呈现不同的导航按钮。

表 3.11 Wizard 控件的常用事件

事 件 名	说 明
ActiveStepChanged	当切换控件中显示的步骤时发生
CancelButtonClick	当单击“取消”按钮时发生
FinishButtonClick	当单击“完成”按钮时发生
NextButtonClick	当单击“下一步”按钮时发生
PreviousButtonClick	当单击“上一步”按钮时发生
SiderBarButtonClick	当单击侧边栏中的按钮时发生

说明：

NextButtonClick、PreviousButtonClick 和 SiderBarButtonClick 事件如不自主添加，控件中也可自动实现“步骤”间的跳转。

【示例 3-7】 在 chapter3 网站根目录下创建名为 example3-7 的网页，页面内包含一个 Wizard 控件和若干其他基本控件，添加 Wizard 的 Steps 属性并进行应用。

(1) 在 chapter3 网站根目录下添加 Wizard 控件，右击“>”，在弹出的快捷菜单中选择“添加/移除 Wizard Steps…”，或者单击“属性”窗口 Wizard Steps 属性后面的“…”图标，出现“WizardStep 集合编辑器”窗口，如图 3.14 所示。

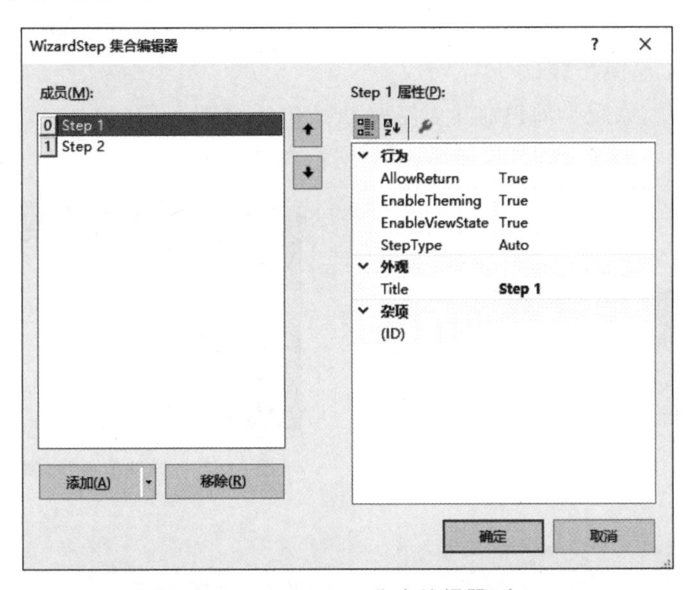

图 3.14 “WizardStep 集合编辑器”窗口

(2) 添加 5 个 WizardStep 默认类型 Step，即 5 个步骤，最后一个“完成”Step 设置其 StepType 属性为 Complete(见图 3.15)，其余均为 Auto 属性。

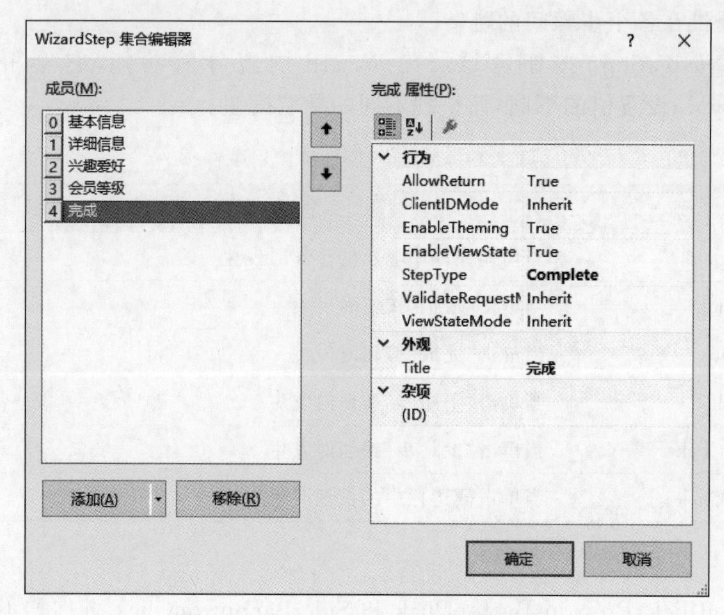

图 3.15　WizardStep 集合编辑器设置

说明：

每个 WizardStep 步骤都有一个 StepType 属性，作用是决定每个步骤中的导航按钮如何显示。StepType 的枚举值如下。

- Start：开始步骤。
- Step：阶段步骤。
- Finish：完成步骤。
- Complete：结束步骤。
- Auto：自动，系统自动识别其为何种 StepType 类型。

(3) 按照每个 Step 的界面，添加相应的控件，具体如图 3.16～图 3.20 所示。

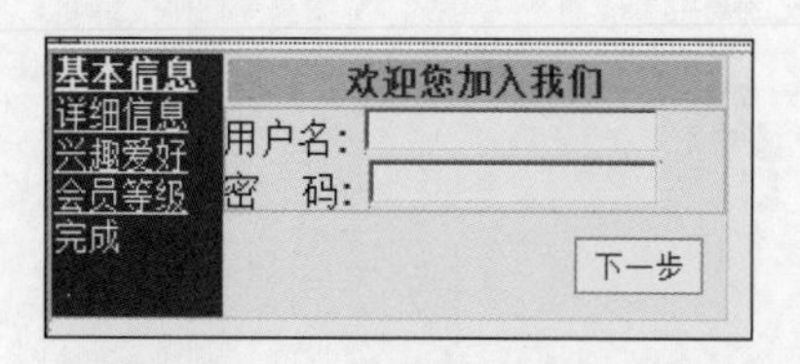

图 3.16　“基本信息”步骤布局

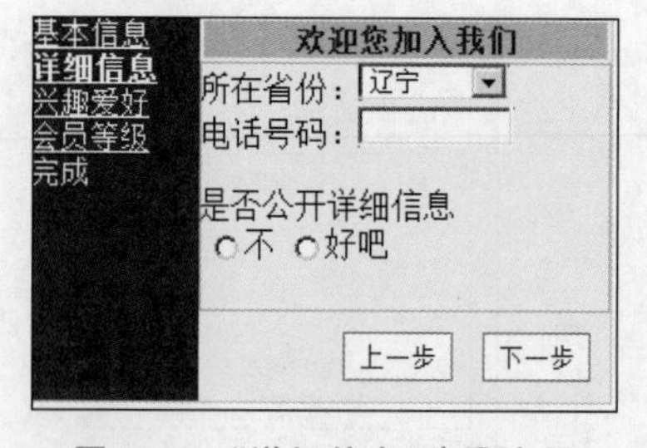

图 3.17　“详细信息”步骤布局

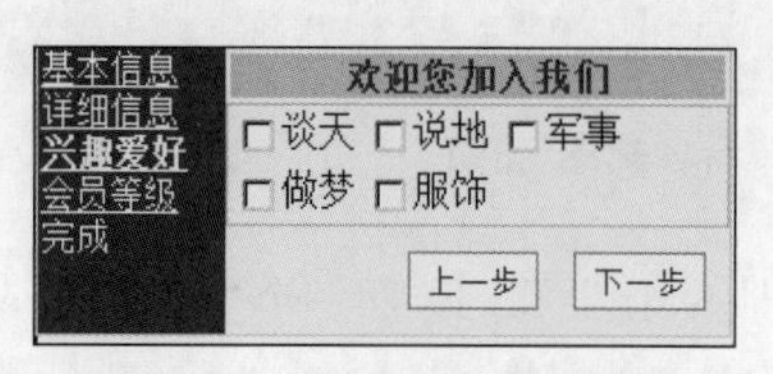

图 3.18　“兴趣爱好”步骤布局

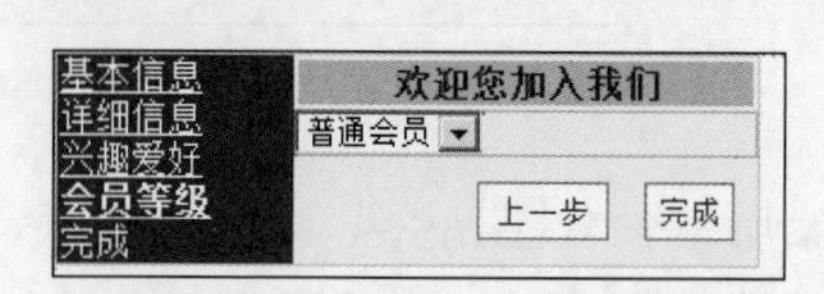

图 3.19　“会员等级”步骤布局

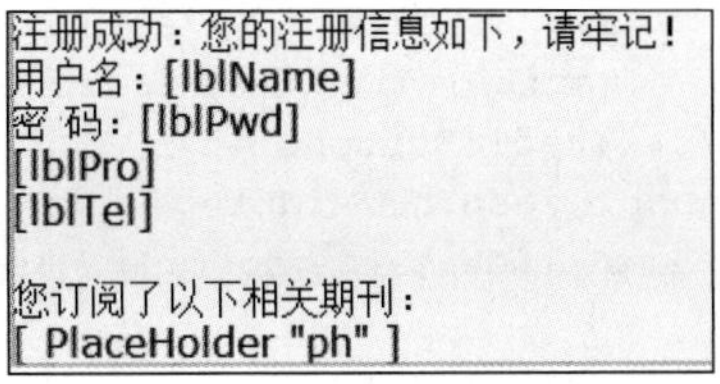

注册成功：您的注册信息如下，请牢记！
用户名：[lblName]
密 码：[lblPwd]
[lblPro]
[lblTel]

您订阅了以下相关期刊：
[PlaceHolder "ph"]

图 3.20 “完成”步骤布局

(4) 设置控件的属性，如源代码所示。

```
<form id="form1" runat="server">
    <div>
        <p>
            <asp:Wizard ID="Wizard2" runat="server" ActiveStepIndex="0"
BackColor="#FFFBD6" BorderColor="#FFDFAD"
                BorderWidth="1px" Font-Names="Verdana" Font-Size="Medium"
OnActiveStepChanged="Wizard2_ActiveStepChanged" Width="295px"
OnFinishButtonClick="Wizard2_FinishButtonClick">
                <SideBarStyle BackColor="#990000" Font-Size="0.9em"
VerticalAlign="Top"/>
                <NavigationButtonStyle BackColor="White" BorderColor="#CC9966"
BorderStyle="Solid"
                BorderWidth="1px" Font-Names="Verdana" Font-Size="0.8em"
ForeColor="#990000"/>
                <WizardSteps>
                    <asp:WizardStep runat="server" Title="基本信息">
                        用户名:
                        <asp:TextBox ID="txtName" runat="server" Width="119px"
TextMode="Password"></asp:TextBox>
                        <br/>
                        密   码:
                        <asp:TextBox ID="txtPwd" runat="server" Width="117px">
</asp:TextBox>
                    </asp:WizardStep>
                    <asp:WizardStep runat="server" Title="详细信息">
                         所在省份:<asp:DropDownList ID="ddlPro" runat="server"
Width="75px">
                            <asp:ListItem>辽宁</asp:ListItem>
                            <asp:ListItem>吉林</asp:ListItem>
                            <asp:ListItem>山东</asp:ListItem>
                            <asp:ListItem>河北</asp:ListItem>
                            <asp:ListItem>上海</asp:ListItem>
                            <asp:ListItem>北京</asp:ListItem>
                        </asp:DropDownList>
                        <br/>
                        电话号码:<asp:TextBox ID="txtTel" runat="server" Width=
"70px"></asp:TextBox>
                        <br/>
                        <br/>
                        是否公开详细信息<br/>
                          <asp:RadioButtonList ID="rbtnlCheck" runat="server"
RepeatDirection="Horizontal">
```

```
                        <asp:ListItem>不</asp:ListItem>
                        <asp:ListItem>好吧</asp:ListItem>
                    </asp:RadioButtonList>
                     </asp:WizardStep>
                <asp:TemplatedWizardStep runat="server" Title="兴趣爱好" ID=
"mb">
                    <ContentTemplate>
                        <asp:CheckBoxList ID="chklLikes" runat="server"
RepeatColumns="3" RepeatDirection="Horizontal">
                            <asp:ListItem>谈天</asp:ListItem>
                            <asp:ListItem>说地</asp:ListItem>
                            <asp:ListItem>军事</asp:ListItem>
                            <asp:ListItem>做梦</asp:ListItem>
                            <asp:ListItem>服饰</asp:ListItem>
                        </asp:CheckBoxList>
                    </ContentTemplate>
                </asp:TemplatedWizardStep>
                <asp:WizardStep runat="server" Title="会员等级">
                    <asp:DropDownList ID="ddlGrade" runat="server">
                        <asp:ListItem>普通会员</asp:ListItem>
                        <asp:ListItem>高级会员</asp:ListItem>
                        <asp:ListItem>VIP 会员</asp:ListItem>
                    </asp:DropDownList>
                </asp:WizardStep>
                <asp:WizardStep runat="server" StepType="Complete" Title=
"完成">
                    注册成功：您的注册信息如下，请牢记!<br/>
                    用户名:<asp:Label ID="lblName" runat="server" Text="">
</asp:Label>
                    <br/>
                    密 码:<asp:Label ID="lblPwd" runat="server" Text=""></
asp:Label>
                    <br/>
                    <asp:Label ID="lblPro" runat="server" Text=""></asp:Label>
                    <br/>
                    <asp:Label ID="lblTel" runat="server" Text=""></asp:Label>
                    <br/>
                    您订阅了以下相关期刊:<br/>
                        <asp:PlaceHolder ID="ph" runat="server"></asp:
PlaceHolder>
                </asp:WizardStep>
            </WizardSteps>
            <SideBarButtonStyle ForeColor="White"/>
                <HeaderStyle BackColor="#FFCC66" BorderColor="#FFFBD6"
BorderStyle="Solid" BorderWidth="2px"
                 Font-Bold="True" Font-Size="0.9em" ForeColor="#333333"
HorizontalAlign="Center"/>
            <HeaderTemplate>
                欢迎您加入我们
            </HeaderTemplate>
        </asp:Wizard>
          </p>
    </form>
```

（5）添加相关事件，对应代码如下。

```
protected void Wizard2_ActiveStepChanged(object sender, EventArgs e)
{
    lblName.Text=txtName.Text;
    lblPwd.Text=txtPwd.Text;
    if (rbtnlCheck.SelectedValue=="不")
    {
      lblPro.Text="用户隐藏了详细信息";
       lblTel.Visible=false;
    }
    else
    {
       lblPro.Text="省 份:"+ddlPro.Text;
       lblTel.Text="电 话:"+txtTel.Text;
    }
    CheckBoxList chkl=(CheckBoxList)mb.ContentTemplateContainer.FindControl
("chklLikes");
    for (int i=0; i<chkl.Items.Count; i++)
    {
       if (chkl.Items[i].Selected)
       {
          Label lbl=new Label();
          lbl.ID="lbl"+i;
          lbl.Text=chkl.Items[i].Text+"<br>";
          ph.Controls.Add(lbl);
       }
    }
}
protected void Wizard2_FinishButtonClick(object sender, WizardNavigationEventArgs e)
{
    string name="尊敬的:"+txtName.Text;
    Response.Write("<script>alert('"+name+"感谢您注册')</script>");
}
```

（6）运行网页，完成会员注册的信息收集功能，执行效果如图 3.21～图 3.26 所示。

图 3.21 “基本信息”步骤演示

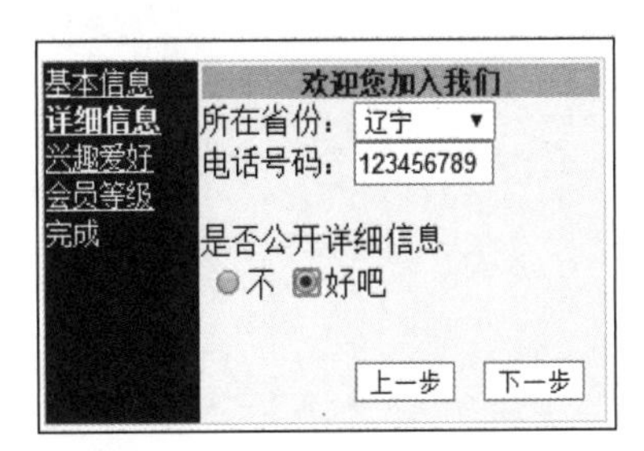

图 3.22 “详细信息”步骤演示

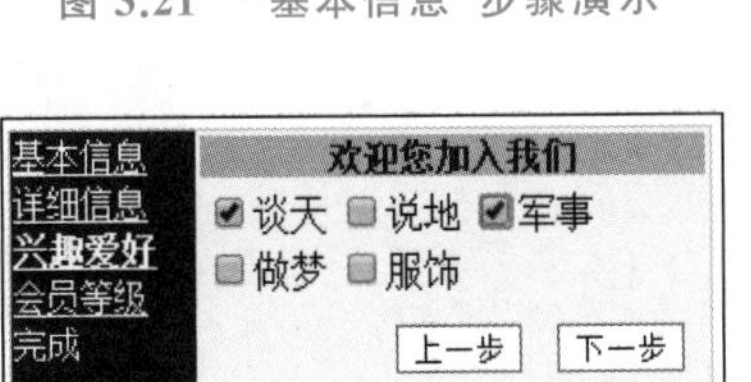

图 3.23 “兴趣爱好”步骤演示

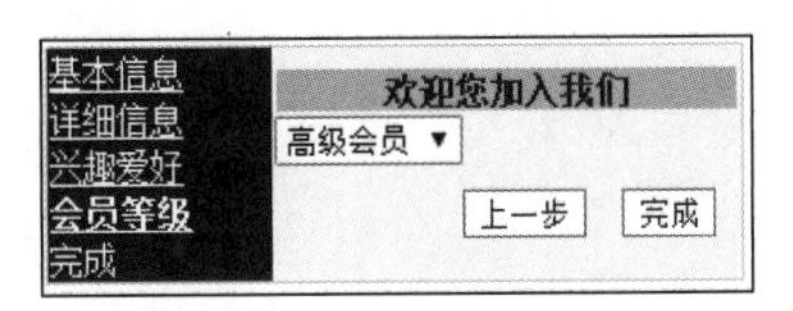

图 3.24 “会员等级”步骤演示

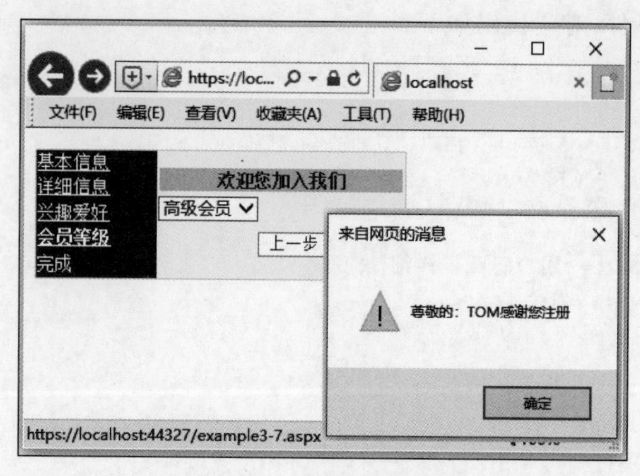

图 3.25 “完成”步骤演示

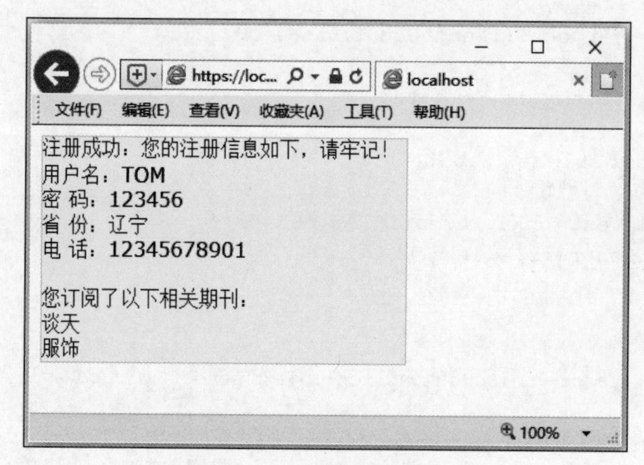

图 3.26 Wizard 控件页面演示

第4章 服务器验证控件

本章学习目标

- 了解客户端验证和服务器端验证
- 熟练掌握 ASP.NET 验证控件的使用方法
- 了解正则表达式
- 熟练掌握验证组的使用方法

本章首先介绍服务器端验证和客户端验证，然后介绍验证控件的基本使用，并详细介绍服务器验证控件的使用，对正则表达式进行了基本讲解，最后讲解验证组的概念及具体应用。

4.1 验证控件介绍

在 Web 应用中，经常利用表单获取用户的信息。用户登录、注册时，如果没有输入信息，页面上会出现“不能为空”提示信息；如果输入的内容不符合标准，则会出现“格式不正确”提示信息。对信息的有效性、合法性等进行验证的控件称为广义上的验证控件，狭义上的验证控件是指在 ASP.NET 网站开发中工具箱中自带的“验证”控件。

4.1.1 服务器端验证与客户端验证

扫一扫

在 ASP.NET Web 开发中，既可以在服务器端进行验证，也可以在客户端进行验证。客户端的验证基本使用脚本实现，如 JavaScript 或 VBScript 等，验证过程中不提交到远程服务器，可为用户提供快速反馈，给人一种运行桌面应用程序的感觉，用户能够及时察觉所填写数据的合法性。客户端验证具有如下的优点和缺点。

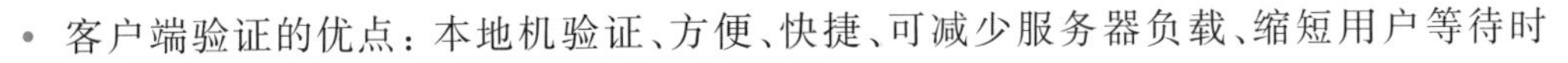

- 客户端验证的优点：本地机验证、方便、快捷、可减少服务器负载、缩短用户等待时

间、用户体验好。

- 客户端验证的缺点：只适用于字符、数字等特定规则的一些简单验证，无法适应复杂的规则，兼容性不好。

服务器端验证基本使用高级语言实现，如 C＃或 VB 等，不论在客户端输入的是什么内容，都将从客户端送往服务器进行数据有效性的验证。服务器端验证具有如下的优缺点。

- 服务器端验证的优点：远程服务器验证，适用于复杂的规则，安全性高，兼容性强。
- 服务器端验证的缺点：服务器负载重，用户等待时间长，用户体验一般。

两种验证各有利弊，应当以适用、高效、快速为标准按具体项目需求选定，如果只进行数字验证、字符检测、简单规则条件、为空判断等验证，通常选择客户端验证，对于涉及数据库、复杂算法、复杂规则条件等的验证，则采用服务器端验证。

【示例 4-1】 在 E 盘 ASP.NET 项目代码目录中创建 chapter4 子目录，将其作为网站根目录，创建一个名为 example4-1 的网页，页面内包含两组注册模块，分别使用 C＃代码与 JavaScript 脚本进行非空验证。

(1) 在页面中添加相应控件，具体如图 4.1 所示。

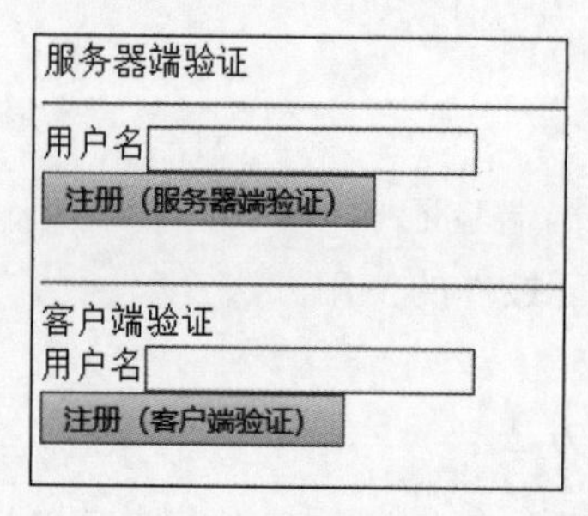

图 4.1 服务器端验证与客户端验证比较页面布局

(2) 设置控件的属性如下。

```
<html>
<head runat="server">
<meta http-equiv="Content-Type" content="text/html; charset=utf-8"/>
    <title>服务器端验证与客户端验证比较</title>
    <script>
        function checknull()
        {
            var txtId=document.getElementById("txtId2").value;
            if (!txtId)
            {
                document.getElementById("lbInfo2").innerHTML="用户名不可为空";
                return false;
            }
            else {
                document.getElementById("lbInfo2").innerHTML="";
                return true;
            }
        }
    </script>
```

```
</head>
<body>
    <form id="form1" runat="server">
            <div>
            服务器端验证<br/>
            <hr/>
            用户名<asp:TextBox ID="txtId" runat="server"></asp:TextBox>
            <asp:Label ID="lbInfo" runat="server" ForeColor="Red"></asp:Label>
            <br/>
             <asp:Button ID="btnReg" runat="server" Text="注册(服务器端验证)"
OnClick="btnReg_Click"/>
            <hr/>
            客户端验证<br/>
            用户名<asp:TextBox ID="txtId2" runat="server"></asp:TextBox>
            <asp:Label ID="lbInfo2" runat="server" ForeColor="Red"></asp:Label>
            <br/>
            <asp:Button ID="btnReg2" runat="server" Text="注册(客户端验证)"
OnClientClick="return checknull();"/>
        </div>
    </form>
</body>
</html>
```

(3) 为控件事件添加 C# 代码如下。

```
protected void btnReg_Click(object sender, EventArgs e)
{
     if (txtId.Text=="")
     {
          lbInfo.Text="用户名不可为空";
     }
     else
     {
         lbInfo.Text="";
     }
}
```

(4) 运行网页,单击"注册(服务器端验证)"按钮时,触发执行按钮的服务器单击事件,回发服务器后执行 C# 代码,实现服务器端验证。单击"注册(客户端验证)"按钮时,触发执行按钮的客户端单击事件,浏览器解析执行 JavaScript 函数,实现客户端验证。未通过非空验证后的效果如图 4.2 所示。

4.1.2 验证控件的使用方法

ASP.NET 为 Web 开发提供了 6 种验证控件,用于对网页上的信息输入进行验证。验证控件的验证方式基本都是客户端验证,在特殊需要时也可以进行服务器端验证。

页面在"回发"至服务器之前,验证控件将检查其所验证的控件内容是否有效,并对应赋值其 IsValid 属性值,如果内容没有通过验证,则属性值为 False。验证控件所在页面的 Page 对象也有 IsValid 属性,只有页面中所有验证控件的 IsValid 属性均为 True,该 Page 对象的 IsValid 属性才为 True,此时页面才可以"回发"至服务器。验证控件可以自动显示

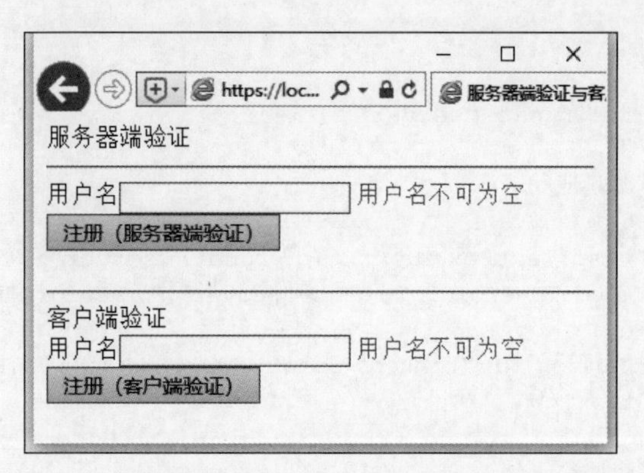

图 4.2　服务器端验证与客户端验证比较页面演示

验证的错误提示，通过验证控件的 ErrorMessage 属性值设置需要显示的提示文本，如果验证未通过，则会显示该错误信息。同时，可以使用 ValidationSummary 控件自动汇总页面中所有未通过验证的 ErrorMessage 属性。

4.1.3　验证控件的公共属性

验证控件都继承自 BaseValidator 类，具有几个相同的公共属性，理解这些属性是使用验证控件的关键。验证控件的公共属性如表 4.1 所示。

表 4.1　验证控件的公共属性

属　性　名	说　　明
ControlToValidate	获取或设置验证控件所验证的控件 ID
Display	获取或设置验证控件的显示形式
EnableClientScript	指示是否启用客户端验证
Enabled	获取或设置是否启用验证控件
ErrorMessage	获取或设置验证失败时显示的错误提示信息
ForeColor	获取或设置错误提示文本的颜色
IsValid	获取或设置 ControlToValidate 属性所指定的输入控件是否通过验证
SetFocusOnError	获取或设置验证失败时焦点是否被设置在 ControlToValidate 属性所指定的控件内
ValidationGroup	获取或设置指定此验证控件所属的验证组名称。

说明：

(1) Display 属性具有如下 3 个枚举值。

- None：指定只在 ValidationSummary 控件中显示错误信息，错误信息不会显示在验证控件中。
- Static：静态显示，指定不希望网页的布局在验证程序控件显示错误信息时改变。显示页面时将在页面上为错误信息分配空间，同一输入控件的多个验证程序将在页面上占据不同的位置。

- Dynamic：流式显示，指定希望在验证失败时在网页上动态放置错误信息。页面上不预先为验证内容分配的空间，页面将动态更改以显示错误信息。多个验证程序可以在页面上共享同一个物理位置。

(2) ErrorMessage 属性不会将特殊字符转换为 HTML 实体。如小于号字符'<'不转换为 <，可以将 HTML 元素(如 <img> 元素等)嵌入该属性的值中。

4.2 常见的验证控件

ASP.NET 中内置的验证控件分别为必填验证控件(RequiredFieldValidator)、范围验证控件(RangeValidator)、比较验证控件(CompareValidator)、正则表达式验证控件(RegularExpressionValidator)、自定义验证控件(CustomValidator)和验证汇总控件(ValidationSummary)，如图 4.3 所示。

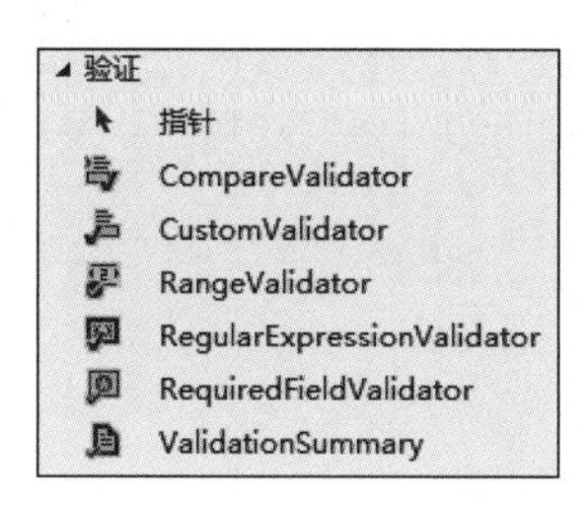

图 4.3 "工具箱"中的所有验证控件

4.2.1 必填验证控件 RequiredFieldValidator

扫一扫

RequiredFieldValidator 控件又称必填验证控件或者非空验证控件，在工具箱中图标为"RequiredFieldValidator"，封装在 System. Web. UI. Control. WebControl 命名空间的 RequiredFieldValidator 类中。RequiredFieldValidator 控件提供了一种验证文本框非空的机制，可确保网页中的某些重要数据必须填写。RequiredFieldValidator 控件除具有公共属性外，特殊属性如表 4.2 所示。

表 4.2 RequiredFieldValidator 控件的特殊属性

属 性 名	说 明
InitialValue	获取或设置输入的初始值

说明：

InitialValue 属性可以起到提示的作用，也具有排除某一特定值的作用，只有当 InitialValue 属性值在页面提交服务器前发生改变才可以通过验证。

【示例 4-2】 在 E 盘 ASP.NET 项目代码目录中创建 chapter4 子目录，将其作为网站根目录，创建一个名为 example4-2 的网页，页面内包含两个必填验证控件，对文本框进行非空验证。

(1) 在页面中添加相应控件，具体如图 4.4 所示。

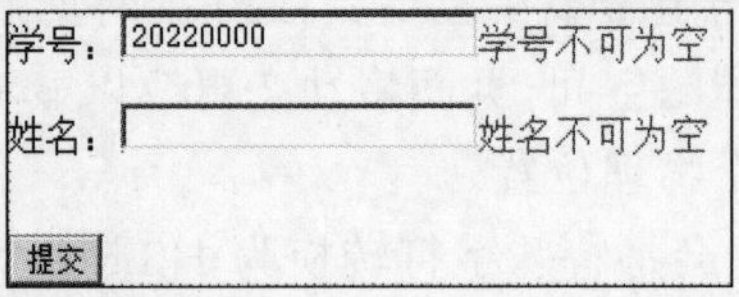

学号：20220000 学号不可为空

姓名： 姓名不可为空

提交

图 4.4 RequiredFieldValidator 控件应用页面布局

(2) 设置相关控件的属性，如下列源代码所示。

```
<form id="form1" runat="server">
        学号:<asp:TextBox ID="txtNum" runat="server">20220000</asp:TextBox>
          <asp:RequiredFieldValidator ID="RequiredFieldValidator1" runat="
server" ControlToValidate="txtNum" ErrorMessage="学号不可为空" ForeColor="Red"
InitialValue="20220000" SetFocusOnError="True"></asp:RequiredFieldValidator>
        <br/>
        姓名:<asp:TextBox ID="txtName" runat="server"></asp:TextBox>
         <asp:RequiredFieldValidator ID="RequiredFieldValidator2" runat="server"
ControlToValidate="txtName" ErrorMessage="姓名不可为空" ForeColor="Red"
SetFocusOnError="True"></asp:RequiredFieldValidator>
        <br/>
     <asp:Button ID="btnSubmit" runat="server" OnClick="btnSubmit_Click" Text="提
交"/>
    </form>
```

(3) 为控件事件添加 C# 代码如下。

```
protected void btnSubmit_Click(object sender, EventArgs e)
{
    Response.Write("<script>alert('学号:"+txtNum.Text+",姓名:"+txtName.Text+"');
</script>");
}
```

(4) 运行页面，当某一文本框未填写，对应必填验证控件会显示错误信息，并自动设置焦点到该文本框，如果学号“20220000”未修改，也不能通过验证，通过验证后的效果如图 4.5 所示。

说明：

如果网页运行中出现如图 4.6 所示的错误信息，是因为在 VS 2019 开发所使用的 Framework 4.7 版本，很多控件默认应用了隐式的验证方式(UnobtrusiveValidationMode)，但并未对其进行赋值，必须手动对其进行设置。

可以对网站的 Web.Config 文件进行手动设置，填写如下标签。

```
<appSettings>
    <add key="ValidationSettings:UnobtrusiveValidationMode" value="None"/>
</appSettings>
```

如果只做验证控件的小测试，也可以简单地将标签中的 targetFramework 值都设置为 4.0。

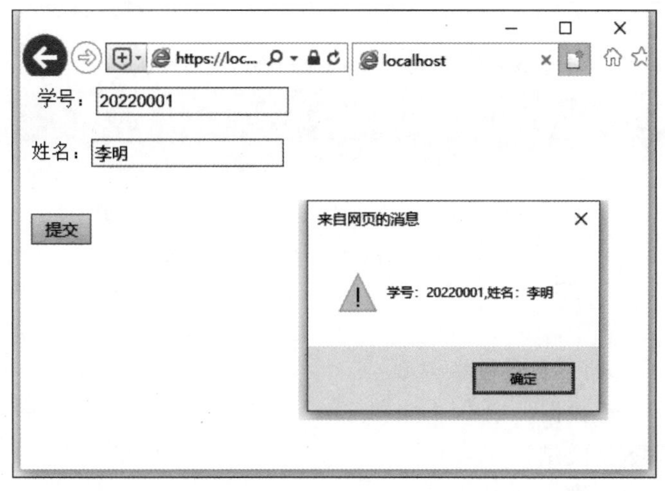

图 4.5　RequiredFieldValidator 控件应用页面演示

图 4.6　RequiredFieldValidator 控件运行错误信息

```
<system.web>
    <compilation debug="false" targetFramework="4.0"/>
    <httpRuntime targetFramework="4.0"/>
</system.web>
```

4.2.2　范围验证控件 RangeValidator

RangeValidator 控件又称范围验证控件，在工具箱中图标为"RangeValidator"，封装在 System. Web. UI. Control. WebControl 命名空间的 RangeValidator 类中。RangeValidator 控件提供指定数据类型输入范围的验证功能，可确保网页中某些重要数据

的上限值和下限值。RangeValidator 控件除具有公共属性外，特殊属性如表 4.3 所示。

表 4.3 RangeValidator 控件的特殊属性

属 性 名	说 明
MaximumValue	获取或设置所验证控件允许输入的最大值
MinimumValue	获取或设置所验证控件允许输入的最小值
Type	获取或设置所验证控件用于比较值的类型

说明：

Type 属性值具有如下几种枚举类型。

- Currency：货币数据类型。
- Date：时间数据类型。
- Double：双精度浮点数据类型。
- Integer：整型数据类型。
- String：字符串数据类型。

【示例 4-3】 在 E 盘 ASP.NET 项目代码的 chapter4 目录下，创建一个名为 example4-3 的网页，使用 RangeValidator 控件进行范围验证。

(1) 在页面中添加相应控件，具体如图 4.7 所示。

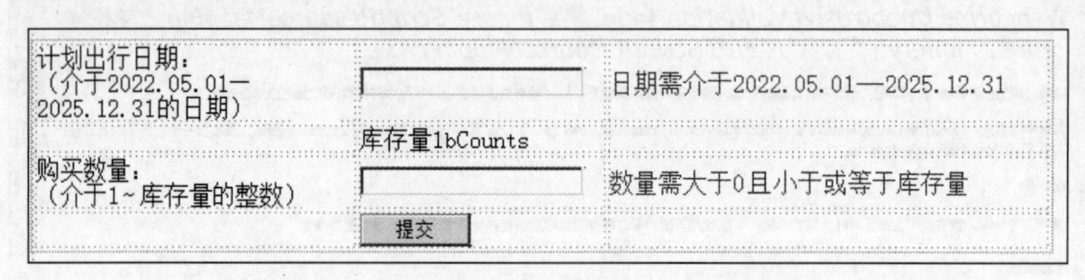

图 4.7 RangeValidator 控件应用页面布局

(2) 按如下源文件设置控件的相关属性值。

```
<form id="form1" runat="server">
        <table>
             <tr>
                 <td>计划出行日期:<br/>
                     (介于 2022.05.01—2025.12.31 的日期)</td>
                 <td>
                     <asp:TextBox ID="txtDate" runat="server"></asp:TextBox>
                 </td>
                 <td >
                     <asp:RangeValidator ID="RangeValidator1" runat="server"
ControlToValidate=" txtDate" ErrorMessage="日期需介于 2022.05.01—2025.12.31"
ForeColor=" Red" MaximumValue=" 2025. 12. 31" MinimumValue=" 2022. 05. 01"
SetFocusOnError="True" Type="Date"></asp:RangeValidator>
                 </td>
             </tr>
             <tr>
                 <td> </td>
```

```
                    <td>库存量<asp:Label ID="lbCount" runat="server" Text="
lbCounts"></asp:Label>
                    </td>
                    <td> </td>
                </tr>
                <tr>
                    <td>购买数量:<br/>
                        (介于 1~库存量的整数)</td>
                    <td>
                        <asp:TextBox ID="txtCounts" runat="server"></asp:TextBox>
                    </td>
                    <td>
                        <asp:RangeValidator ID="RangeValidator2" runat="server"
ControlToValidate="txtCounts" ErrorMessage="数量需大于 0 且小于或等于库存量"
ForeColor="Red" MinimumValue="1" SetFocusOnError="True" Type="Integer"></
asp:RangeValidator>
                    </td>
                </tr>
                <tr>
                    <td> </td>
                    <td>
        <asp:Button ID="btnSubmit" runat="server" Text="提交" OnClick="
btnSubmit_Click" Width="73px"/>
                    </td>
                    <td>
                    </td>
                </tr>
            </table>
    </form>
```

(3) 为控件添加事件,并编辑代码如下。

```
protected void Page_Load(object sender, EventArgs e)
{
    Random random=new Random();
    lbCount.Text=random.Next(1, 1000).ToString();
    RangeValidator2.MaximumValue=lbCount.Text;
}
protected void btnSubmit_Click(object sender, EventArgs e)
{
    Response.Write("<script>alert('日期:"+txtDate.Text+",数量:"+txtCounts.
Text+"');</script>");
}
```

(4) 运行页面,日期范围必须在 2022 年 5 月 1 日—2025 年 12 月 31 日,购买数量的数值范围必须为 1～22,无法通过验证的效果如图 4.8 所示。

说明:

(1) 如果输入值无法转换为指定的数据类型,验证也会失败。

(2) 如果 MaximumValue 属性值小于 MinimumValue 属性值,则程序会直接报错。

(3) String 类型的范围比较规则为:首先比较字符串首字母所对应的 ASCII 码,如果

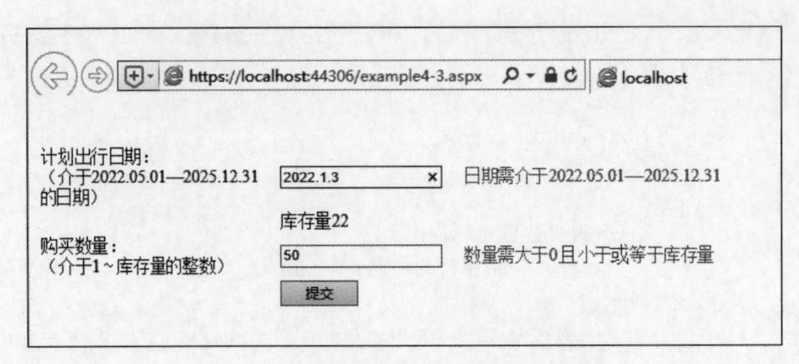

图 4.8 RangeValidator 控件应用页面演示

相同，则依次向后比较，直到出现不同字符，之后比较它们的大小关系。

（4）库存量通常取自数据库，此处以 1～1000 的随机数字进行模拟测试。

扫一扫

4.2.3 比较验证控件 CompareValidator

CompareValidator 控件又称比较验证控件，在工具箱中图标为"CompareValidator"，封装在 System. Web. UI. Control. WebControl 命名空间的 CompareValidator 类中。CompareValidator 控件可实现网页中的某些重要数据与固定值或其他控件值的某种特定比较关系。CompareValidator 控件除具有公共属性外，特殊属性如表 4.4 所示。

表 4.4 CompareValidator 控件的特殊属性

属　性　名	说　　明
ControlToCompare	获取或设置与所验证的输入控件进行比较的另一个输入控件
ValueToCompare	获取或设置与所验证的输入控件进行比较的常数值
Type	获取或设置所验证控件用于比较值的类型
Operator	获取或设置要执行的操作

说明：

Operator 属性具有如下几个枚举值。

- Equal：等于。
- GreaterThan：大于。
- GreaterThanEqual：大于或等于。
- LessThan：小于。
- LessThanEqual：小于或等于。
- NotEqual：不等于。
- DataTypeCheck：数据类型检测，当选取该枚举值时，ControlToCompare 和 ValueToCompare 属性值可不设置，将根据 Type 属性值实现类型检查。

【示例 4-4】 在 E 盘 ASP.NET 项目代码的 chapter4 目录下，创建一个名为 example4-4 的网页，使用 CompareValidator 控件进行范围验证及类型检查。

（1）在页面中添加相应控件，具体如图 4.9 所示。

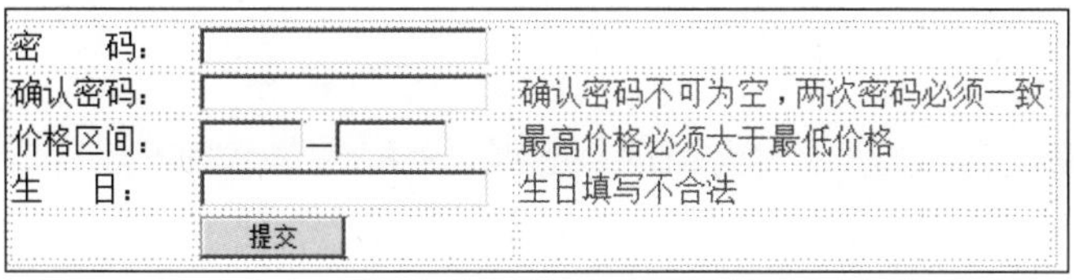

密　　码:	[ ]	
确认密码:	[ ]	确认密码不可为空，两次密码必须一致
价格区间:	[ ] —[ ]	最高价格必须大于最低价格
生　　日:	[ ]	生日填写不合法
	提交	

图 4.9　CompareValidator 控件应用页面布局

（2）按如下源文件设置控件的相关属性值。

```
<form id="form1" runat="server">
        <div>
        <br/>
              <table>
                  <tr>
                      <td>密   码:</td>
                       <td><asp:TextBox ID="txtPwd" runat="server" TextMode=
"Password"></asp:TextBox>
                      </td>
                      <td> </td>
                  </tr>
                  <tr>
                      <td>确认密码:</td>
                      <td><asp:TextBox ID="txtRePwd" runat="server" TextMode=
"Password"></asp:TextBox>
                      </td>
                      <td>
          < asp: RequiredFieldValidator ID =" RequiredFieldValidator1" runat =
"server" ErrorMessage="确认密码不可为空，" ForeColor="Red" ControlToValidate=
"txtRePwd" Display="Dynamic"></asp:RequiredFieldValidator>
            < asp: CompareValidator ID =" CompareValidator3" runat =" server"
ErrorMessage="两次密码必须一致" ForeColor="Red" ControlToCompare="txtPwd"
ControlToValidate="txtRePwd" Display="Dynamic"></asp:CompareValidator>
                      </td>
                  </tr>
                  <tr>
                      <td>价格区间:</td>
                      <td><asp:TextBox ID="txtMinPrice" runat="server" Width="
45px"></asp:TextBox>
                          ~<asp:TextBox ID="txtMaxPrice" runat="server" Width="
50px"></asp:TextBox>
                      </td>
                      <td>
            < asp: CompareValidator ID =" CompareValidator2" runat =" server"
ControlToCompare="txtMinPrice" ControlToValidate="txtMaxPrice" ErrorMessage="最高
价格必须大于最低价格" ForeColor="Red" Operator="GreaterThan" SetFocusOnError=
"True" Type="Integer"></asp:CompareValidator>
                      </td>
                  </tr>
                  <tr>
                      <td>生   日:</td>
```

```
                    <td><asp:TextBox ID="txtBirthDay" runat="server"></asp:
TextBox>
                    </td>
                    <td>
          < asp: CompareValidator  ID =" CompareValidator4 "  runat =" server "
ErrorMessage="生日填写不合法" ForeColor="Red" ControlToValidate="txtBirthDay"
Operator="DataTypeCheck" Type="Date"></asp:CompareValidator>
                    </td>
                </tr>
                <tr>
                    <td> </td>
                    <td>
        < asp: Button ID =" btnSubmit " runat =" server " Text =" 提 交 " OnClick ="
btnSubmit_Click" Width="74px"/>
                    </td>
                    <td> </td>
                </tr>
            </table>

        </div>
    </form>
```

(3) 为控件添加事件,并编辑代码如下。

```
protected void btnSubmit_Click(object sender, EventArgs e)
{
    Response.Write("<script>alert('通过验证');</script>");
}
```

(4) 运行页面,验证确认密码与密码的一致性,最高价格必须大于最低价格,生日输入必须是日期类型,未通过验证时的效果如图 4.10 所示。

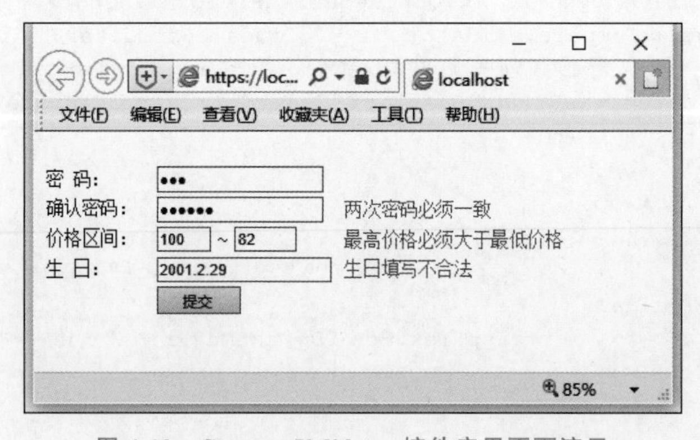

图 4.10 CompareValidator 控件应用页面演示

扫一扫

4.2.4 正则表达式验证控件 RegularExpressionValidator

RegularExpressionValidator 控件又称正则表达式验证控件,在工具箱中图标为"RegularExpressionValidator",封装在 System.Web.UI.Control.WebControl 命名空间的 RegularExpressValidator 类中。RegularExpressionValidator 控件可使用系统提供的正

则表达式完成某些特定验证，也可以使用自定义正则表达式实现各种特定的验证。该控件除具有公共属性外，最重要的属性是 ValidationExpression 属性，该属性由正则表达式字符组成。正则表达式字符及说明如表 4.5 所示。常用的正则表达式字符如表 4.6 所示。

表 4.5　正则表达式字符及说明

正则表达式字符	说　明
\	将下一个字符标记为一个特殊字符、一个原义字符、一个向后引用或一个八进制转义符
^	匹配输入字符串的开始位置
$	匹配输入字符串的结束位置
*	匹配前面的子表达式 0 次或多次
+	匹配前面的子表达式一次或多次
{n}	n 是一个非负整数，匹配次数为 n 次
{n,}	n 是一个非负整数，至少匹配 n 次
{n,m}	m 和 n 均为非负整数，其中 n≤m，最少匹配 n 次且最多匹配 m 次
?	匹配前面表达式 0 次或 1 次
x|y	匹配 x 或 y
[xyz]	字符集合，匹配所包含的任意一个字符
[^xyz]	负值字符集合，匹配未包含的任意字符
[a-z]	字符范围，匹配指定范围内的任意字符
[^a-z]	负值字符范围，匹配不在指定范围内的任意字符
\d	匹配一个数字字符，等价于[0-9]
\D	匹配一个非数字字符，等价于[^0-9]
\f	匹配一个换页符，等价于\x0c 和\cL
\n	匹配一个换行符，等价于\x0a 和\cJ
\r	匹配一个回车符，等价于\x0d 和\cM

表 4.6　常用的正则表达式字符

正则表达式字符	说　明
^[A-Z]+$	匹配由 26 个大写英文字母组成的字符串
^[a-z]+$	匹配由 26 个小写英文字母组成的字符串
^[A-Za-z]+$	匹配由 26 个英文字母组成的字符串
^[A-Za-z0-9]+$	匹配英文和数字组成的字符串
^[\u4e00-\u9fa5]{0,}$	匹配由汉字组成的字符串
^[\u4e00-\u9fa5A-Za-z0-9_]+$	匹配由中文、英文、数字(包括下画线)组成的字符串
^([1-9][0-9]*){1,3}$	匹配非零的正整数

续表

<table>
<tr><th>正则表达式字符</th><th>说　明</th></tr>
<tr><td>^(\-)? \d+(\.\d{1,2})? $</td><td>匹配带 1～2 位小数的正数或负数</td></tr>
<tr><td>^\d{4}-\d{1,2}-\d{1,2}</td><td>匹配 yyyy-mm-dd 格式的日期</td></tr>
<tr><td>^(0? [1-9]|1[0-2]) $</td><td>匹配一年的 12 个月</td></tr>
<tr><td>^((0? [1-9])|((1|2)[0-9])|30|31) $</td><td>匹配一个月的 31 天</td></tr>
<tr><td>^[a-zA-Z][a-zA-Z0-9_]{4,15} $</td><td>匹配以字母开头，由字母、数字、下画线组成，长度为 5～16 的账号</td></tr>
<tr><td>^[a-zA-Z]\w{5,17} $</td><td>匹配以字母开头，由字母、数字和下画线组成，长度为 6～18 的密码</td></tr>
<tr><td>^(? =.*\d)(? =.*[a-z])(? =.*[A-Z]).{8,10} $</td><td>匹配必须包含大小写字母和数字，无特殊字符，长度为 8～10 的强密码</td></tr>
<tr><td>^http://([\w-]+\.)+[\w-]+(/[\w-./?%&=]*)? $</td><td>匹配网址 URL 字符串</td></tr>
<tr><td>[1-9][0-9]{4,8}</td><td>匹配从 10000 开始至多 9 位数的腾讯 QQ 号</td></tr>
<tr><td>\d+\.\d+\.\d+\.\d+</td><td>匹配 IP 地址字符串</td></tr>
</table>

【示例 4-5】 在 E 盘 ASP.NET 项目代码的 chapter4 目录下，创建一个名为 example4-5 的网页，使用 RegularExpressionValidator 控件进行正则验证。

（1）在页面中添加相应控件，具体如图 4.11 所示。

（2）设置 RegularExpressionValidator 控件的 ValidationExpression 属性，如图 4.12 所示。单击“...”图标，在“正则表达式编辑器”内选择系统预定的正则表达式或者设置自定义的 Custom 表达式，各控件的相关属性设置如下。

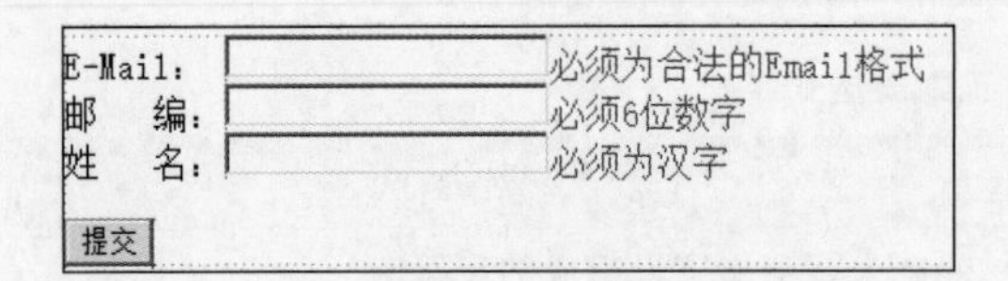

图 4.11　RegularExpressionValidator 控件应用页面布局

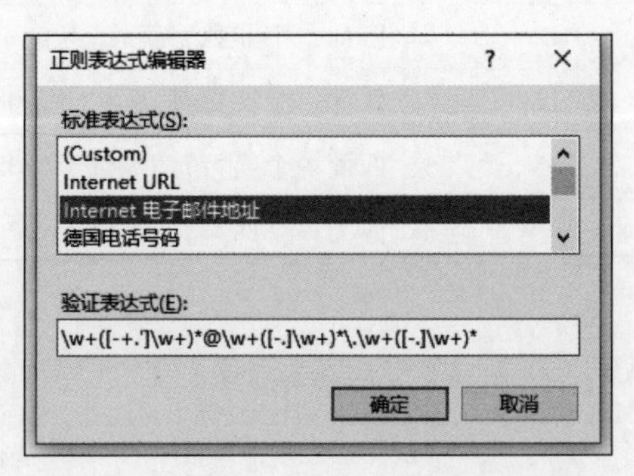

图 4.12　正则表达式编辑器

```
<form id="form1" runat="server">
        E-Mail:<asp:TextBox ID="txtEmail" runat="server"></asp:TextBox>
        <asp:RegularExpressionValidator ID="RegularExpressionValidator1" runat="
server" ErrorMessage="必须为合法的 Email 格式" ControlToValidate="txtEmail"
ForeColor="Red" ValidationExpression="\w+([-+.']\w+)*@\w+([-.]\w+)*\.\w+([-.]\
w+)*"></asp:RegularExpressionValidator>
```

```
        <br/>
        邮   编:<asp:TextBox ID="txtNode" runat="server"></asp:TextBox>
          < asp: RegularExpressionValidator ID =" RegularExpressionValidator2 "
runat="server" ErrorMessage="必须 6 位数字" ControlToValidate="txtNode" ForeColor=
"Red" ValidationExpression="\d{6}"></asp:RegularExpressionValidator>
        <br/>
        姓   名:<asp:TextBox ID="txtName" runat="server"></asp:TextBox>
         < asp: RegularExpressionValidator ID =" RegularExpressionValidator3" runat =
"server" ErrorMessage =" 必 须 为 汉 字 " ControlToValidate =" txtName " ForeColor =" Red"
ValidationExpression="^[\u4e00-\u9fa5]{0,}$ "></asp:RegularExpressionValidator>
        <br/>
     <asp:Button ID="btnSubmit" runat="server" Text="提交" OnClick="btnSubmit_
Click"/>
    </form>
```

(3) 为控件添加事件,并编辑代码如下。

```
protected void btnSubmit_Click(object sender, EventArgs e)
{
    Response.Write("<script>alert('通过验证');</script>");
}
```

(4) 运行页面,相关控件未通过验证时效果如图 4.13 所示。

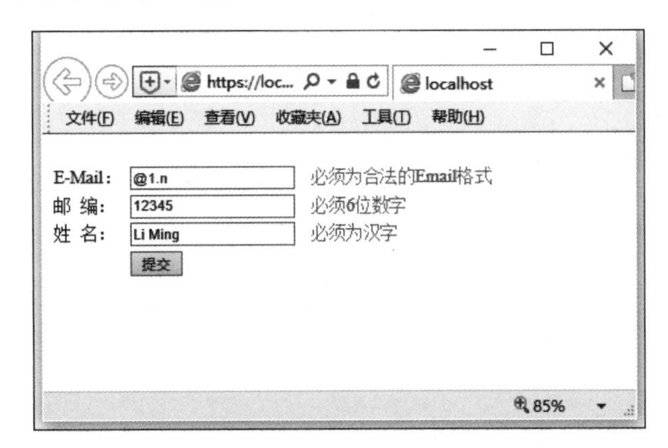

图 4.13 RegularExpressionValidator 控件应用页面演示

4.2.5 自定义验证控件 CustomValidator

CustomValidator 控件又称自定义验证控件,在工具箱中图标为"CustomValidator",封装在 System. Web. UI. Control. WebControl 命名空间的 CustomValidator 类中。CustomValidator 控件允许创建自定义的验证规则完成某些特殊的验证。该控件除具有公共属性外,其他特殊的属性和事件如表 4.7 和表 4.8 所示。

表 4.7 CustomValidator 控件的特殊属性

属性名	说明
ClientValidationFunction	获取或设置脚本函数,可实现 CustomValidator 控件的客户端验证

表 4.8 CustomValidator 控件的特殊事件

事 件 名	说 明
OnServerValidate	当页面提交时触发服务器验证

【示例 4-6】 在 E 盘 ASP.NET 项目代码的 chapter4 目录下，创建一个名为 example4-6 的网页，使用 CustomValidator 控件进行客户端验证和服务器端验证。

(1) 在页面中添加相应控件，具体如图 4.14 所示。

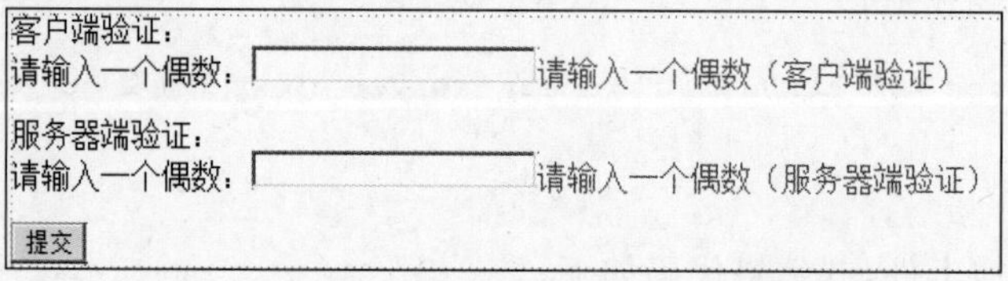

客户端验证：
请输入一个偶数： 请输入一个偶数（客户端验证）
服务器端验证：
请输入一个偶数： 请输入一个偶数（服务器端验证）
提交

图 4.14 CustomValidator 控件应用页面布局

(2) 按如下源文件设置控件的相关属性值。

```
<form id="form1" runat="server">
        客户端验证:<br/>
        请输入一个偶数:<asp:TextBox ID="txtNum1" runat="server"></asp:TextBox>
        <asp:CustomValidator ID="CustomValidator1" runat="server"  ErrorMessage="请
输入一个偶数(客户端验证)" ClientValidationFunction="IsEven" ControlToValidate=
"txtNum1" ForeColor="Red"></asp:CustomValidator>
        <br/>
        服务器端验证:<br/>
        请输入一个偶数:<asp:TextBox ID="txtNum2" runat="server"></asp:TextBox>
        <asp:CustomValidator ID="CustomValidator2" runat="server"  ErrorMessage="请
输入一个偶数(服务器端验证)" OnServerValidate="CustomValidator1_ServerValidate"
ControlToValidate="txtNum2" ForeColor="Red"></asp:CustomValidator>
      <br/>
    <asp:Button ID="btnSubmit" runat="server" Text="提交" OnClick="btnSubmit_
Click"/>
   </form>
```

(3) 在源文件的<head>标签内构建 JavaScript 脚本函数，对应代码如下。

```
<head runat="server">
<meta http-equiv="Content-Type" content="text/html; charset=utf-8"/>
    <title></title>
    <script type="text/javascript">
        function IsEven(source, args)
        {
            if (args.Value %2==0)
            {
                arg.IsValid=true;
            }
            else
            {
                args.IsValid=false;
```

```
            }
        }
    </script>
</head>
```

(4) 为控件添加事件,并编辑代码如下。

```
protected void CustomValidator1_ServerValidate(object source, ServerValida-
teEventArgs args)
{
    args.IsValid=(int.Parse(args.Value) %2==0);
}
protected void btnSubmit_Click(object sender, EventArgs e)
{
    if (CustomValidator2.IsValid)
    {
          Response.Write("<script>alert('通过验证');</script>");
      }
      else
      {
          Response.Write("<script>alert('未通过验证');</script>");
      }
}
```

(5) 网页运行后,第一个文本框内的数值进行客户端验证,不需要提交服务器,只有该值通过验证才将页面整体提交至服务器,否则页面不提交至服务器,如图 4.15 所示。第二个控件内的数值进行服务器端验证,执行服务器端 C# 代码完成验证操作,执行效果如图 4.16 所示。

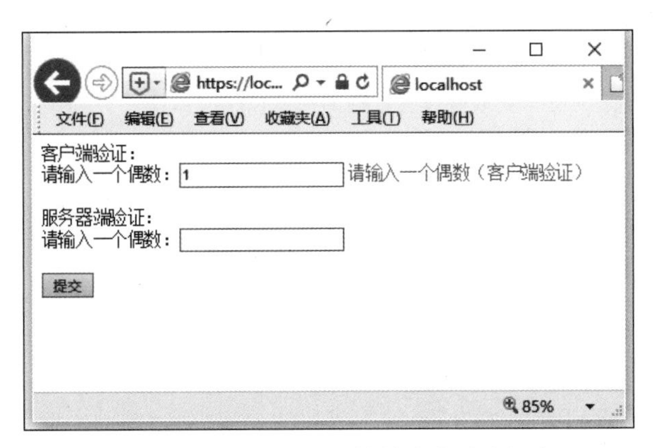

图 4.15 CustomValidator 控件客户端验证演示

说明:

(1) function IsEven(source,args)函数中的 args 变量具有 IsValid 和 Value 两个属性。其中 IsValid 属性对应验证控件是否通过验证,True 为通过验证;Value 属性表示被该控件所验证的控件内的文本值。

(2) 关于客户端验证和服务器端验证,前文已经详细讲解,客户端验证可以实现本地验证,减少用户的等待时间,客户体验比较友好。服务器端验证更安全,代码在客户端是看不

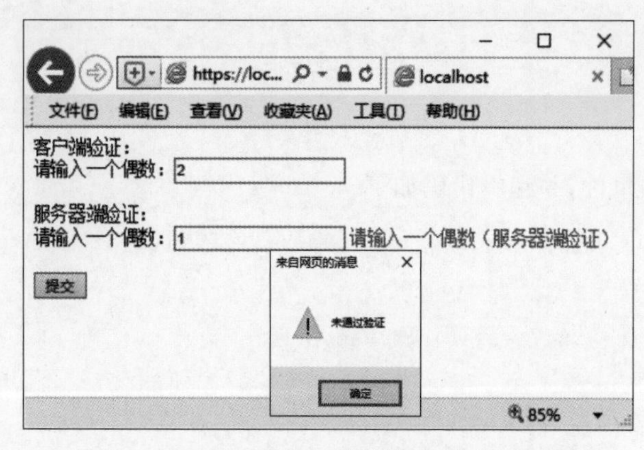

图 4.16　CustomValidator 控件服务器端验证演示

到的，而客户端验证可以通过“源文件”进行查看。

扫一扫

4.2.6　验证汇总控件 ValidationSummary

ValidationSummary 控件又称验证汇总控件，在工具箱中图标为“ValidationSummary”，封装在 System. Web. UI. Control. WebControl 命名空间的 ValidationSummary 类中。ValidationSummary 控件用于在网页、消息框或在这两者中内联显示所有验证错误的摘要，其属性如表 4.9 所示。

表 4.9　ValidationSummary 控件属性

属 性 名	说　明
DisplayMode	获取或设置显示模式
ForeColor	获取或设置控件的前景色
HeaderText	获取或设置控件中的标题文本
ShowMessageBox	获取或设置是否在消息框中显示验证摘要
ShowSummary	获取或设置是否显示验证摘要

说明：

(1) DisplayMode 属性具有如下几个枚举值。

- BulletList：子弹列表形式显示。
- List：普通列表形式显示。
- SingleParagraph：按行文字形式显示。

(2) 控件中显示的错误消息是由每个验证控件的 ErrorMessage 属性值设定的，若未设置验证控件的 ErrorMessage 属性，则不会显示该验证控件的错误消息。

【示例 4-7】 在 E 盘 ASP.NET 项目代码的 chapter4 目录下，创建一个名为 example4-7 的网页，使用 ValidationSummary 控件的验证信息汇总功能。

(1) 按图 4.17 所示添加相应控件。

(2) 按如下源文件设置控件的相关属性值。

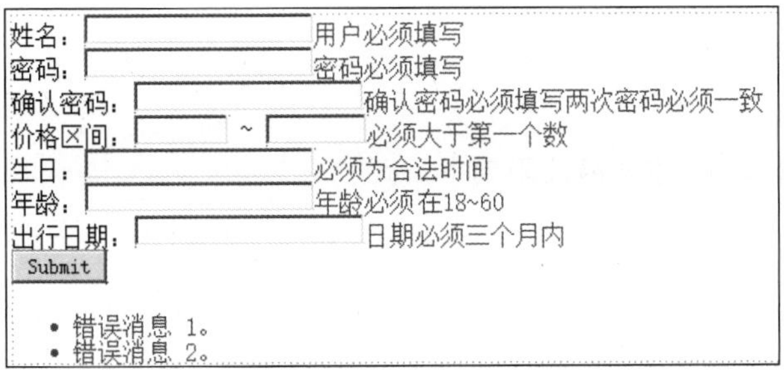

图 4.17 ValidationSummary 控件应用页面布局

```
<form id="form1" runat="server">
    姓名:<asp:TextBox ID="txtName" runat="server"></asp:TextBox>
    <asp:RequiredFieldValidator ID="RequiredFieldValidator1" runat="server"
ControlToValidate="txtName" ErrorMessage="用户必须填写" ForeColor="Red"
SetFocusOnError="True"></asp:RequiredFieldValidator><br/>
    密码:<asp:TextBox ID="txtPwd" runat="server"></asp:TextBox>
    <asp:RequiredFieldValidator ID="RequiredFieldValidator2" runat="server"
ControlToValidate="txtPwd" ErrorMessage="密码必须填写" ForeColor="Red"
SetFocusOnError="True"></asp:RequiredFieldValidator><br/>
    确认密码:<asp:TextBox ID="txtRePwd" runat="server"></asp:TextBox>
      <asp:RequiredFieldValidator ID="RequiredFieldValidator3" runat="server"
ControlToValidate="txtRePwd" Display="Dynamic" ErrorMessage="确认密码必须填写"
ForeColor="Red" SetFocusOnError="True"></asp:RequiredFieldValidator>
        < asp: CompareValidator ID =" CompareValidator1 " runat =" server "
ControlToCompare =" txtPwd " ControlToValidate =" txtRePwd " Display =" Dynamic"
ErrorMessage="两次密码必须一致" ForeColor="Red" SetFocusOnError="True"></asp:
CompareValidator><br/>
    价格区间:<asp:TextBox ID="txtMin" runat="server" Width="53px"></asp:
TextBox>
                ~<asp:TextBox ID="txtMax" runat="server" Width="57px"></
asp:TextBox>
       < asp: CompareValidator ID =" CompareValidator2 " runat =" server "
ControlToCompare="txtMin" ControlToValidate="txtMax" ErrorMessage="必须大于第
一个数" ForeColor="Red" Operator="GreaterThan" SetFocusOnError="True" Type=
"Integer"></asp:CompareValidator><br/>
    生日:<asp:TextBox ID="txtBirthday" runat="server"></asp:TextBox>
      < asp: CompareValidator  ID =" CompareValidator3 " runat =" server "
ControlToValidate="txtBirthday" ErrorMessage="必须为合法时间" ForeColor="Red"
Operator="DataTypeCheck" Type="Date"></asp:CompareValidator><br/>
    年龄:<asp:TextBox ID="txtAge" runat="server"></asp:TextBox>
    <asp:RangeValidator ID="RangeValidator1" runat="server" ControlToValidate="
txtAge" ErrorMessage="年龄必须在 18~60" ForeColor="Red" MaximumValue="60"
MinimumValue="18" Type="Integer"></asp:RangeValidator><br/>
    出行日期:<asp:TextBox ID="txtDate" runat="server"></asp:TextBox>
      < asp: RangeValidator  ID ="  RangeValidator2 "  runat  ="  server "
ControlToValidate="txtDate" ErrorMessage="日期必须三个月内" ForeColor="Red"
Type="Date"></asp:RangeValidator><br/>
    <asp:Button ID="btnSubmit" runat="server" Text="Submit"/>  <br/>
```

```
        < asp: ValidationSummary  ID =" ValidationSummary1 "  runat =" server "
DisplayMode=" BulletList" ForeColor="Red" ShowMessageBox="True"/>
</form>
```

(3) 为控件添加事件,并编辑代码如下。

```
protected void Page_Load(object sender, EventArgs e)
{
    RangeValidator2.MinimumValue=DateTime.Now.ToShortDateString();
    RangeValidator2.MaximumValue=DateTime.Now.AddMonths(3).ToShortDateString();
}
```

(4) 运行网站,执行效果如图 4.18 所示。

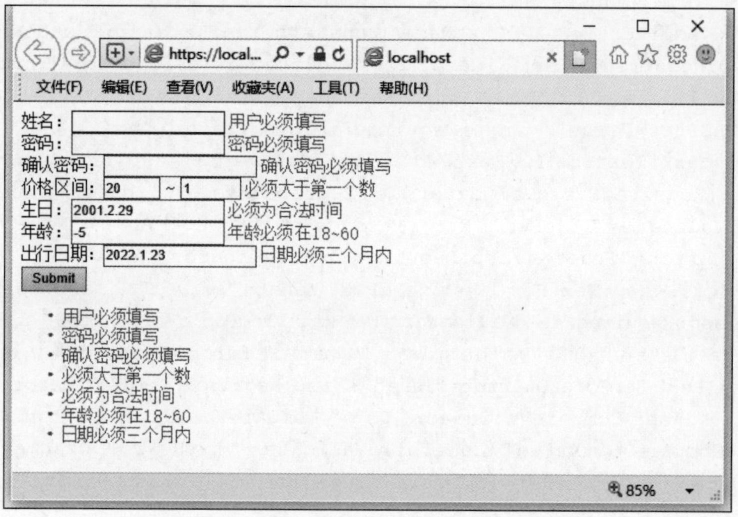

图 4.18　ValidationSummary 控件应用页面演示

4.3 验证控件组的使用

验证控件组也称为验证组,它不是一个单独的控件,而是验证控件的一种分组划分。实际应用中通常包含若干输入文本框,如果在提交表单时对所有控件都进行验证,则会降低网页运行效率,将验证控件归为若干组,对每个验证组分别执行验证,每一组验证控件称为一个验证组。

将一组控件的 ValidationGroup 属性设置为同一个名称(字符串)即创建了一个验证组,验证组名可以设置为任何名称,但必须对该组的所有成员使用相同的名称。然后对应设置按钮、单选按钮、复选框等触发验证的控件的 ValidationGroup 属性值,指定“回发”时将要进行验证的验证组,只要该验证组内的所有验证均通过就将“回发”至服务器,而不对其他验证组进行验证。

【示例 4-8】 在 E 盘 ASP.NET 项目代码的 chapter4 目录下,创建一个名为 example4-8 的网页,使用验证组进行分组验证。

（1）按图 4.19 所示添加相应控件。

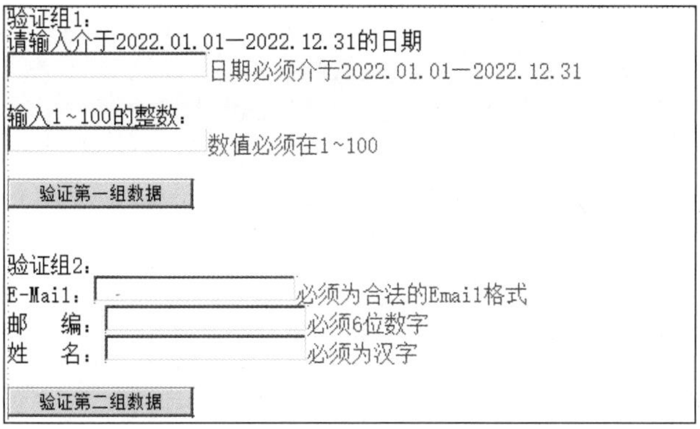

验证组1：
请输入介于2022.01.01—2022.12.31的日期
日期必须介于2022.01.01—2022.12.31
输入1～100的整数：
数值必须在1～100
验证第一组数据
验证组2：
E-Mail：必须为合法的Email格式
邮　编：必须6位数字
姓　名：必须为汉字
验证第二组数据

图 4.19　验证控件组应用页面布局

（2）按如下源文件设置控件的相关属性值。

```
<form id="form1" runat="server">
     验证组 1:<br/>
    请输入介于 2022.01.01—2022.12.31 的日期<br/>
    <asp:TextBox ID="txtDate" runat="server"></asp:TextBox>
    <asp:RangeValidator ID="RangeValidator1" runat="server" ErrorMessage
="日期必须介于 2022.01.01—2022.12.31" MaximumValue="2022-12-31" MinimumValue=
"2022 - 01 - 01" Type =" Date" ControlToValidate =" txtDate" ForeColor =" Red"
ValidationGroup="Valid1"></asp:RangeValidator>
     <br/>
    输入 1~100 的整数:<br/>
    <asp:TextBox ID="txtNums" runat="server"></asp:TextBox>
    <asp:RangeValidator ID="RangeValidator2" runat="server" ErrorMessage
="数值必须在 1~100" ControlToValidate="txtNums" ForeColor="Red" MaximumValue="
100" MinimumValue =" 1" Type =" Integer" ValidationGroup =" Valid1" > </asp:
RangeValidator>
     <br/>
    <asp:Button ID="btn1" runat="server" OnClick="btn1_Click" Text="验证第
一组数据" ValidationGroup="Valid1"/>
    <br/>
    验证组 2:<br/>
    E-Mail:<asp:TextBox ID="txtEmail" runat="server"></asp:TextBox>
    <asp:RegularExpressionValidator ID="RegularExpressionValidator1" runat="
server" ErrorMessage="必须为合法的 Email 格式" ControlToValidate="txtEmail"
ForeColor="Red" ValidationExpression="\w+([-+.']\w+)*@\w+([-.]\w+)*\.\w+([-.]\
w+)*" ValidationGroup="Valid2"></asp:RegularExpressionValidator>
      邮   编:<asp:TextBox ID="txtNode" runat="server"></asp:TextBox>
      < asp: RegularExpressionValidator ID =" RegularExpressionValidator2"
runat=" server" ErrorMessage =" 必须 6 位数字" ControlToValidate =" txtNode"
ForeColor="Red" ValidationExpression="\d{6}" ValidationGroup="Valid2"></asp:
RegularExpressionValidator>
    <br/>
    姓   名:<asp:TextBox ID="txtName" runat="server"></asp:TextBox>
```

```
        <asp:RegularExpressionValidator ID="RegularExpressionValidator3" runat="server" ErrorMessage="必须为汉字" ControlToValidate="txtName" ForeColor="Red" ValidationExpression="^[\u4e00-\u9fa5]{0,}$ " ValidationGroup="Valid2"></asp:RegularExpressionValidator>
        <br/>
        <asp:Button ID="btn2" runat="server" Text="验证第二组数据" OnClick="btn2_Click" ValidationGroup="Valid2"/>
</form>
```

(3) 为控件添加事件,并编辑代码如下。

```
protected void btn1_Click(object sender, EventArgs e)
{
    Response.Write("<script>alert('第一个验证组通过验证');</script>");
}
protected void btn2_Click(object sender, EventArgs e)
{
    Response.Write("<script>alert('第二个验证组通过验证');</script>");
}
```

(4) 运行页面,单击"验证第一组数据"按钮时只对 ValidationGroup 属性值为 Valid1 的控件进行验证,具体如图 4.20 所示。单击"验证第二组数据"按钮时只对 ValidationGroup 属性值为 Valid2 的控件进行验证。

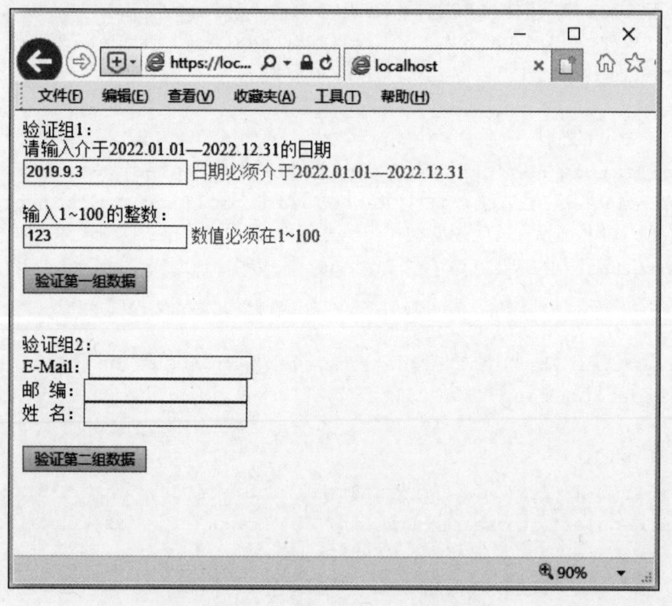

图 4.20 验证控件组应用页面演示

说明:

如果触发验证的控件未设置 ValidationGroup 属性值,则将只触发验证同样未设置 ValidationGroup 属性值的验证控件;如果希望某一控件对所有验证控件均不触发验证,可设置其 CausesValidation 属性值为 False。

综合实验四 注册模块数据验证

主要任务

创建 ASP.NET 应用程序,并创建 Web 窗体,实现用户注册模块,相关输入验证功能见表 4.10。

表 4.10 注册模块输入约束

输入内容	控件 ID	输入约束	验证控件类型
用户名	txtID	非空且用户名必须在 3~10 字节	必填、正则表达式验证控件
用户密码	txtPassword	非空且密码不少于 6 位	必填、正则表达式验证控件
重复密码	txtComparePassword	非空且需与用户密码相同	必填、比较验证控件
密码提示问题	txtQuestion	非空且密码提示问题不少于 6 个字	必填、正则表达式验证控件
密码提示答案	txtAnswer	非空且密码提示答案不少于 6 个字	必填、正则表达式验证控件
真实姓名	txtName	非空且姓名需为中文	必填、正则表达式验证控件
身份证号	txtCardID	非空且需 18 位身份证号	必填、正则表达式验证控件
学生证号	txtStudentID	非空且需以 16 或 17 开头的 4 位数字	必填、自定义验证控件
电子邮箱	txtEmail	非空且需满足邮箱格式	必填、正则表达式验证控件
联系 QQ	txtQQ	非空且需 5~9 位数字	必填、正则表达式验证控件

实验步骤

步骤 1:在 VS 2019 菜单上选择"文件"→"新建"→"项目"命令。

步骤 2:创建 ASP.NET 应用程序,命名为"综合实验四"。

步骤 3:在网站的根目录上右击,创建"综合实验四.aspx"窗体,添加相应控件,具体如图 4.21 所示。

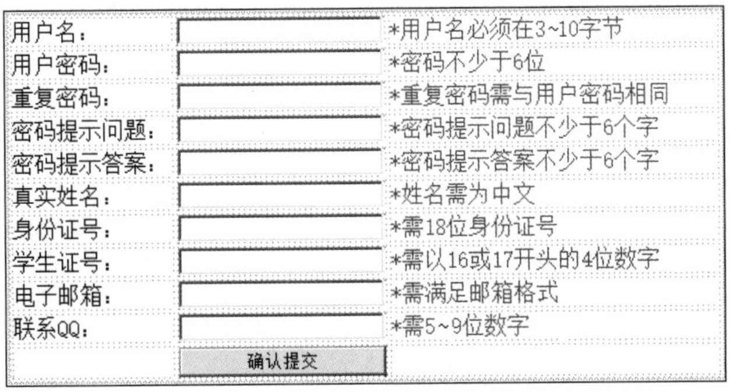

用户名: *用户名必须在3~10字节
用户密码: *密码不少于6位
重复密码: *重复密码需与用户密码相同
密码提示问题: *密码提示问题不少于6个字
密码提示答案: *密码提示答案不少于6个字
真实姓名: *姓名需为中文
身份证号: *需18位身份证号
学生证号: *需以16或17开头的4位数字
电子邮箱: *需满足邮箱格式
联系QQ: *需5~9位数字
确认提交

图 4.21 注册模块页面布局

步骤 4：按如下源文件设置控件的相关属性值。

```
<form id="form1" runat="server">
        <div>
            <table>
                <tr>
                    <td>用户名:</td>
                    <td>
                        <asp:TextBox ID="txtID" runat="server"></asp:TextBox>
                    </td>
                    <td>
                         <asp:RequiredFieldValidator ID="RequiredFieldValidator1"
runat="server" ErrorMessage="*" ForeColor="Red" SetFocusOnError="true" ControlTo-
Validate="txtID"></asp:RequiredFieldValidator>
                        <asp:RegularExpressionValidator ID="RegularExpressionVal-
idator6" runat="server" ErrorMessage="用户名必须在 3~10 字节" ControlToValidate=
"txtID" ValidationExpression="^.{3,10}$ " ForeColor="Red" SetFocusOnError="True">
</asp:RegularExpressionValidator>
                    </td>
                </tr>
                <tr>
                    <td>用户密码:</td>
                    <td>
                        <asp:TextBox ID="txtPassword" runat="server" TextMode="
Password"></asp:TextBox>
                    </td>
                    <td>
                         <asp:RequiredFieldValidator ID="RequiredFieldValidator2"
runat="server" ErrorMessage="*" ForeColor="Red" SetFocusOnError="true" ControlTo-
Validate="txtPassword"></asp:RequiredFieldValidator>
                        <asp:RegularExpressionValidator ID="RegularExpressionVal-
idator7" runat="server" ErrorMessage="密码不少于 6 位" ControlToValidate="txtPass
word" ValidationExpression="^.{6,}$ " ForeColor="Red" SetFocusOnError=
"True"></asp:RegularExpressionValidator>
                    </td>
                </tr>
                <tr>
                    <td>重复密码:</td>
                    <td>
                          <asp:TextBox ID="txtComparePassword" runat="server"
TextMode="Password"></asp:TextBox>
                    </td>
                    <td><asp:RequiredFieldValidator ID="RequiredFieldValidator10"
runat="server" ErrorMessage="*" ForeColor="Red" SetFocusOnError="true"
ControlToValidate="txtComparePassword"></asp:RequiredFieldValidator>
                        <asp:CompareValidator ID="CompareValidator1" runat=
"server" ControlToCompare="txtPassword" ControlToValidate="txtComparePassword"
ErrorMessage="重复密码需与用户密码相同" ForeColor="Red" SetFocusOnError=
"True"></asp:CompareValidator>
                    </td>
                </tr>
                <tr>
                    <td>密码提示问题:</td>
```

```
                    <td>
                        <asp:TextBox ID="txtPwdQuestion" runat="server"></asp:
TextBox>
                    </td>
                    <td>
                        <asp:RequiredFieldValidator ID="RequiredFieldValidator3"
runat="server" ErrorMessage="*" ForeColor="Red" SetFocusOnError="true" Control-
ToValidate="txtPwdQuestion"></asp:RequiredFieldValidator>
                        <asp:RegularExpressionValidator ID="RegularExpression-
Validator8" runat="server" ErrorMessage="密码提示问题不少于 6 个字" ControlTo-
Validate="txtPwdQuestion" ValidationExpression="^.{6,}$ " ForeColor="Red"
SetFocusOnError="True"></asp:RegularExpressionValidator>
                    </td>
                </tr>
                <tr>
                    <td>密码提示答案:</td>
                    <td>
                        <asp:TextBox ID="txtPwdAnswer" runat="server"></asp:
TextBox>
                    </td>
                    <td>
                        <asp:RequiredFieldValidator ID="RequiredFieldValidator4"
runat="server" ErrorMessage="*" ForeColor="Red" SetFocusOnError="true" Control-
ToValidate="txtPwdAnswer"></asp:RequiredFieldValidator>
                        <asp:RegularExpressionValidator ID="RegularExpression-
Validator9" runat="server" ErrorMessage="密码提示答案不少于 6 个字" ControlTo-
Validate="txtPwdAnswer" ValidationExpression="^.{6,}$" ForeColor="Red" SetFo-
cusOnError="True"></asp:RegularExpressionValidator>
                    </td>
                </tr>
                <tr>
                    <td>真实姓名:</td>
                    <td>
                            <asp:TextBox ID="txtName" runat="server"></asp:
TextBox>
                    </td>
                    <td>
                        <asp:RequiredFieldValidator ID="RequiredFieldValidator5"
runat="server" ErrorMessage="*" ForeColor="Red" SetFocusOnError="true" Control-
ToValidate="txtName"></asp:RequiredFieldValidator>
                        <asp:RegularExpressionValidator ID="RegularExpression-
Validator10" runat="server" ControlToValidate="txtName" ErrorMessage="姓名需
为中文" ForeColor="Red" SetFocusOnError="True" ValidationExpression="^[\u4e00
-\u9fa5]{0,}$ "></asp:RegularExpressionValidator>
                    </td>
                </tr>
                <tr>
                    <td>身份证号:</td>
                    <td>
                          <asp:TextBox ID="txtCardID" runat="server"></asp:
TextBox>
                    </td>
                    <td>
```

```
                    <asp:RequiredFieldValidator ID="RequiredFieldValidator6"
runat="server" ErrorMessage="*" ForeColor="Red" SetFocusOnError="true" Control-
ToValidate="txtCardID"></asp:RequiredFieldValidator>
                    <asp:RegularExpressionValidator ID="RegularExpression-
Validator11" runat="server" ControlToValidate="txtCardID" ErrorMessage="需 18
位身份证号" ForeColor="Red" SetFocusOnError="True" ValidationExpression="\d
{17}[\d|X]"></asp:RegularExpressionValidator>
                </td>
            </tr>
            <tr>
                <td>学生证号:</td>
                <td>
                    <asp:TextBox ID="txtStudentID" runat="server"></asp:
TextBox>
                </td>
                <td>
                    <asp:RequiredFieldValidator ID="RequiredFieldValidator7"
runat="server" ErrorMessage="*" ForeColor="Red" SetFocusOnError="true" Control-
ToValidate="txtStudentID"></asp:RequiredFieldValidator>
                    <asp:CustomValidator ID="CustomValidator1" runat="serv-
er"  ClientValidationFunction="checkstudentId" ControlToValidate="txtStuden-
tID" ErrorMessage="需以 16 或 17 开头的 4 位数字" ForeColor="Red" SetFocusOnError
="True"></asp:CustomValidator>
                </td>
            </tr>
            <tr>
                <td>电子邮箱:</td>
                <td>
                    <asp:TextBox ID="txtEmail" runat="server"></asp:
TextBox>
                </td>
                <td>
                    <asp:RequiredFieldValidator ID="RequiredFieldValidator8"
runat="server" ErrorMessage="*" ForeColor="Red" SetFocusOnError="true" Control-
ToValidate="txtEmail"></asp:RequiredFieldValidator>
                    <asp:RegularExpressionValidator ID="RegularExpression-
Validator12" runat="server" ControlToValidate="txtEmail" ErrorMessage="需满足
邮箱格式" ForeColor="Red" SetFocusOnError="True" ValidationExpression="\w+([-
+.']\w+)*@\w+([-.]\w+)*\.\w+([-.]\w+)*"></asp:RegularExpressionValidator>
                </td>
            </tr>
            <tr>
                <td>联系 QQ:</td>
                <td>
                    <asp:TextBox ID="txtQQ" runat="server"></asp:TextBox>
                </td>
                <td>
                    <asp:RequiredFieldValidator ID="RequiredFieldValidator9"
runat="server" ErrorMessage="*" ForeColor="Red" SetFocusOnError="true" Control-
ToValidate="txtQQ"></asp:RequiredFieldValidator>
                    <asp:RegularExpressionValidator ID="RegularExpression-
Validator13" runat="server" ControlToValidate="txtQQ" ErrorMessage="需 5~9 位
数字" ForeColor="Red" SetFocusOnError="True" ValidationExpression="[1-9]\d{4,
8}"></asp:RegularExpressionValidator>
```

```
                    </td>
                </tr>
                <tr>
                    <td> </td>
                    <td>
                        <asp:Button ID="btnSubmit" runat="server" Text="确认提
交" Width="149px"/>
                    </td>
                    <td> </td>
                </tr>
            </table
        </div>
    </form>
```

步骤5：在综合实验四源代码的<head>标签内添加JavaScript脚本，编辑代码如下。

```
<script>
        function checkstudentId(source, args)
        {
            if ((args.Value.substring(0, 2)=="16" || args.Value.substring(0, 2)==
"17") && args.Value.length==4)
                args.IsValid=true;
            else
                args.IsValid=false;
        }
</script>
```

步骤6：注册模块页面运行测试如图4.22所示。

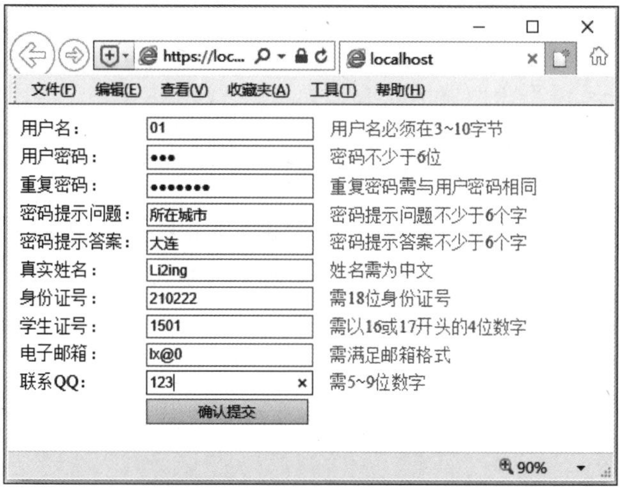

图4.22　注册模块页面运行测试

第5章 ASP.NET内置对象

本章学习目标

- 了解内置对象的基本原理
- 熟练掌握 Page 对象的使用方法
- 熟练掌握 Response 对象的使用方法
- 熟练掌握 Request 对象的使用方法
- 熟练掌握 Server 对象的使用方法
- 熟练掌握 Application 对象的使用方法
- 熟练掌握 Session 对象的使用方法
- 了解 Cookie 对象的使用方法
- 熟练掌握全局应用程序类中的事件

本章首先介绍内置对象的作用，然后对 ASP.NET 提供的 Page、Request、Response、Application、Session、Server、Cookie 等内置对象进行详细讲解，最后添加全局应用程序类，并对其中的事件进行应用测试。

5.1 Page 对象

扫一扫

5.1.1 Page 对象的属性和方法

Page 对象是 System.Web.UI 命名空间 Page 类的一个实例，是网页中所有服务器控件的容器，在 ASP.NET 中每个页面都派生自 Page 类，并继承该类所有公开的方法和属性。Page 对象的主要属性如表 5.1 所示，主要事件如表 5.2 所示。

表 5.1 Page 对象的主要属性

属 性 名	说 明
IsPostBack	获取一个值,指示页面是首次加载还是回发加载
IsValid	获取一个值,指示页面验证是否成功

说明：

1. IsPostBack 属性可以检查.aspx 页是否为"回发",常用于判断页面是否为首次加载,若为首次加载,则该值为 False。

2. IsValid 属性可以判断页面中输入的所有内容是否通过验证,使用服务器端验证时常使用该属性。

表 5.2 Page 对象的主要事件

事 件 名	说 明
PreInit	当页面初始化时触发
PreLoad	在页面 Load 事件之前触发
Load	当服务器加载到 Page 对象时触发
Init	当服务器控件初始化时触发
PreRender	在加载控件之后、呈现控件之前触发
Unload	当服务器控件从内存中卸载时触发
InitComplete	当页初始化完成时触发
LoadComplete	当页加载结束时触发

5.1.2 Page 对象的应用

网页 C#代码中默认提供了 Page.Load 事件所对应的 Page_Load()方法,而其他的事件则需要编写其对应的方法,不需要如控件一样在源中声明。各事件触发执行的先后顺序为：Page. PreInit、Page. Init、Page. InitComplete、Page. PreLoad、Page. Load、Page. LoadComplete、Page.PreRender 和 Page.Unload。

【示例 5-1】 在 E 盘 ASP.NET 项目代码目录中创建 chapter5 子目录,将其作为网站根目录,创建名为 example5-1 的网页,演示加载页面时 Page 对象的各种事件的触发执行顺序。

(1) 在页面中添加相应控件,具体如图 5.1 所示。

Page对象事件的触发顺序
[lblInfo] 提交

图 5.1 Page 对象应用页面设计

(2) 设置相关控件的属性,如下列源代码所示。

```
<form id="form1" runat="server">
    <span>Page 对象事件的触发顺序</span><hr>
    <asp:Label ID="lblInfo" runat="server"></asp:Label>
    <br/>
    <asp:Button ID="btnSubmit" runat="server" OnClick="btnSubmit_Click" Text="提
交"/>
</form>
```

（3）为事件添加 C#代码如下。

```
protected void Page_Load(object sender, EventArgs e)
{
    if (! IsPostBack)
    {
        lblInfo.Text+="页面第一次加载";
    }
    else
    {
        lblInfo.Text+="页面第二次或第二次以上加载";
    }
    lblInfo.Text+="触发了 Page 对象的 Load 事件<br>";
}
protected void Page_Init(object sender, EventArgs e)
{
    lblInfo.Text+="触发了 Page 对象的 Init 事件<br>";
}
protected void Page_PreInit(object sender, EventArgs e)
{
    lblInfo.Text+="触发了 Page 对象的 PreInit 事件<br>";
}
protected void Page_InitComplete(object sender, EventArgs e)
{
    lblInfo.Text+="触发了 Page 对象的 InitComplete 事件<br>";
}
protected void Page_PreLoad(object sender, EventArgs e)
{
    lblInfo.Text+="触发了 Page 对象的 PreLoad 事件<br>";
}
protected void Page_LoadComplete(object sender, EventArgs e)
{
    lblInfo.Text+="触发了 Page 对象的 LoadComplete 事件<br>";
}
protected void Page_PreRender(object sender, EventArgs e)
{
    lblInfo.Text+="触发了 Page 对象的 PreRender 事件<br>";
}
protected void Page_Unload(object sender, EventArgs e)
{
    lblInfo.Text+="触发了 Page 对象的 Unload 事件<br>";
}
protected void btnSubmit_Click(object sender, EventArgs e)
{
```

```
        lblInfo.Text+="触发了按钮的 Click 事件<br>";
    }
```

（4）运行页面，页面显示相关事件执行顺序，如图 5.2 所示。单击“提交”按钮，页面显示“回发”时相关事件执行的顺序，如图 5.3 所示。

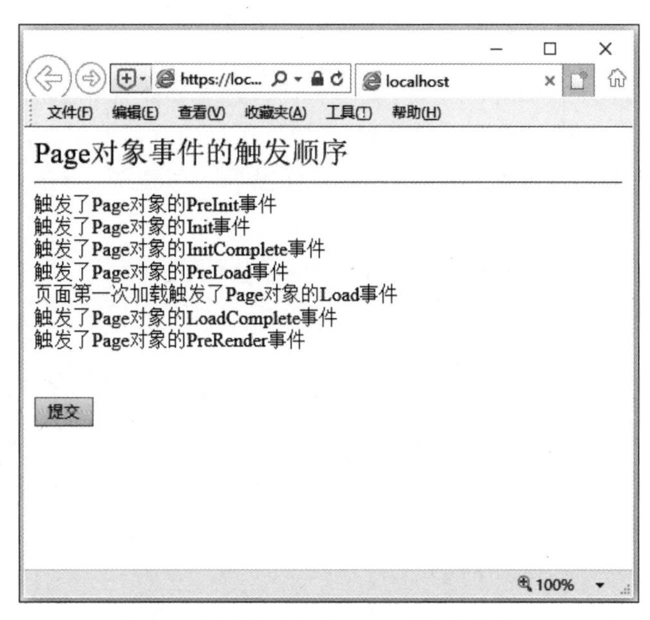

图 5.2　页面初次加载 Page 对象事件演示

图 5.3　页面“回发”时 Page 对象事件演示

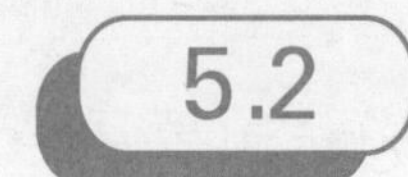

5.2 Response 对象

5.2.1 Response 对象的属性和方法

扫一扫

Response 对象是 HttpResponse 类的一个实例，封装来自 ASP.NET 操作的 HTTP 响应信息，提供对当前页的输出流访问，控制服务器发送给浏览器的信息，包括直接发送信息给浏览器、重新定向浏览器到另一个 URL，以及设置 Cookie 值等。Response 对象的主要属性如表 5.3 所示，主要方法如表 5.4 所示。

表 5.3 Response 对象的主要属性

属 性 名	说 明
Buffer	获取或设置一个值，该值指示是否缓冲输出，并在完成处理整个响应之后将其发送
BufferOutput	获取或设置一个值，该值指示是否缓冲输出，并在完成处理整个页之后将其发送
Cache	获取 Web 页的缓存策略，如过期时间、保密性、变化子句等
Charset	获取或设置输出流的 HTTP 字符集
Cookie	获取响应的 Cookie 集合
Expires	获取或设置在浏览器上缓存的页过期前的分钟数

表 5.4 Response 对象的主要方法

方 法 名	说 明
AppendCookie()	向响应对象的 Cookie 集合中增加一个 Cookie 对象
Clear()	清空缓冲区中的所有内容输出
Close()	关闭当前服务器到客户端的连接
End()	终止响应，并且将缓冲区中的输出发送到客户端
Redirect()	重定向当前请求
Write()	将信息写入 HTTP 的响应输出流
WriteFile()	将指定的文件直接写入 HTTP 的响应输出流

5.2.2 Response 对象的应用

【示例 5-2】 在 E 盘 ASP.NET 项目代码的 chapter5 目录下，添加一个文本文件，命名为“1.txt”，编写文本内容为“This is a text”，创建名为 example5-2 的网页，练习使用 Response 对象的各种属性和方法。

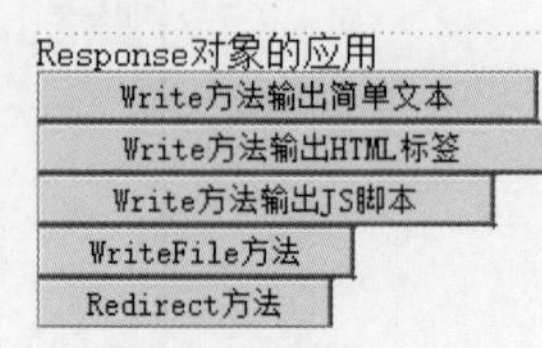

图 5.4 Response 对象应用页面设计

(1) 在页面中添加相应控件，具体如图 5.4 所示。

(2) 设置相关控件的属性，如下列源代码所示。

```
<form id="form1" runat="server">
    <div>
        Response 对象的应用<br/>
        <asp:Button ID="btnWriteTxt" runat="server"  Text="Write 方法输出简单文
本" OnClick="btnWriteTxt_Click"/>
        <br/>
        <asp:Button ID="btnWriteHTML" runat="server"  Text="Write 方法输出 HTML
标签" OnClick="btnWriteHTML_Click"/>
        <br/>
        <asp:Button ID="btnWriteJS" runat="server"  Text="Write 方法输出 JS 脚
本" OnClick="btnWriteJS_Click"/>
        <br/>
        <asp:Button ID="btnWriteFile" runat="server"  Text="WriteFile 方法"
OnClick="btnWriteFile_Click"/>
        <br/>
        <asp:Button ID="btnRedirect" runat="server"  Text="Redirect 方法"
OnClick="btnRedirect_Click"/>
      </div>
    </form>
```

(3) 为事件添加 C#代码如下。

```
protected void btnWriteTxt_Click(object sender, EventArgs e)
{
    Response.Write("普通文字输出");
}
protected void btnWriteHTML_Click(object sender, EventArgs e)
{
    Response.Write("<a href=123>包含 html 标签的文本</a>");
}
protected void btnWriteJS_Click(object sender, EventArgs e)
{
    Response.Write("<script>alert('包含脚本文本');</script>");
}
protected void btnWriteFile_Click(object sender, EventArgs e)
{
    Response.WriteFile(@"E:\ASP.NET 项目代码\chapter51.txt");
}
protected void btnRedirect_Click(object sender, EventArgs e)
{
    Response.Redirect("example5-1.aspx");
}
```

(4) 运行页面,单击按钮,可获取 Response 对象各方法和属性的基本功能,单击"WriteFile 方法"按钮,运行时效果如图 5.5 所示。

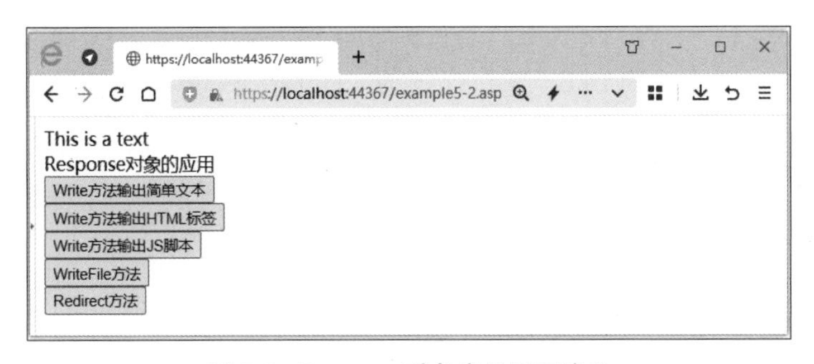

图 5.5　Response 对象应用页面演示

5.3 Request 对象

扫一扫

5.3.1 Request 对象的属性和方法

Request 对象是 System.Web.HttpRequest 类的实例，当客户请求 ASP.NET 页面时，所有的请求信息，包括请求报头、请求方法、客户端基本信息等被封装在 Request 对象中，利用 Request 对象可以读取这些请求信息。Request 对象的主要属性如表 5.5 所示，主要方法如表 5.6 所示。

表 5.5 Request 对象的主要属性

属 性 名	说 明
Browser	获取正在请求的客户端浏览器功能的信息
Cookies	获取客户端发送的 Cookie 的集合
Form	获取表单变量的集合
FilePath	获取当前请求的虚拟路径
Param	获取地址栏中的参数集合
QueryString	获取 HTTP 查询字符串变量集合
UserHostAddress	获取远程客户端的 IP 主机地址
Url	获取有关当前请求的 URL 信息
UserHostName	获取远程客户端的 DNS 名称

表 5.6 Request 对象的主要方法

方 法 名	说 明
BinaryRead()	执行对当前输入流进行指定字节数的二进制读取
MapPath()	将请求的 URL 中的虚拟路径映射到服务器上的物理路径
SaveAs()	将 HTTP 请求保存到文件中

5.3.2 Request 对象的应用

【示例 5-3】 在 E 盘 ASP.NET 项目代码的 chapter5 目录下，创建名为 example5-3 的网页，练习使用 Request 对象的各种属性。

(1) 在页面中添加相应控件，具体如图 5.6 所示。

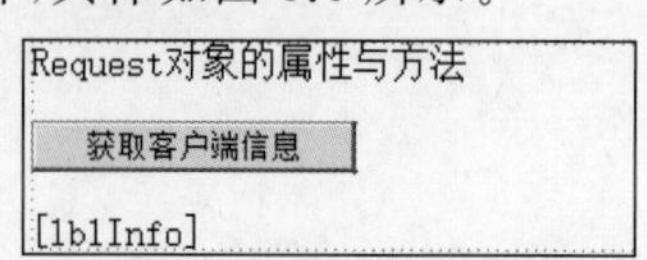

图 5.6 Request 对象应用页面设计

(2) 设置相关控件的属性，如下列源代码所示。

```
<form id="form1" runat="server">
    Request 对象的属性与方法<br/>
    <br/>
    <asp:Button ID="btnGetInfo" runat="server" OnClick="btnGetInfo_Click"
Text="获取客户端信息"/>
    <br/>
    <asp:Label ID="lblInfo" runat="server"></asp:Label>
</form>
```

(3) 为事件添加 C# 代码如下。

```
protected void btnGetInfo_Click(object sender, EventArgs e)
{
    lblInfo.Text="<br>客户端浏览器名称:"+Request.Browser.Type
        +"<br>版本号:"+Request.Browser.Version
        +"<br>客户端使用的操作系统:"+Request.Browser.Platform
        +"<br>客户端 IP 地址:"+Request.UserHostAddress
        +"<br>当前请求的 URL:"+Request.Url
        +"<br>当前请求的虚拟路径:"+Request.Path
        +"<br>当前请求的物理路径:"+Request.PhysicalPath;
}
```

(4) 运行页面，单击相关按钮，运行结果如图 5.7 所示。

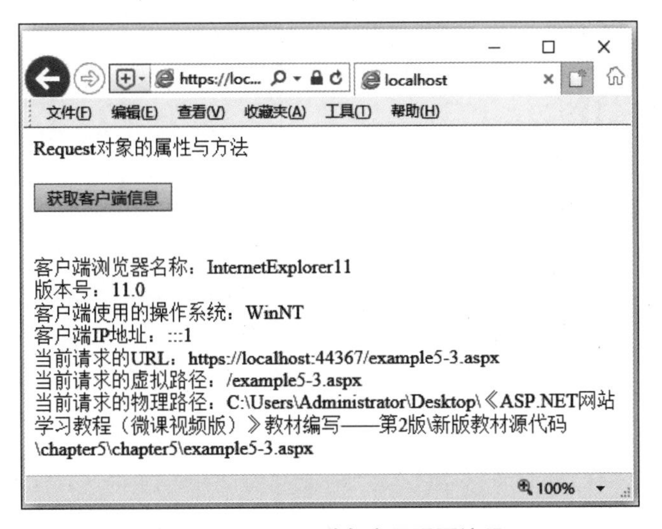

图 5.7　Request 对象应用页面演示

【示例 5-4】 在 E 盘 ASP.NET 项目代码的 chapter5 目录下，创建 example5-4 页面和 example5-4-2 页面，练习使用 Request 对象和 Response 对象进行地址栏传值。

(1) 在 example5-4 网页和 example5-4-2 页面中添加相应控件，具体如图 5.8 和图 5.9 所示。

example5-4 页面：

图 5.8　Request 对象应用的网页设计 1

example5-4-2 页面：

欢迎登录
ID:[lblID]
Name:[lblName]

图 5.9　Request 对象应用的网页设计 2

（2）设置相关控件的属性，如下列源代码所示。

```
example5-4.aspx 页:
<form id="form1" runat="server">
        用户 ID:<asp:TextBox ID="txtID" runat="server"></asp:TextBox>
        <br/>
        昵 称:<asp:TextBox ID="txtName" runat="server"></asp:TextBox>
        <br/>
         <asp:Button ID="btnRedirect" runat="server" onclick=" btnRedirect_
Click " Text="页码跳转并地址栏传值"/>
</form>
example5-4-2.aspx 页:
<form id="form1" runat="server">
        欢迎登录<br/>
        ID:<asp:Label ID="lblID" runat="server"></asp:Label>
        <br/>
        Name:<asp:Label ID="lblName" runat="server"></asp:Label>
</form>
```

（3）为事件添加 C# 代码如下。

```
example5-4.cs:
protected void btnRedirect_Click(object sender, EventArgs e)
{
    Response.Redirect("example5-4-2.aspx? id="+txtID.Text+"&name="+txtName.
Text);
}
example5-4-2.cs:
protected void Page_Load(object sender, EventArgs e)
{
    if (Request.Params["id"] !=null && Request["name"] !=null)
    {
         lblID.Text=Request.Params["id"].ToString();
         lblName.Text=Request.Params["name"].ToString();
    }
    else
    {
       Response.Redirect("example5-4.aspx");
    }
}
```

(4) 运行页面,在 example5-4 页面输入用户 ID、昵称,单击“页面跳转并地址栏传值”按钮提交服务器,如图 5.10 所示。在 example5-4-2 页面获取地址栏中的参数值,显示在页面中,如图 5.11 所示。

图 5.10 Request 对象应用页面演示 2

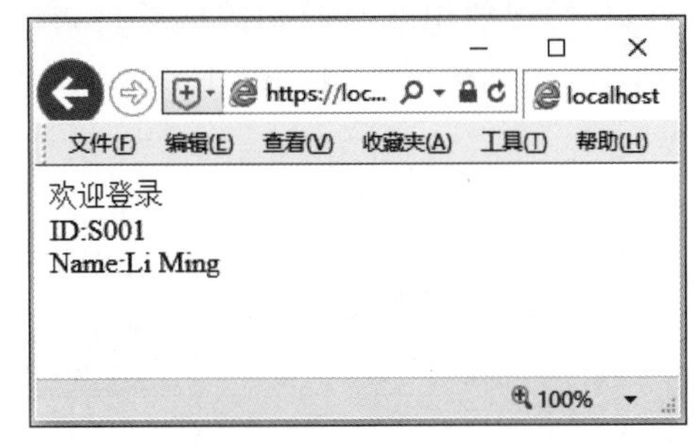

图 5.11 Request 对象应用页面演示 3

5.4 Server 对象

5.4.1 Server 对象的属性和方法

扫一扫

Server 对象是 HttpServerUtility 类的一个实例,提供对服务器属性和方法的访问功能,可以处理页面请求时所需的功能,如建立 COM 对象、字符串的编译码等工作。Server 对象的主要属性如表 5.7 所示,主要方法如表 5.8 所示。

表 5.7 Server 对象的主要属性

属 性 名	说 明
MachineName	获取服务器的名称
ScriptTimeOut	获取或设置请求的超时值(单位为秒)

表 5.8 Server 对象的主要方法

方 法 名	说 明
Execute()	执行指定的资源,并且在执行完之后再执行本页的代码
HtmlDecode()	对 HTML 编码的字符串进行解码
HtmlEncode()	对要在浏览器中显示的字符串进行 HTML 编码
MapPath()	获取指定相对路径在服务器上的物理路径
Transfer()	停止执行当前程序,执行指定的资源
UrlDecode()	对已被编码的 URL 字符串进行解码
UrlEncode()	对 URL 的字符串进行编码,通过 URL 从 Web 服务器到客户端进行 HTTP 传输

5.4.2 Server 对象的应用

【示例 5-5】 在 E 盘 ASP.NET 项目代码的 chapter5 目录下，创建一个名为 example5-5 的网页，练习使用 Server 对象的各种属性。

（1）在页面中添加相应控件，具体如图 5.12 所示。

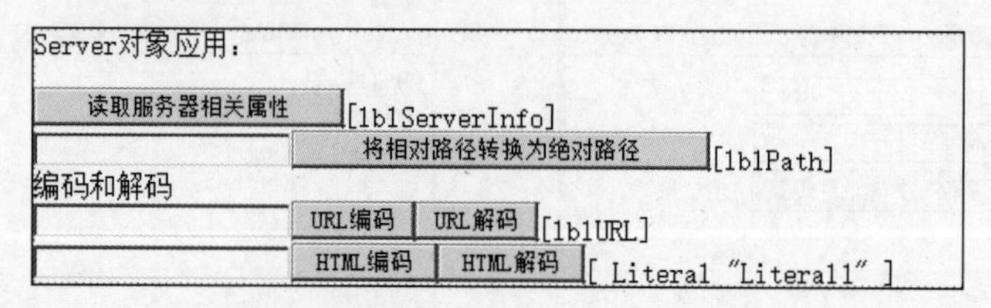

图 5.12 Server 对象应用页面设计

（2）设置相关控件的属性，如下列源代码所示。

```
<form id="form1" runat="server">
    <div>
        Server 对象应用:<br/>
        <br/>
        <asp:Button ID="btnServerInfo" runat="server"  Text="读取服务器相关属性" OnClick="btnServerInfo_Click" />
        <asp:Label ID="lblServerInfo" runat="server"></asp:Label>
        <br/>
        <asp:TextBox ID="txtPath" runat="server"></asp:TextBox>
        <asp:Button ID="btnPath" runat="server" Text="将相对路径转换为绝对路径" OnClick="btnPath_Click"/>
        <asp:Label ID="lblPath" runat="server"></asp:Label>
        <br/>
        编码和解码<br/>
        <asp:TextBox ID="txtURL" runat="server"></asp:TextBox>
        <asp:Button ID="btnURlEncode" runat="server"  Text="URL 编码" OnClick="btnURlEncode_Click"/>
        <asp:Button ID="btnURLDecode" runat="server"  Text="URL 解码" OnClick="btnURLDecode_Click"/>
        <asp:Label ID="lblURL" runat="server"></asp:Label>
        <br/>
        <asp:TextBox ID="txtHtml" runat="server"></asp:TextBox>
        <asp:Button ID="btnHTMLEncode" runat="server"  Text="HTML 编码" OnClick="btnHTMLEncode_Click"/>
        <asp:Button ID="btnHTMLDecode" runat="server"  Text="HTML 解码" OnClick="btnHTMLDecode_Click"/>
        <asp:Literal ID="Literal1" runat="server" Mode="Encode"></asp:Literal>
    </div>
</form>
```

（3）为事件添加 C# 代码如下。

```
protected void btnPath_Click(object sender, EventArgs e)
{
    lblPath.Text=Server.MapPath(txtPath.Text);
}
protected void btnServerInfo_Click(object sender, EventArgs e)
{
    lblServerInfo.Text="<br>服务器计算机名:"+Server.MachineName
            +"<br>页面请求超时时间:"+Server.ScriptTimeout.ToString()+"秒";
}
protected void btnURlEncode_Click(object sender, EventArgs e)
{
    lblURL.Text=Server.UrlEncode(txtURL.Text);
}
protected void btnURLDecode_Click(object sender, EventArgs e)
{
    lblURL.Text=Server.UrlDecode(txtURL.Text);
}
protected void btnHTMLEncode_Click(object sender, EventArgs e)
{
    Literal1.Text=Server.HtmlEncode(txtHtml.Text);
}
protected void btnHTMLDecode_Click(object sender, EventArgs e)
{
    Literal1.Text=Server.HtmlDecode(txtHtml.Text);
}
```

(4) 运行页面,单击相关按钮,可获取服务器相关属性,实现 URL 编码与解码,HTML 编码与解码,具体如图 5.13 所示。

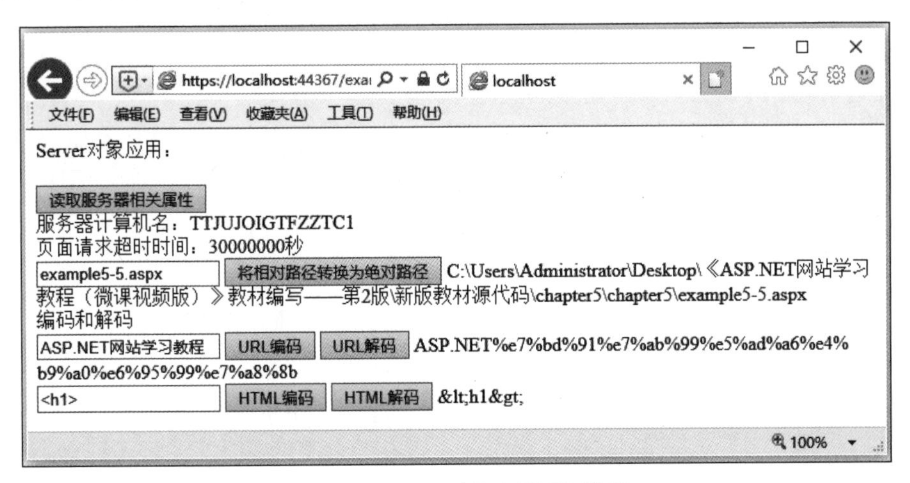

图 5.13 Server 对象应用页面演示

说明:

页面运行时会出现如图 5.14 所示的错误,这时需要在对应.aspx 源文件的＜Page＞标签内添加 validateRequest="false"属性。

```
<%@Page .... validateRequest="false" %>
```

"/"应用程序中的服务器错误。

从客户端(txtHtml="<br/>")中检测到有潜在危险的 Request.Form 值。

图 5.14　Server 对象应用时错误信息

5.5　Application 对象

扫一扫

5.5.1　Application 对象的属性和方法

Application 对象是 HttpApplicationState 类的一个实例，可以在多个请求、连接之间共享公用信息，也可以在各个请求连接之间充当信息传递的管道，使用 Application 对象保存希望传递的变量。由于在整个应用程序生存周期中 Application 对象都是有效的，因此在不同的页面中都可以对它进行存取，就如同 C 语言中的全局变量一样方便。Application 对象的主要属性如表 5.9 所示，主要方法如表 5.10 所示。

表 5.9　Application 对象的主要属性

属 性 名	说　明
AllKeys	获取 HttpApplicationState 集合中的访问键
Count	获取 HttpApplicationState 集合中的对象数

表 5.10　Application 对象的主要方法

方 法 名	说　明
Add()	新增一个新的 Application 对象变量
Clear()	清除全部的 Application 对象变量
GetKey()	按索引关键字获取变量名称
Get()	按索引关键字或变数名称得到变量值
Remove()	按变量名称删除一个 Application 对象
RemoveAt()	按索引名称删除一个 Application 对象
RemoveAll()	删除所有 Application 对象
Lock()	锁定全部的 Application 变量
Set()	使用变量名更新一个 Application 对象变量的内容
UnLock()	解除锁定的 Application 变量

5.5.2 Application 对象的应用

【示例 5-6】 在 E 盘 ASP.NET 项目代码的 chapter5 目录下，创建名为 example5-6 的网页，创建简单聊天室练习 Application 对象的使用。

（1）在页面中添加相应控件，具体如图 5.15 所示。

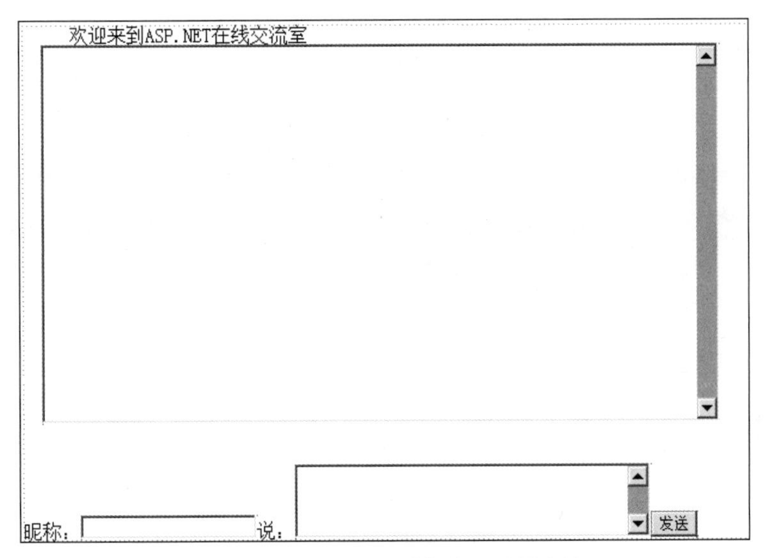

图 5.15 Application 对象应用页面设计

（2）在网站根目录添加全局应用程序文件 Global.asax，在 Application_Start 事件下添加代码，在网站运行时初始化聊天内容。关于 Global.asax 的具体讲解见 5.8 节。

```
void Application_Start(object sender, EventArgs e)
{
    Application["content"]="";
}
```

（3）设置相关控件的属性，如下列源代码所示。

```
<form id="form1" runat="server">
           欢迎来到 ASP.NET 在线交流室<br/>
 <asp:TextBox ID="txtShow" runat="server" Height="302px" TextMode="
MultiLine" Width="561px"></asp:TextBox>
            <br/>
            昵称:<asp:TextBox ID="txtName" runat="server"></asp:TextBox>
            说:<asp:TextBox ID="txtChart" runat="server" Height="54px" TextMode="
MultiLine" Width="290px"></asp:TextBox>
          <asp:Button ID="btnSend" runat="server" Text="发送" OnClick="btnSend
_Click"/>
</form>
```

（4）为事件添加 C#代码如下。

```
protected void Page_Load(object sender, EventArgs e)
{
```

```
    txtShow.Text=Application["content"].ToString();
}
protected void btnSend_Click(object sender, EventArgs e)
{
    Application["content"]=Application["content"] .ToString()+txtName.Text+"
说:"+txtChart.Text+"("+DateTime.Now.ToString()+")\n";
    txtShow.Text=Application["content"].ToString();
}
```

(5) 运行页面,输入昵称和聊天内容,单击“发送”按钮,可显示不同用户之间的聊天,运行效果如图 5.16 所示。

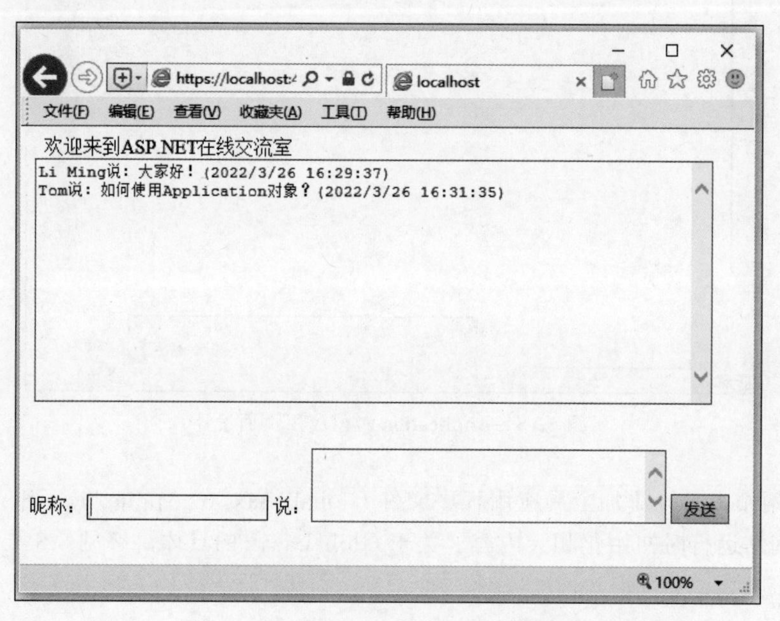

图 5.16 Application 对象应用页面演示

5.6 Session 对象

5.6.1 Session 对象的属性和方法

Session 对象是 HttpSessionState 的一个实例,为当前用户会话提供信息,可用于存储会话范围的访问以及管理会话的方法。Session 对象的变量只对单一用户有效,不同用户的会话信息用不同的 Session 对象变量存储。在网络环境下 Session 对象是有生命周期的,如果规定的时间未对 Session 对象的变量刷新,系统会终止这些变量。

当用户第一次请求.aspx 文件时,ASP.NET 将生成一个 SessionID。SessionID 是由一个复杂算法生成的号码,它唯一标识每个用户会话。在新会话开始时,服务器将 SessionID 作为一个 Cookie 存储在用户的 Web 浏览器中。Session 对象的主要属性如表 5.11 所示,主要方法如表 5.12 所示。

表 5.11　Session 对象的主要属性

属 性 名	说　明
Count	获取会话状态集合中 Session 对象的个数
TimeOut	获取或设置在会话状态提供程序终止会话之前各请求之间所允许的超时期限(单位为分钟)
SessionID	获取用于标识会话的唯一会话 ID

表 5.12　Session 对象的主要方法

方 法 名	说　明
Add()	新增一个 Session 对象变量
Clear()	清除全部的 Session 对象变量
CopyTo()	将 Session 对象复制到一维数组中
Get()	按索引关键字或变数名称得到变量值
Remove()	按变量名称删除一个 Session 对象
RemoveAt()	按索引名称删除一个 Session 对象
RemoveAll()	删除所有 Session 对象

5.6.2　Session 对象的应用

【示例 5-7】　在 E 盘 ASP.NET 项目代码的 chapter5 目录下，创建名为 example5-7 和 example5-7-2 的页面，练习 Session 对象各种属性的使用。

(1) 在页面中添加相应控件，具体如图 5.17 和图 5.18 所示。

example5-7 页面：

图 5.17　Session 对象应用页面设计 1

example5-7-2 页面：

欢迎[lblName]进入聊天室

您的SessionID号为：[lblSessionID]

图 5.18　Session 对象应用页面设计 2

(2) 设置页面源代码如下。

example5-7 页面：

```
<form id="form1" runat="server">
        Session 对象页面间传值<br/>
        用户名:<asp:TextBox ID="txtName" runat="server"></asp:TextBox>
```

```
        <asp:Button ID="btnLogin" runat="server"   Text="进入聊天室" OnClick="
btnLogin_Click"/>
</form>
```

example5-7-2 页面：

```
<form id="form1" runat="server">
    欢迎<asp:Label ID="lblName" runat="server"></asp:Label>进入聊天室<br/>
    您的 SessionID 号为:<asp:Label ID="lblSessionID" runat="server"></asp:
Label>
</form>
```

(3) 为事件添加 C#代码如下。

example5-7.cs：

```
protected void btnLogin_Click(object sender, EventArgs e)
{
    Session["name"]=txtName.Text;
    Response.Redirect("example5-7-2.aspx");
}
```

example5-7-2.cs：

```
protected void Page_Load(object sender, EventArgs e)
{
    //判断若 Session["name"]值为 null,则跳转回登录页
    if (Session["name"] !=null)
    {
        lblName.Text=Session["name"].ToString();
        lblSessionID.Text=Session.SessionID;
    }
    else
    {
        Response.Redirect("example5-7.aspx");
    }
}
```

(4) 运行页面，填写用户名，单击“进入聊天室”按钮后进入聊天室页面，不同会话对应不同 SessionID 值，Session["name"]值不共享，运行效果如图 5.19 和图 5.20 所示。

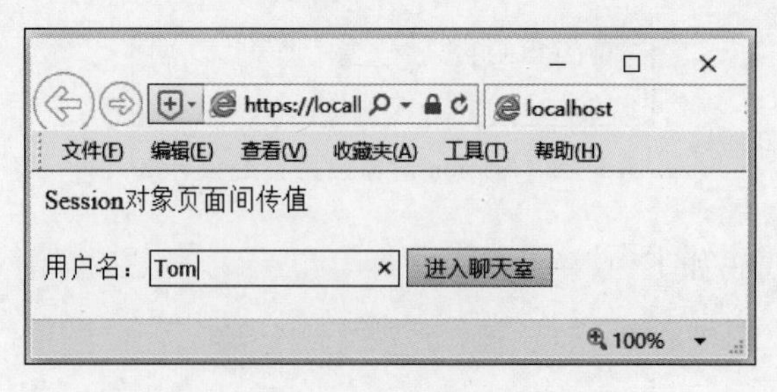

图 5.19 Session 对象页面演示 1

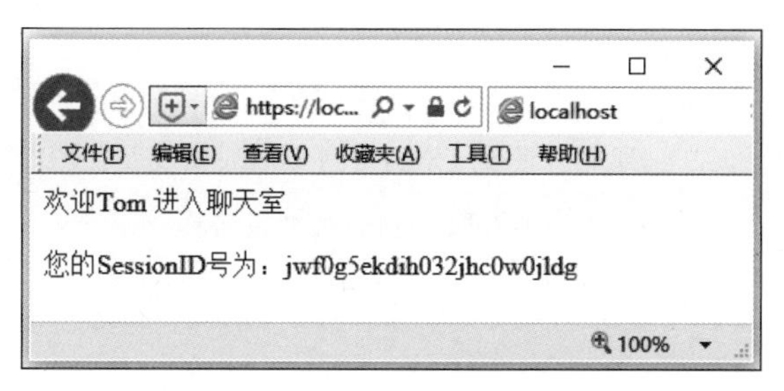

图 5.20　Session 对象页面演示 2

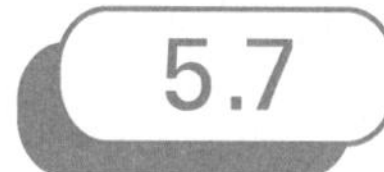

5.7 Cookie 对象

5.7.1 Cookie 对象的属性和方法

扫一扫

Cookie 对象是 HttpCookie 类的对象，是保存在客户端的一小段文本信息(4kb 左右)，可以保存少量数据，伴随着用户请求在 Web 服务器和浏览器之间传递，用户每次访问站点时，Web 应用程序都可以读取 Cookie 信息。

Cookie 对象与 Session、Application 类似，也用来保存相关信息，但它和其他对象最大的不同是 Cookie 将信息保存在客户端，而 Session 和 Application 将信息保存在服务器端。无论何时用户连接到服务器，Web 站点都可以访问 Cookie 信息，既方便用户的使用，也方便网站对用户的管理。

可以通过 HttpRequest 的 Cookies 集合访问客户端的 Cookie 文件，通过 HttpResponse 的 Cookies 集合创建新的文件并传输保存到客户端。Cookie 对象的主要属性如表 5.13 所示，主要方法如表 5.14 所示。

表 5.13　Cookie 对象的主要属性

属 性 名	说 明
Name	获取或设置 Cookie 的名称
Value	获取或设置 Cookie 的值
Values	获取在单个 Cookie 对象中包含的键值对集合
Expires	获取或设置 Cookie 的过期日期和时间
Version	获取或设置 Cookie 的 HTTP 版本

表 5.14　Cookie 对象的主要方法

方 法 名	说 明
Add()	新增一个 Cookie 变量
Clea()	清除 Cookies 集合内的变量

续表

方法名	说明
Get()	通过变量名或索引得到 Cookie 的变量值
GetKey()	以索引值获取 Cookie 的变量名称
Remove()	通过 Cookie 变量名删除 Cookie 变量

5.7.2 Cookie 对象的应用

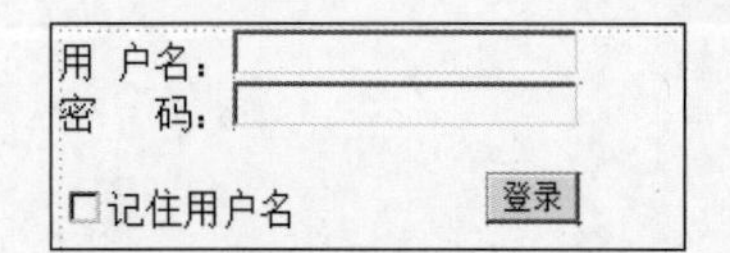

图 5.21 Cookie 对象应用页面设计

【示例 5-8】 在 E 盘 ASP.NET 项目代码的 chapter5 目录下，创建一个名为 example5-8 的网页，练习使用 Request 对象的各种属性。

(1) 在页面中添加相应控件，具体如图 5.21 所示。

(2) 设置相关控件的属性，如下列源代码所示。

```
<form id="form1" runat="server">
        用 户名:<asp:TextBox ID="txtID" runat="server"></asp:TextBox>
        <br/>
         密   码: < asp: TextBox ID = " txtPwd" runat = " server" TextMode = "
Password"></asp:TextBox><br/>
        <asp:CheckBox ID="chkRem" runat="server" Text="记住用户名"/>
     
       <asp:Button ID="btnLogin" runat="server" OnClick="btnLogin_Click" Text
="登录"/>
          <br/>
</form>
```

(3) 为事件添加 C#代码如下。

```
protected void Page_Load(object sender, EventArgs e)
{
     if (Request.Cookies["login"] !=null)
    {
        txtID.Text=Request.Cookies["login"].Values["id"];
    }
}
protected void btnLogin_Click(object sender, EventArgs e)
{
    if (chkRem.Checked)
    {
        HttpCookie cookie=new HttpCookie("login");
        cookie.Values.Add("id", txtID.Text);
        cookie.Expires=DateTime.Now.AddMonths(1);
        Response.Cookies.Add(cookie);
    }
    Response.Redirect("example5-8.aspx");
}
```

(4) 运行页面，勾选“记住用户名”复选框，可将用户名存入 Cookie 文件，下次登录时从客户端的 Cookie 文件中自动读取该用户信息，如图 5.22 所示。

图 5.22 Cookie 对象应用页面演示

说明：

由于浏览器不会传递 Cookies 过期时间，而且无法删除客户端的文件，因此无论是修改还是删除 Cookies，通常都是创建一个新的 Cookies 去覆盖原有的 Cookies 以达到该效果。

5.8 全局应用程序类 Global.asax 文件

Global.asax 文件也称作 ASP.NET 应用程序文件，位于应用程序根目录下，是一个可选文件，提供了一种响应应用程序级事件的方法。ASP.NET 页面框架能够自动识别出对 Global.asax 文件所做的任何更改，在 Global.asax 被更改后 ASP.NET 页面框架会重新启动应用程序，关闭所有的浏览器会话，去除所有状态信息，并重新启动应用程序域。

扫一扫

默认 Global.asax 文件的基本模板代码如下。

```
<%@Application Language="C#" %>
<script runat="server">
    void Application_Start(object sender, EventArgs e)
    {
        // 在应用程序启动时运行的代码
    }
    void Application_End(object sender, EventArgs e)
    {
        //  在应用程序关闭时运行的代码
    }
     void Application_Error(object sender, EventArgs e)
    {
        // 在出现未处理的错误时运行的代码
    }
    void Session_Start(object sender, EventArgs e)
    {
        // 在新会话启动时运行的代码
    }
    void Session_End(object sender, EventArgs e)
    {
        // 在会话结束时运行的代码。
        // 注意：只有在 Web.config 文件中的 sessionstate 模式设置为
```

```
        // InProc 时,才会引发 Session_End 事件。如果会话模式设置为 StateServer
        // 或 SQLServer,则不引发该事件。
    }
</script>
```

Global.asax 中处理的事件如表 5.15 所示。

表 5.15　Global.asax 中处理的事件

事 件 名	说 明
Application_Start	在应用程序(网站)第一次运行时触发执行
Session_Start	在每个会话第一次访问应用程序时触发执行
Application_Error	在应用程序的用户抛出一个错误时触发,适合于提供应用程序级的错误处理,或者把错误记录到服务器的事件日志中
Session_End	在会话或者超时失效时触发执行
Application_End	在应用程序(网站)结束时触发。因为 ASP.NET 内有很好的垃圾处理机制,可以有效地完成关闭和清理剩余对象的任务,所以通常该事件使用较少

【示例 5-9】 在 E 盘 ASP.NET 项目代码的 chapter5 目录下,创建名为 example5-9 的网页,在 Global.asax 文件中添加相关事件,综合使用 Session 对象和 Application 对象。

(1) 在页面中添加相应控件,具体如图 5.23 所示。

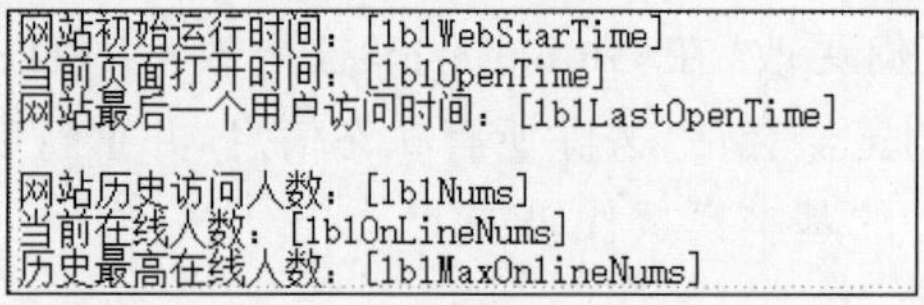

网站初始运行时间：[lblWebStarTime]
当前页面打开时间：[lblOpenTime]
网站最后一个用户访问时间：[lblLastOpenTime]

网站历史访问人数：[lblNums]
当前在线人数：[lblOnLineNums]
历史最高在线人数：[lblMaxOnlineNums]

图 5.23　Global.asax 文件应用页面设计

(2) 设置相关控件的属性,如下列源代码所示。

```
<form id="form1" runat="server">
        网站初始运行时间:<asp:Label ID="lblWebStarTime" runat="server"></asp:
Label><br/>
        当前页面打开时间:<asp:Label ID="lblOpenTime" runat="server"></asp:Label>
<br/>
        网站最后一个用户访问时间:<asp:Label ID="lblLastOpenTime" runat="server">
</asp:Label><br/>
        网站历史访问人数:<asp:Label ID="lblNums" runat="server"></asp:Label>
<br/>
        当前在线人数:<asp:Label ID="lblOnLineNums" runat="server" Text=""></asp:
Label><br/>
        历史最高在线人数:<asp:Label ID="lblMaxOnlineNums" runat="server" Text="">
</asp:Label>
</form>
```

(3) 为事件添加 C# 代码如下。

```
example5-9.cs:
protected void Page_Load(object sender, EventArgs e)
```

```
{
    lblWebStarTime.Text=Application["startTime"].ToString();
    lblOpenTime.Text=Session["openTime"].ToString();
    lblLastOpenTime.Text=Application["lastOpenTime"].ToString();
    lblNums.Text=Application["nums"].ToString();
    lblOnLineNums.Text=Application["online"].ToString();
    lblMaxOnlineNums.Text=Application["maxOnline"].ToString();
}
Global.asax:
void Application_Start(object sender, EventArgs e)
{
    Application["content"]="";
    //在应用程序启动时运行的代码
    //设置网站初始运行时间
    Application["startTime"]=DateTime.Now.ToString();
    //设置网站访问总人数为 0
    Application["nums"]=0;
    //设置网站在线人数为 0
    Application["online"]=0;
    //设置历史最高在线人数为 0
    Application["maxOnline"]=0;
}
void Session_Start(object sender, EventArgs e)
{
    //设置 Session 失效时间为 1 分钟
    Session.Timeout=1;
    //设置当前会话打开时间
    Session["openTime"]=DateTime.Now.ToString();
    Application.Lock();
    //设置整个网站内最新会话打开时间
    Application["lastOpenTime"]=DateTime.Now.ToString();
    //新会话打开,设置访问人数加 1
    Application["nums"]=(int)Application["nums"]+1;
    //新会话打开,设置在线人数加 1
    Application["online"]=(int)Application["online"]+1;
    //若当前在线人数大于历史最高在线人数,则重新赋值历史最高在线人数
    if ((int)Application["online"] >(int)Application["maxOnline"])
    {
        Application["maxOnline"]=Application["online"];
    }
    Application.UnLock();
}
void Session_End(object sender, EventArgs e)
{
    Application.Lock();
    //会话关闭,设置在线人数减 1
    Application["online"]=(int)Application["online"]-1;
    Application.UnLock();
}
```

（4）运行页面，打开多个浏览器，输入网址 http://localhost:24292/example5-9.aspx，模拟多用户访问网站，分析各事件的触发执行以及 Application 对象和 Session 对象的作用范围，运行效果如图 5.24 所示。

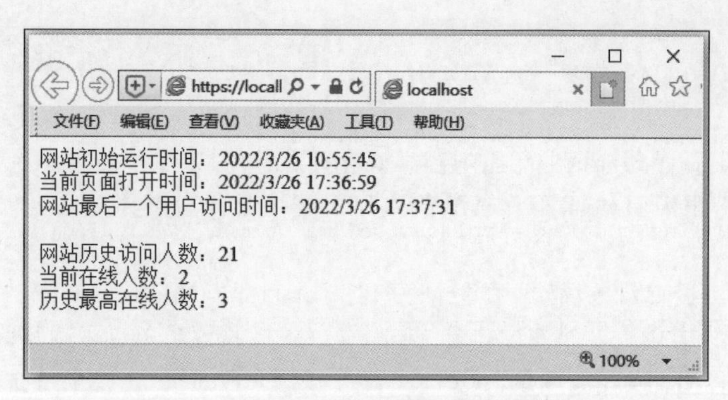

图 5.24　Global.asax 文件应用页面演示

说明：

(1) Session.Timeout 属性表示 Session 失效时间，默认值为 20 分钟，页面处于非活动状态或者关闭后 20 分钟后触发 Session_End 事件。Session.Timeout 的最小值为 1(1 分钟)，最大值为 1440(24 小时)。该数值越小越早触发 Seeion_End 事件，但短有效期会让登录等操作很快失效；该数值越大有效期越长，但较晚触发 Seeion_End 事件也会导致同步性较差。

(2) Application 对象的值为网站中所有会话共享，可能出现两个或者多个用户同时对这一变量进行操作而产生冲突的情况，Application.Lock()和 Application.Unlock()就是为了解决这一问题的，使用 Lock()就能确保在某一时段所有连接到服务器的用户中只有一个用户能获得存取或修改该 Application 变量的权限，即对该公共变量进行锁定操作，其他用户只能等当前权限用户执行 UnLock()方法结束其锁定或者当前 ASP.NET 程序终止执行才可继续访问。

(3) 本示例只做了最基本的网站人数统计，网站的实际应用中可在 Application_End 事件内添加代码，将网站关闭时的统计信息存入数据库或文件，在 Application_Start 事件内添加代码，在网站再次运行时读取前期统计信息。

综合实验五　简易购物车

主要任务

创建 ASP.NET 应用程序，添加 Web 窗体实现基于 Session 内置对象的简易购物车模块，其具有的主要功能见表 5.16。

表 5.16　简易购物车功能介绍

功 能 名	对 象 页 面	功 能 描 述
用户登录	Login.aspx	对用户名进行非空验证，将用户输入的用户名存入 Session 对象，进行页面跳转
显示商品列表	综合实验五.aspx	在复选框列表中显示商品的信息，包括名称及图片

续表

功能名	对象页面	功能描述
添加购物车	综合实验五.aspx	根据复选框列表中商品的选中状态,将其存入 Session 对象
查看购物车	ShoppingCart.aspx	从 Session 对象中读出购物车信息,并将其显示在复选框列表中

实验步骤

步骤 1：在 VS 2019 菜单上选择“文件”→“新建”→“项目”命令。

步骤 2：创建 ASP.NET 应用程序,命名为“综合实验五”。

步骤 3：在网站的根目录上右击,创建“综合实验五.aspx”窗体,添加相应控件,具体如图 5.25 所示。

图 5.25 注册模块页面布局

步骤 4：按如下源文件设置控件的相关属性值。

```
<form id="form1" runat="server">
        <div>
            <table>
                <tr>
                    <td> </td>
                    <td>当前账号:<asp:Label ID="lbID" runat="server" Text=""></
asp:Label>
                    </td>
                    <td>
                        <asp:LinkButton ID="lbtnMyShoppingCart" runat="server"
  PostBackUrl="~/ShoppingCart.aspx">我的购物车</asp:LinkButton>
                    </td>
                    <td> </td>
                    <td> </td>
                </tr>
                <tr>
                    <td> </td>
                    <td colspan="2" rowspan="2">
                          <asp:CheckBoxList ID="chklGoodsList" runat="server"
RepeatDirection="Horizontal" TextAlign="Left">
                        </asp:CheckBoxList>
                    </td>
                    <td> </td>
                    <td> </td>
                </tr>
                <tr>
                    <td> </td>
                    <td> </td>
                    <td> </td>
                </tr>
                <tr>
                    <td> </td>
                    <td> </td>
                    <td> </td>
       <td> </td>
```

```
                    <td> </td>
                </tr>
                <tr>
                    <td> </td>
                    <td>
                         <asp:Button ID="btnAddShoppingCart" runat="server" Text=
"加入购物车" OnClick="btnAddShoppingCart_Click"/>
                    </td>
                    <td>
                     <asp:Button ID="btnCloseWebsite" runat="server" Text="关闭
当前页"/>
                    </td>
                    <td> </td>
                    <td> </td>
                </tr>
            </table>
```

步骤 5：在综合实验五.aspx.cs 文件中添加 System.Collections 命名空间引用，并编辑代码如下。

```
using System.Collections;
protected void Page_Init(object sender, EventArgs e)
{
    ListItem[] listItems=new ListItem[]
    {
        new ListItem(@"<img src='image\beverage.png' width='100' height='100'>",
"饮料汽水"),
        new ListItem(@"<img src='image\candy.png' width='100' height='100'>","糖果
巧克力"),
        new ListItem(@"<img src='image\food.png' width='100' height='100'>","休闲
食品"),
        new ListItem(@"<img src='image\fruit.png' width='100' height='100'>","新鲜
水果"),
        new ListItem(@"<img src='image\pickles.png' width='100' height='100'>","酱
菜辅料"),
    };
    foreach (ListItem li in listItems)
    {
        chklGoodsList.Items.Add(li);
    }
}
protected void Page_Load(object sender, EventArgs e)
{
    if (Session["id"] !=null)
    {
         lbID.Text=Session["id"].ToString();
         lbtnMyShoppingCart.Enabled=true;
     }
     else
     {
         lbtnMyShoppingCart.Enabled=false;
     }
```

```
    }
    protected void btnAddShoppingCart_Click(object sender, EventArgs e)
    {
        if (Session["id"]==null)
        {
            Response.Write("<script>alert('请先登录');location='Login.aspx';
</script>");
        }
        else if(chklGoodsList.SelectedIndex!=-1)
        {
            ArrayList arrayList=new ArrayList();
            foreach (ListItem li in chklGoodsList.Items)
            {
                if(li.Selected)
                {
                    arrayList.Add(li);
                }
            }
            Session["shoppingcart"]=arrayList;
            Response.Write("<script>alert('添加购物车成功!');</script>");
        }
        else
        {
            Response.Write("<script>alert('请选择要添加的商品!');</script>");
        }
    }
```

步骤 6：在网站的根目录上右击，创建 Login.aspx 窗体，添加相应控件，具体如图 5.26 所示。

图 5.26　登录页面测试

步骤 7：按如下源文件设置 Login.aspx 窗体控件的相关属性值。

```
<form id="form1" runat="server">
        <table>
                <tr>
                    <td> </td>
                    <td colspan="2">XX 购物网站</td>
                    <td> </td>
                </tr>
                <tr>
                    <td> </td>
                    <td>登录账号</td>
                    <td>
                        <asp:TextBox ID="txtID" runat="server"></asp:TextBox>
                         <asp:RequiredFieldValidator ID="RequiredFieldValidator1"
runat=" server" ControlToValidate =" txtID" ErrorMessage =" *" ForeColor =" Red"
SetFocusOnError="True"></asp:RequiredFieldValidator>
```

```
                </td>
                <td> </td>
            </tr>
            <tr>
                <td> </td>
                <td> </td>
                <td>
                      < asp:Button ID="btnLogin" runat="server" OnClick="
btnLogin_Click" Text="登录" Width="65px"/>
 
                             < asp: Button ID = " btnClose " runat = " server "
CausesValidation="False" OnClientClick="window.close();" Text="关闭" Width="
65px"/>
                </td>
                <td> </td>
            </tr>
            <tr>
                <td> </td>
                <td> </td>
                <td> </td>
                <td> </td>
            </tr>
        </table>
    </form>
```

步骤 8：编辑 Login.aspx.aspx.cs 文件代码如下。

```
protected void btnLogin_Click(object sender, EventArgs e)
{
    Session["id"]=txtID.Text;
    Response.Redirect("综合实验五.aspx");
}
```

步骤 9：在网站的根目录上右击，创建 ShoppingCart.aspx 窗体，添加相应控件，具体如图 5.27 所示。

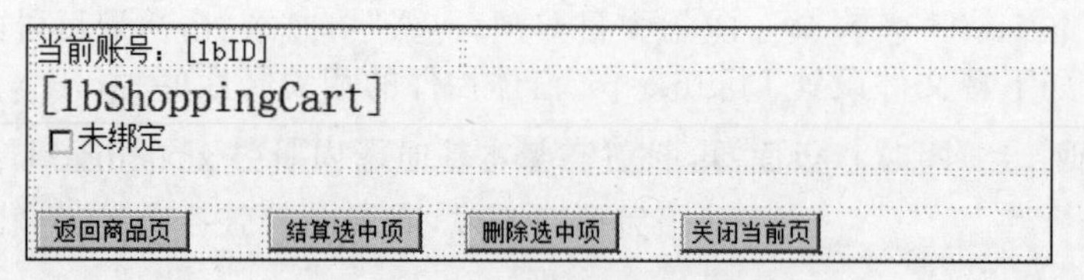

图 5.27　购物车页面运行测试

步骤 10：按如下源文件设置 ShoppingCart.aspx 窗体控件的相关属性值。

```
<form id="form1" runat="server">
    <table>
            <tr>
                <td> </td>
                  <td colspan="2">当前账号:<asp:Label ID="lbID" runat="
server" Text=""></asp:Label>
                </td>
```

```
                <td colspan="2"> </td>
                <td>
                     </td>
                <td> </td>
                <td> </td>
            </tr>
            <tr>
                <td> </td>
                <td colspan="5" rowspan="2">
                    <asp:Label ID="lbShoppingCart" runat="server" Font-Size
="X-Large" ForeColor="#3333CC"></asp:Label>
                    <asp:CheckBoxList ID="chklShoppingCart" runat="server"
RepeatDirection="Horizontal">
                    </asp:CheckBoxList>
                </td>
                <td> </td>
                <td> </td>
            </tr>
            <tr>
                <td> </td>
                <td> </td>
                <td> </td>
            </tr>
            <tr>
                <td> </td>
                <td colspan="4">
                     </td>
                <td>
                     </td>
                <td> </td>
                <td> </td>
            </tr>
            <tr>
                <td> </td>
                <td>
                    <asp:Button ID="btnBacktoGoodsList" runat="server" Text
="返回商品页"  PostBackUrl="~/综合实验五.aspx" Width="84px"/>
                </td>
                <td>
                    <asp:Button ID="btnBalance" runat="server" Text="结算选
中项" Width="84px"/>
                </td>
                <td>
                    <asp:Button ID="btnDelete" runat="server" Text="删除选中
项" Width="84px"/>
                </td>
                <td>
                    <asp:Button ID="btnCloseWebsite" runat="server" Text="
关闭当前页" CausesValidation="False" OnClientClick="window.close();" Width="
84px"/>
                </td>
                <td>
                     </td>
                <td> </td>
                <td> </td>
```

```
                </tr>
                <tr>
                    <td> </td>
                    <td colspan="4"> </td>
                    <td> </td>
                    <td> </td>
                    <td> </td>
                </tr>
        </table>
</form>
```

步骤 11：在 ShoppingCart.aspx.cs 文件中添加 System.Collections 命名空间的引用，并编辑代码如下。

```
using System.Collections;
protected void Page_Load(object sender, EventArgs e)
{
    if (Session["id"] !=null)
    {
        lbID.Text=Session["id"].ToString();
        if (Session["shoppingcart"]==null)
        {
            lbShoppingCart.Text="购物车中 0 件商品";
        }
        else
        {
            ArrayList arrayList=(ArrayList)Session["shoppingcart"];
            lbShoppingCart.Text=string.Format("购物车中{0}件商品", arrayList.
Count);
            foreach (object obj in arrayList)
            {
                chklShoppingCart.Items.Add((ListItem)obj);
            }
        }
    }
    else
    {
        Response.Write("<script>alert('请先登录');location='Login.aspx';
</script>");
    }
}
```

步骤 12：注册模块页面运行测试如图 5.28 所示，添加购物车如图 5.29 所示，查看购物车如图 5.30 所示。

图 5.28　注册模块页面运行测试

图 5.29 添加购物车

图 5.30 查看购物车

第6章 主题、母版页与用户控件

本章学习目标

- 了解主题的基本使用方法
- 熟练掌握母版页的使用方法
- 熟练掌握用户控件的使用方法

本章首先介绍主题的概念，创建主题并进行主题的动态选择应用，然后讲解母版页的基本原理及应用，最后讲解自定义控件的应用以及 Web 窗体与自定义控件的转换。

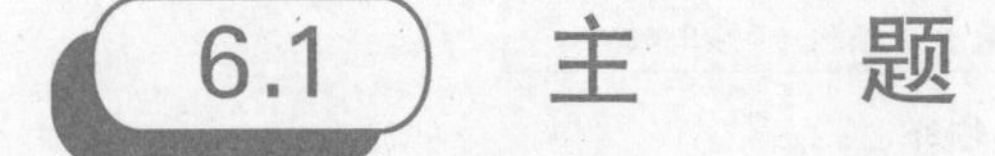

6.1 主　　题

扫一扫

6.1.1 主题的简单应用

在 Web 应用程序开发中，良好的 Web 应用界面能够让访问者耳目一新。网站的美观主要涉及页面布局和控件的属性设置，具体实现中通过 CSS 可以控制静态页面布局以及各标签的样式，但服务器控件的大部分属性无法通过 CSS 样式表进行控制，为了解决这一问题，ASP.NET 提供了主题这一新外观设置方式。

主题(Theme)是控件属性设置的集合，包括一系列元素，如皮肤(.skin 文件)、样式表(.css 文件)、图像等资源，通过主题的设置能够定义页面和控件的样式。主题文件的创建方法如图 6.1 所示。主题文件通常保存在 Web 应用程序的特殊目录下，创建外观文件时 VS 2019 会提示是否将文件存放到特定目录，如图 6.2 所示。

单击“是”按钮后，主题文件会存放在 App_Themes 文件夹中，可以根据需要在外观模板中分别设置网站默认外观标记和包含皮肤 SkinId 属性的命名外观标记，主题提示文本如下。

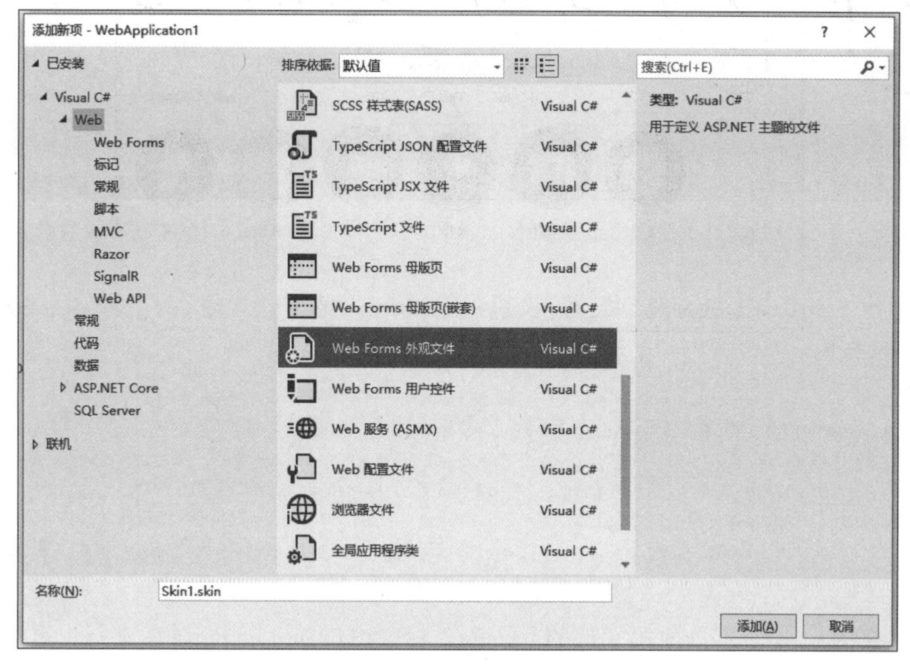

图 6.1　创建新外观文件

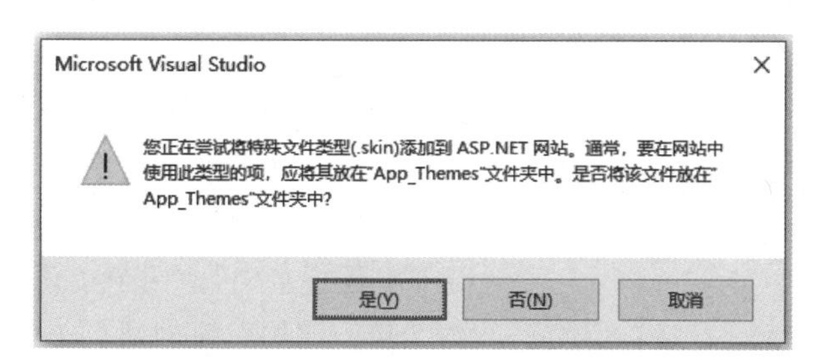

图 6.2　将主题文件存放在系统文件夹

```
<%--
默认的外观模板。以下外观仅作为示例提供。
1. 命名的控件外观。SkinId 的定义应唯一，因为在同一主题中不允许一个控件类型有重复的
SkinId。
<asp:GridView runat="server" SkinId="gridviewSkin" BackColor="White" >
    <AlternatingRowStyle BackColor="Blue" />
</asp:GridView>
2. 默认外观。未定义 SkinId。在同一主题中每个控件类型只允许有一个默认的控件外观。
<asp:Image runat="server" ImageUrl="~/images/image1.jpg" />
--%>
```

【示例 6-1】　在 E 盘 ASP.NET 项目代码目录中创建 chapter6 子目录，将其作为网站根目录，创建名为 example6-1 的网页。

（1）在 example6-1 页面中添加若干 Button 按钮，如图 6.3 所示。

（2）在网站根目录下添加名为 Skin1.skin 的皮肤文件，存放在 App_Themes 文件夹的子文件夹 Skin1 内，具体如图 6.4 所示。

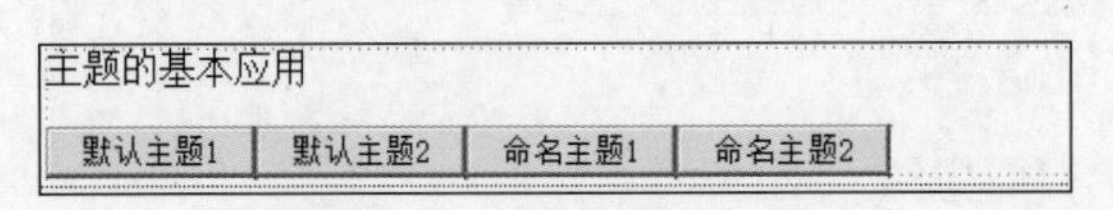

图 6.3　主题应用页面设计

图 6.4　主题存放目录

(3) 对 Skin1.skin 文件进行配置,设置一个 Button 的默认主题,两个命名主题 red 和 blue,代码如下。

```
<asp:Button  runat="server"  Font-Bold="True" Font-Names="Algerian"
ForeColor="#669999" Height="30px" />
<asp:Button SkinID="blue" runat="server"  BackColor="#3333CC" BorderStyle="
Dashed" BorderWidth="1px" Font-Size="30pt" ForeColor="#FFFF99" />
<asp:Button SkinID="red" runat="server"  BackColor="Red" Font-Bold="True"
Font-Italic="True" Font-Overline="True" Font-Size="25pt" Font-Strikeout="
True" Font-Underline="True" />
```

(4) 在 example6-1 页面源文件的＜Page＞标签内添加 Theme 属性进行主题声明,如果不声明主题,则页面无法找到该主题,示例代码如下。

```
<%@Page Language="C#" AutoEventWireup="true" CodeFile="example6-1.aspx.cs"
Inherits="example6_1" Theme="Skin1" %>
```

(5) 设置"命名主题 1"和"命名主题 2"按钮的 SkinID 属性值,分别赋值为 red 和 blue。可以在源文件中直接设置,也可以在属性窗口中选择,具体如图 6.5 所示。

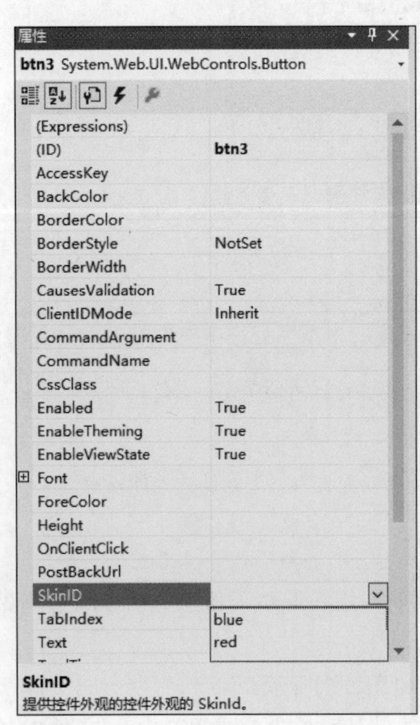

图 6.5　控件主题 ID 属性设置

(6) 运行页面,前两个按钮直接应用默认主题,后两个按钮应用对应的命名主题,显示外观如图 6.6 所示。

说明:

控件皮肤标签内的属性与控件声明标签内的属性基本相同,构建皮肤标签比较简单的方法是直接将已经设置好外观的控件标签粘贴到皮肤文件中,然后去掉唯一标识 ID 属性和呈现文本外观的 Text 等属性即可,其中 ID 属性是必须去掉的。

6.1.2　页面主题和全局主题

主题文件可以应用于网站的一个或多个页面,这种主题文件称为"页面主题"。主题文件也可应用于网站的每个页面,使得每个页面都使用默认的主题,这种主题文件称为"全局主题"。页面主题可以灵活地设置每个页面的风格,上述示例 6-1 使用的就是页面主题;而全局主题可以使整个网站中的页

图 6.6　主题应用页面演示

面保持一致的风格，大型网站通常采用全局主题。

使用全局主题时，需要修改 Web.config 配置文件中的<pages>配置进行主题的全局设定，添加代码如下。

```
<system.web>
    <pages theme="Skin1">
    </pages>
</system.web>
```

当一个控件使用主题后，再对页面中该控件的属性进行设置是没有任何效果的，只要应用了默认主题或者命名主题，就会服从主题属性设置。对于页面而言，如果既存在页面主题，也存在全局主题，页面主题的属性将会被全局属性更改，全局属性中没有设置的属性将继续保留。

如果控件或页面已经定义了外观，不希望主题将其属性进行重写和覆盖，就可以禁用主题，对页面可以用声明的方法进行禁用，示例代码如下。

```
<%@Page Language="C#" AutoEventWireup="true" EnableTheming="false" %>
```

当页面的某个控件不使用主题时，将该控件的 EnableTheming 属性值赋值为 False，即可实现该控件不应用主题。

6.1.3　主题的动态选择

扫一扫

当主题制作完成后，通过编写 C＃代码更改页面的 StyleSheetTheme 属性就能对页面的主题进行更改，同样可以更改控件的主题，达到动态更改控件主题的效果。StyleSheetTheme 属性的更改代码只能编写在 Page 对象的 PreInit 事件中。

【示例 6-2】　在 chapter6 网站根目录下创建名为 example6-2 和 example6-2-2 的网页，练习主题的动态选择。

(1) 在 example6-2 和 example6-2-2 页面分别添加若干控件，具体如图 6.7 和图 6.8 所示。

example6-2 页面：

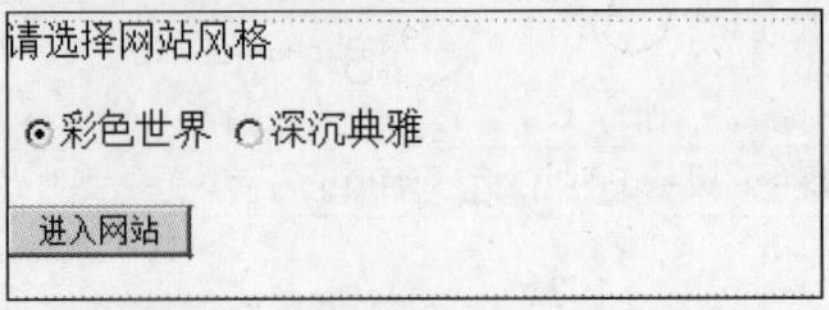

请选择网站风格

⊙彩色世界 ○深沉典雅

进入网站

图 6.7　动态主题应用页面设计 1

example6-2-2 页面：

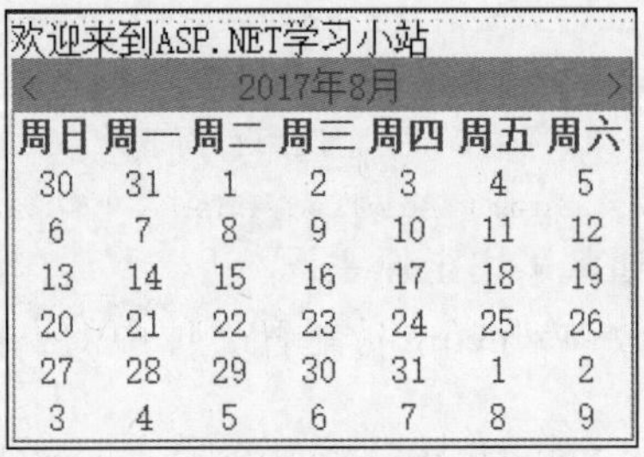

欢迎来到ASP.NET学习小站

<			2017年8月			>
周日	周一	周二	周三	周四	周五	周六
30	31	1	2	3	4	5
6	7	8	9	10	11	12
13	14	15	16	17	18	19
20	21	22	23	24	25	26
27	28	29	30	31	1	2
3	4	5	6	7	8	9

图 6.8　动态主题应用页面设计 2

(2) 在“属性窗口”或“源”视图中修改控件属性如下。

example6-2 页面：

```
<form id="form1" runat="server">
    <div>
        请选择网站风格<br />
        <asp:RadioButtonList ID="rbtnlTheme" runat="server" RepeatDirection=
"Horizontal">
            <asp:ListItem Selected="True" Value="Skin2">彩色世界</asp:ListItem>
            <asp:ListItem Value="Skin3">深沉典雅</asp:ListItem>
            </asp:RadioButtonList>
        <br />
      <asp:Button ID="btnLogin" runat="server" OnClick="btnLogin_Click" Text=
"进入网站" />
         </div>
    </form>
```

example6-2-2 页面：

```
<form id="form1" runat="server">
    <div>
       <asp:Label ID="lblInfo" runat="server" Text="欢迎来到 ASP.NET 学习小站">
</asp:Label>
        <br />
        <asp:Calendar ID="Calendar1" runat="server" ForeColor="Gray"></asp:
Calendar>
    </div>
</form>
```

(3) 为页面添加相关事件,并编辑代码如下。

```
    example6-2.cs:
protected void btnLogin_Click(object sender, EventArgs e)
{
    Session["theme"]=rbtnlTheme.SelectedValue;
    Response.Redirect("example6-2-2.aspx");
}
    example6-2-2.cs:
protected void Page_PreInit(object sender, EventArgs e)
{
    //若未选择主题,则跳转页面
    if (Session["theme"]==null)
    {
        Response.Redirect("example6-2.aspx");
    }
    else
    {
       //设置页面应用的主题
        Page.Theme=Session["theme"].ToString();
    }
}
```

(4) 在网站根目录下添加名为 Theme1 和 Theme2 的文件夹以及 Skin2.skin 和 Skin3.skin 的皮肤文件,具体结构如图 6.9 所示。

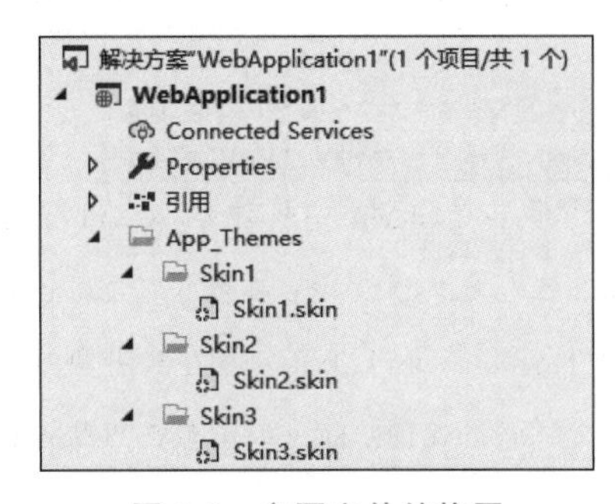

图 6.9　主题文件结构图

(5) 在 Skin2.skin 和 Skin3.skin 中设置主题属性,代码如下。

```
Skin2.skin:
<asp: Calendar runat =" server" BackColor =" # FFFFCC" BorderColor =" # FFCC66"
BorderWidth="1px" DayNameFormat="Shortest" Font-Names="Verdana"
         Font-Size="8pt" ForeColor="#663399" Height="200px" ShowGridLines="
True" Width="220px">
         <DayHeaderStyle BackColor="#FFCC66" Font-Bold="True" Height="1px" />
         <NextPrevStyle Font-Size="9pt" ForeColor="#FFFFCC" />
         <OtherMonthDayStyle ForeColor="#CC9966" />
         <SelectedDayStyle BackColor="#CCCCFF" Font-Bold="True" />
         <SelectorStyle BackColor="#FFCC66" />
         <TitleStyle BackColor="#990000" Font-Bold="True" Font-Size="9pt"
ForeColor="#FFFFCC" />
         <TodayDayStyle BackColor="#FFCC66" ForeColor="White" />
      </asp:Calendar>
<asp:Label runat="server" Font-Bold="True" Font-Italic="True" Font-Size="
30pt" ForeColor="Red" ></asp:Label>
```

```
Skin3.skin:
  < asp: Calendar   runat = " server " BackColor = " White " BorderColor = " Black " DayNameFormat="Shortest"
            Font-Names="Times New Roman" Font-Size="10pt" ForeColor="Black" Height="220px"
          NextPrevFormat="FullMonth" TitleFormat="Month" Width="400px">
          <DayHeaderStyle BackColor="#CCCCCC" Font-Bold="True" Font-Size="7pt" ForeColor="#333333" Height="10pt" />
          <DayStyle Width="14%" />
          <NextPrevStyle Font-Size="8pt" ForeColor="White" />
          <OtherMonthDayStyle ForeColor="#999999" />
          <SelectedDayStyle BackColor="#CC3333" ForeColor="White" />
          <SelectorStyle BackColor="#CCCCCC" Font-Bold="True" Font-Names="Verdana" Font-Size="8pt" ForeColor="#333333" Width="1%" />
          <TitleStyle BackColor="Black" Font-Bold="True" Font-Size="13pt" ForeColor="White" Height="14pt" />
          <TodayDayStyle BackColor="#CCCC99" />
      </asp:Calendar>
  <asp:Label  runat="server" Font-Bold="True"  Font-Size="30pt" ForeColor="Gray" ></asp:Label>
```

(6) 运行 example6-2 页面，选择网站风格后，单击“进入网站”按钮，如图 6.10 所示。选择“深沉典雅”主题后，example6-2-2 页面运行效果如图 6.11 所示。

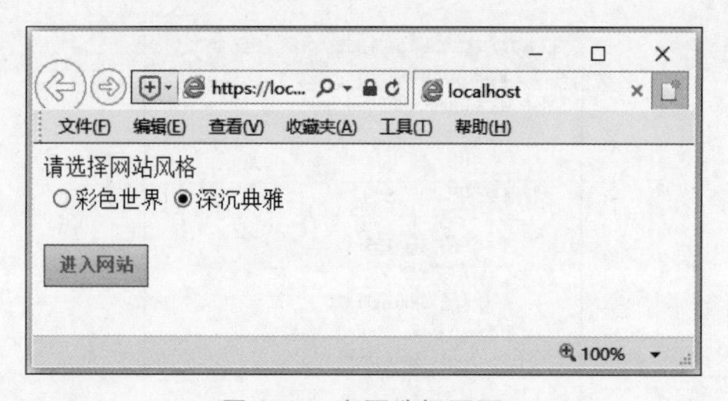

图 6.10 主题选择页面

图 6.11 “深沉典雅”主题页面演示

6.2 母版页

在 Web 应用开发过程中,很多元素在页面的布局及基本内容都是相同的,如导航栏、版权信息等。可以使用 CSS 和主题实现多页面布局,同时 ASP.NET 4.7 还提供了更加方便实用、健壮的母版页技术。

6.2.1 母版页基础

母版页是用来设置页面公共部分外观和功能的模板,是一种特殊的 ASP.NET 网页文件。使用母版页可以定义某一组页面的呈现样式,从而实现定义整个网站页面的呈现样式。

母版页的扩展名以.master 结尾,可以在页面内放置 HTML 控件和 Web 控件等。母版页仅是一个页面模板,不能被浏览器直接查看,必须被其他页面使用后才能显示。母版页的使用与普通页的使用一样,可以可视化地设计,也可以编写后置代码,与普通页不一样的是,它包含 ContentPlaceHolder 控件。ContentPlaceHolder 控件是预留显示内容的页面区域。

母版页和内容页有着严格的对应关系,母版页中包含多少个 ContentPlaceHolder 控件,内容页中也必须设置多少个与其相对应的 Content 控件,当客户端浏览器向服务器发出请求,要求浏览某个内容页面时,引擎将同时执行内容页和母版页的代码,并将最终结果发送给客户端浏览器,具体如图 6.12 所示。

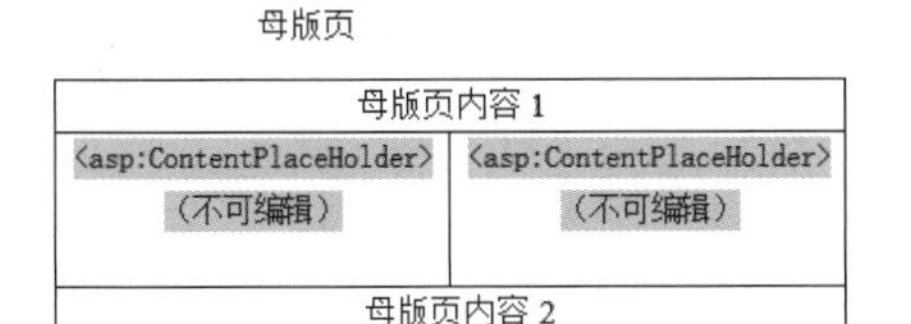

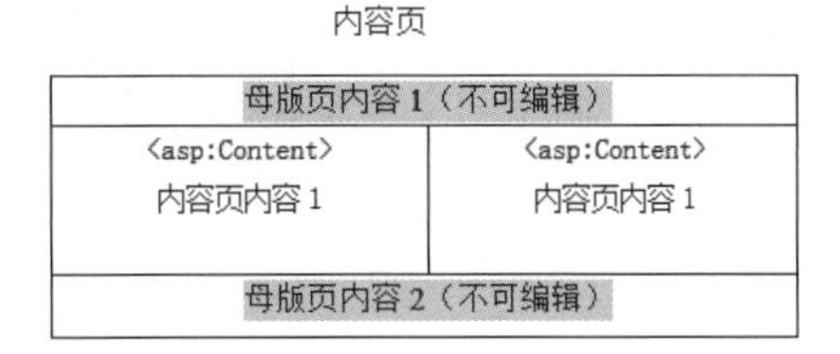

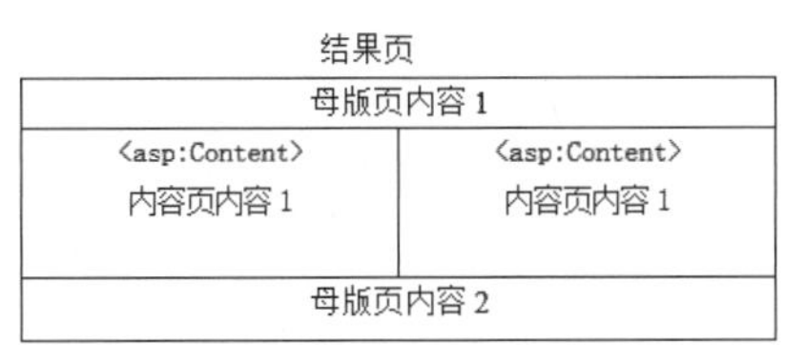

图 6.12 母版页和内容窗体

在母版页运行后,内容窗体中的 Content 控件会被映射到母版页的 ContentPlaceHolder 控件,并向母版页中的 ContentPlaceHolder 控件填充自定义控件。母版页和内容窗体将会整合形成结果页面,然后呈现给用户的浏览器。母版页运行的具体步骤如下。

(1) 通过 URL 指令加载内容页面。

(2) 处理页面指令。

(3) 将更新过内容的母版页合并到内容页面的控件集合里。

(4) 将单独的 ContentPlaceHolder 控件的内容合并到相对的母版页中。

(5) 加载合并的页面并显示给浏览器。

扫一扫

6.2.2 母版页的应用

从用户的角度来说，母版页和内容窗体的运行并没有本质的区别，运行过程中URL是唯一的。而从开发人员的角度来说，母版页和内容窗体是单独而离散的页面，分别进行各自的工作，在运行后合并生成相应的结果页面呈现给用户。下面以具体示例讲解母版页的应用。

【示例6-3】 在chapter6网站根目录下创建一个母版页，简单设置母版页，并添加内容页example6-3使其继承自该母版页。

(1) 在网站根目录右击"添加项"选项，选择"母版页"项目，向项目中添加一个母版页，如图6.13所示。

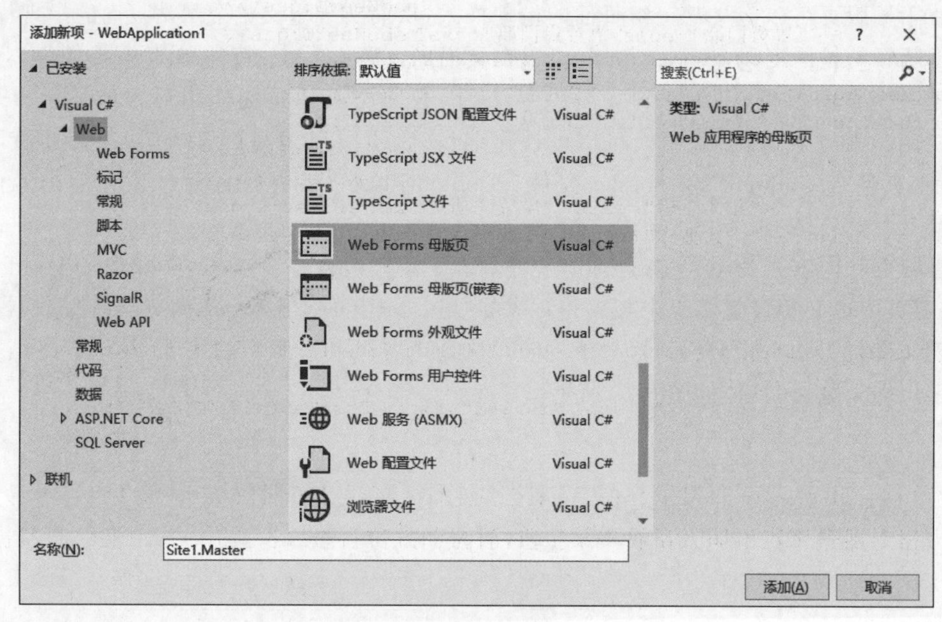

图6.13 添加母版页

(2) 对MasterPage.master母版页进行页面布局，拆分为如图6.14所示的5部分。母版页最终页面设计如图6.15所示。

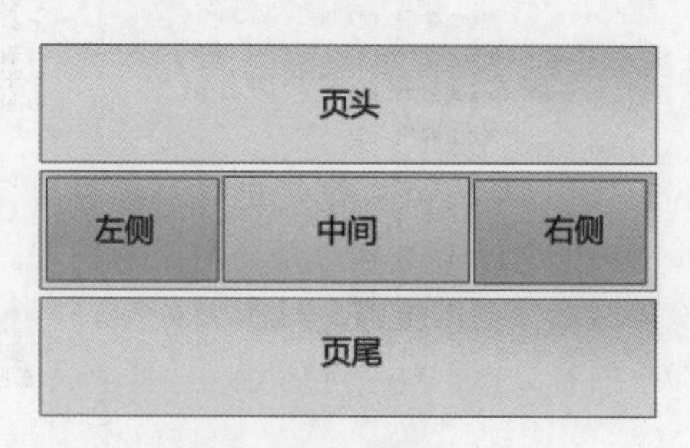

图6.14 母版页面布局

(3) 在页头、页尾分别设置文本内容，对中部内容项添加ContentPlaceHolder控件，设

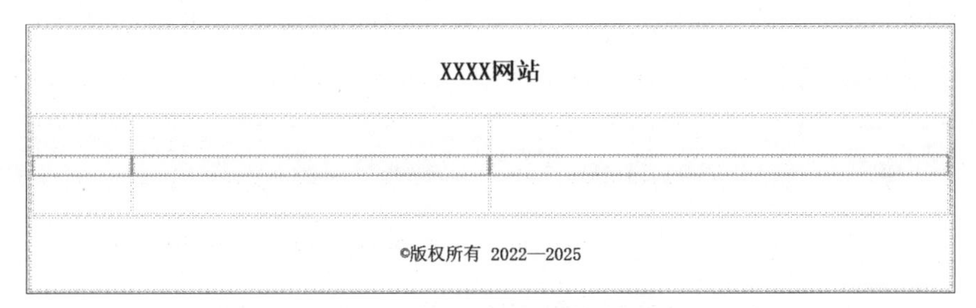

图 6.15　母版页最终页面效果

置源文件如下。

```
<form id="form1" runat="server">
        <div>
      <table>
          <tr>
              <td>
                  XXXX 网站</td>
          </tr>
          <tr>
              <td>
                  <table>
                      <tr>
                          <td>
                              <asp: ContentPlaceHolder id=" ContentPlaceHolder1"
runat="server">
                               </asp:ContentPlaceHolder>
                          </td>
                          <td>
                            <asp: ContentPlaceHolder  id =" ContentPlaceHolder2"
runat="server">
                          </asp:ContentPlaceHolder></td>
                          <td>
                              <asp: ContentPlaceHolder id=" ContentPlaceHolder3"
runat="server">
                            </asp:ContentPlaceHolder>
                          </td>
                      </tr>
                  </table>
              </td>
          </tr>
          <tr>
              <td>
              &copy;版权所有 2022—2025
              </td>
          </tr>
      </table>
    </div>
</form>
```

(4) 添加内容页 example6-3，在创建 Web 窗体时选择“包含母版页的 Web 窗体”，如图 6.16 所示。单击“添加”按钮，系统会提示选择相应的母版页，选择相应的母版页后，单击

"确定"按钮即可创建内容页窗体,如图 6.17 所示。

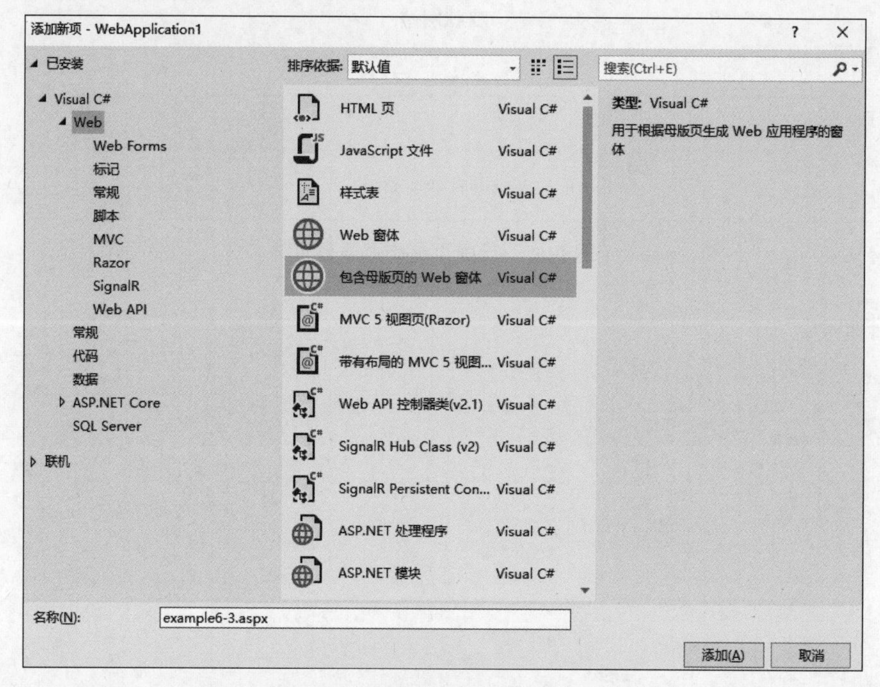

图 6.16 新建包含母版页的内容页

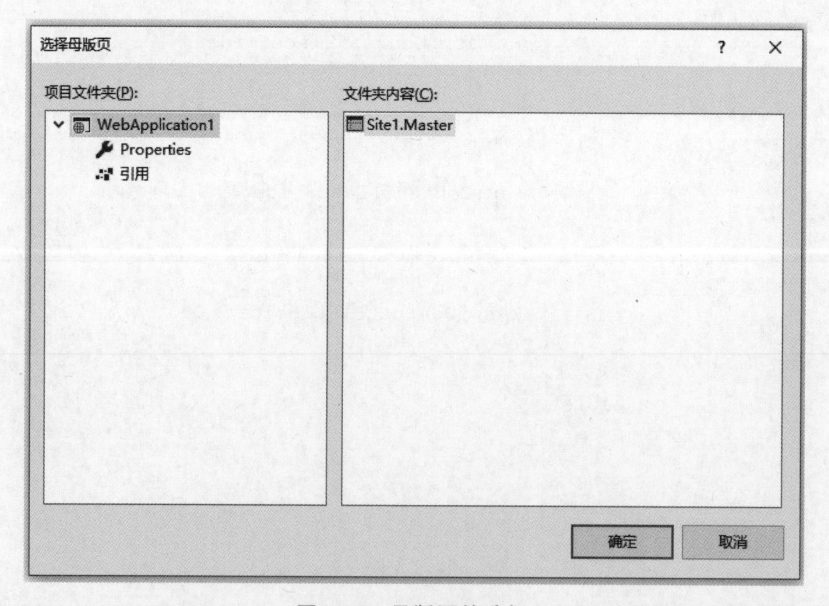

图 6.17 母版页的选择

(5) 内容页如图 6.18 所示,继承了母版页的基本页面布局,对应母版页的三个 ContentPlaceHolder 控件,自动创建三个对应的 Content 控件区,可在该内容标签内增加控件或自定义内容。

(6) 向内容页的三个占位符添加简单的文本,运行页面,最终效果如图 6.19 所示。

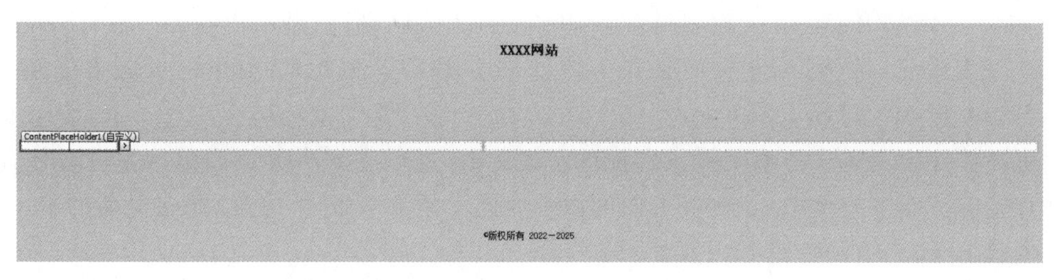

图 6.18　内容页

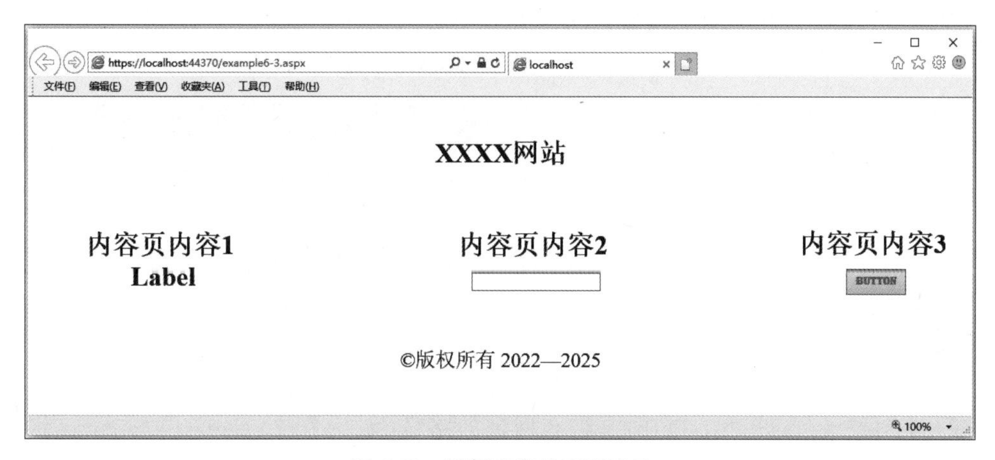

图 6.19　母版页应用页面演示

当编写 Web 应用时，可以使用母版页设计较大型的框架布局，对一个页面进行整体样式的控制，可以使用母版页与母版页嵌套，对细节的地方进行细分。

6.3　用户控件

通过本章前两节的学习，使用主题可以增强属性设置的可复用性，使用母版页可以增强页面设计的可复用性，本节将讲解如何使用自定义控件提升控件的可复用性。在 ASP 编程中，开发人员经常使用 Include 方式包含其他文件从而简化编程过程，而在 ASP.NET 中，使用控件能够提高应用程序中控件的复用性。在 ASP.NET 中提供的用户控件，允许用户进行控件组合并编写代码实现某些特定功能。

6.3.1　用户控件基础

根据应用程序的需求，方便地定义和编写用户控件，所使用的编程技术与编写 Web 窗体的技术相同，并且当对控件进行修改时，可同步更改所有使用该控件的页面。

用户控件创建完毕后，会生成一个.ascx 页面，源文件如下。

```
<%@Control Language="C#" AutoEventWireup="true" CodeFile="UserLogin.ascx.cs"
Inherits="UserLogin" %>
```

.ascx 页面结构同.aspx 页面编辑区域基本没有太大区别，只是用户控件中没有<html><body>等标记，因为.ascx 页面是作为控件被引用到其他页面，引用的页面中已经包含<body><html>等标记。

用户控件使用方法与系统提供的控件基本一致，可以在解决方案内鼠标左键按住.ascx 文件直接向页面中拖曳添加，对应页面的源中会同步添加<ucl>标签，并在页面的 Page 标签下生成 Register 标签，基本内容如下。

```
<%@Register Src="~/UserLogin.ascx" TagPrefix="uc1" TagName="UserLogin" %>
```

在标签内声明了用户控件的引用，其主要属性如表 6.1 所示。

表 6.1 Register 标签的主要属性

属 性 名	说 明
TagPrefix	定义控件位置的命名空间，从而在同一页面中使用不同功能
TagName	指向所用控件的名字
Src	用户控件的文件路径，可以为相对路径或绝对路径，但不能使用物理路径

说明：

(1) 物理路径(Physical Path)是指硬盘上文件的路径，如 d:\asp\html\a.html。

(2) 绝对路径(Absolute Path)是带有网址的路径，如一个域名 www.asp.com，其域名指向 d:\asp，那么文件就可以表示为 http://www.asp.com/html/a.html。

6.3.2 用户控件的应用

在网站程序开发中，导航控件栏、登录框等经常使用用户控件实现，当功能发生改变时只修改用户控件即可，用户控件的使用如同函数调用一样方便。

【示例 6-4】 在 chapter6 网站根目录下创建一个用户控件，使用基本控件组合一个简单登录功能用户控件，添加内容页 example6-4 并在该页中使用登录用户控件。

(1) 在网站根目录右击“添加项”选项，向项目中添加一个名为 UserLogin.ascx 的 Web 用户控件，如图 6.20 所示。

(2) 向 UserLogin.ascx 控件内添加基本动态控件，页面设计如图 6.21 所示。

(3) 设置用户控件内各控件的属性，如下列源代码所示。

```
<table>
    <tr>
        <td colspan="3">用户登录</td>
    </tr>
    <tr>
        <td>ID:</td>
        <td colspan="2">
            <asp:TextBox ID="txtID" runat="server"></asp:TextBox>
                <asp:RequiredFieldValidator ID="RequiredFieldValidator1" runat
="server" ErrorMessage="用户名不可为空" ControlToValidate="txtID" ForeColor="
Red" SetFocusOnError="True"></asp:RequiredFieldValidator>
        </td>
```

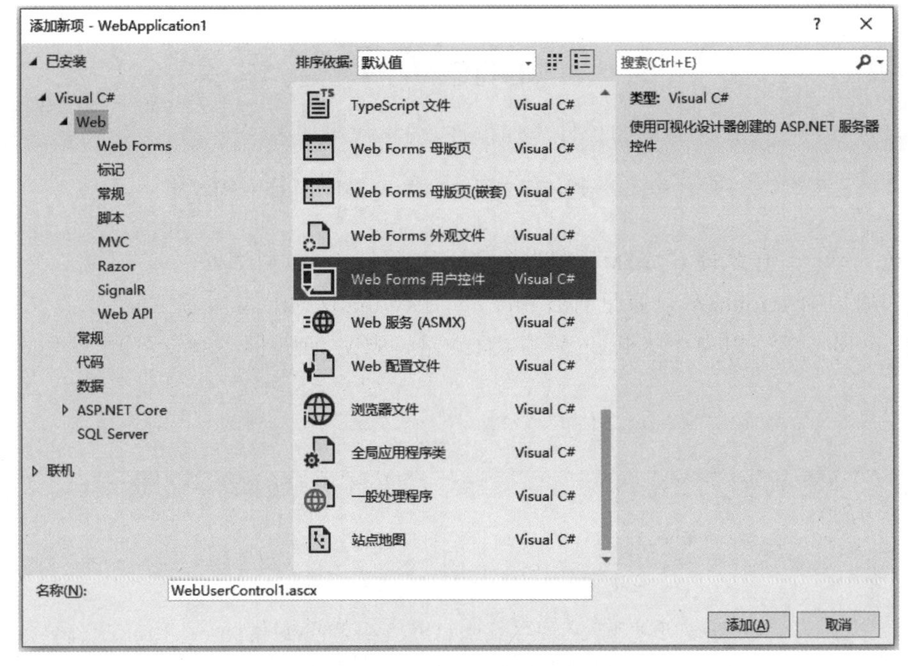

图 6.20 创建用户控件

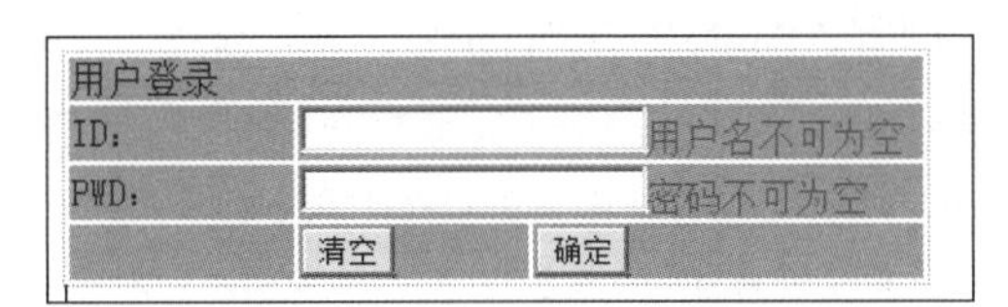

图 6.21 用户控件页面设计

```
        </tr>
            <tr>
            <td>PWD:</td>
            <td colspan="2">
            <asp:TextBox ID="txtPassword" runat="server" TextMode="Password">
</asp:TextBox>
            <asp:RequiredFieldValidator ID="RequiredFieldValidator2" runat="
server" ErrorMessage="密码不可为空" ControlToValidate="txtPassword" ForeColor
="Red" SetFocusOnError="True"></asp:RequiredFieldValidator>
        </td>
    </tr>
    <tr>
        <td> </td>
        <td>
                <asp:Button ID="btnClear" runat="server" OnClick="btnClear_
Click" Text="清空" />
        </td>
        <td>
            <asp:Button ID="btnLogin" runat="server" OnClick="btnLogin_Click"
Text="确定" />
```

```
        </td>
    </tr>
</table>
```

（4）为 UserLogin.ascx 内的控件添加事件，编辑代码如下。

```
protected void btnClear_Click(object sender, EventArgs e)
{
    txtID.Text="";
    txtPassword.Text ="";
    txtID.Focus();
}
protected void btnLogin_Click(object sender, EventArgs e)
{
    if (txtID.Text=="admin" && txtPassword.Text=="123456")
    {
        Response.Write("<script>alert('登录成功!');location='admin.aspx';</
script>");
    }
    else
    {
        Response.Write("<script>alert('用户名或者密码错误!');</script>");
    }
}
```

（5）在网站根目录添加页面 example6-4.aspx。

（6）在解决方案中鼠标左键按住 UserLogin.ascx 控件的图标“UserLogin.ascx”，拖曳添加到 example6-4.aspx 页面中。

（7）运行 example6-4 页面，可以直接使用 UserLogin.ascx 用户控件的所有功能，具体如图 6.22 所示。

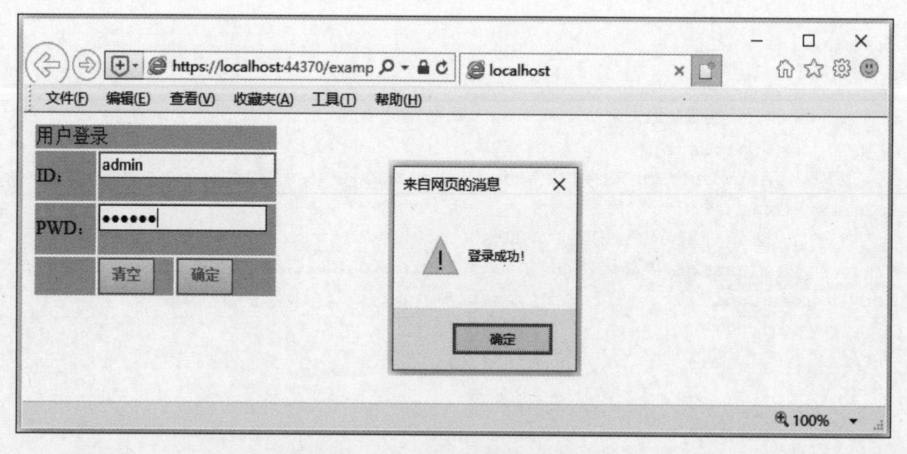

图 6.22　用户控件应用页面演示

6.3.3　将 Web 窗体转换成用户控件

在编写用户控件时，会发现 Web 窗体的结构和用户控件的结构基本相同，如果已经设计好 Web 窗体，并在应用程序全局中多次使用此 Web 窗体，就可以将 Web 窗体改成用户控件。

首先，对比 Web 窗体和用户控件的区别如下。

- Web 窗体中有<body><html><head>等标记，而用户控件没有。
- Web 窗体和用户控件所声明的方法不同。

了解以上区别后，可以很快将 Web 窗体转换成用户控件。首先需要删除<body><html><head>等标记。删除标记后，对窗体的声明方式进行更改，对于 Web 窗体，其标记代码如下。

```
<%@ Page Language="C#" AutoEventWireup="true" CodeBehind="Default.aspx.cs"
Inherits="example6-5.Default" %>
```

而对于用户控件，声明代码如下。

```
<%@Control Language="C#" AutoEventWireup="true" CodeBehind="mycontrol.ascx.
cs" Inherits="example6-5.mycontrol" %>
```

在将 Web 窗体更改为用户控件时，将 Page 标签更改为 Control 标签，就可以完成从 Web 窗体向用户控件的转换。

综合实验六　购物网站导航条

主要任务

创建 ASP.NET 应用程序，添加 Web 用户控件制作购物网站导航条。

实验步骤

步骤 1：在 VS 2019 菜单上选择“文件”→“新建”→“项目”命令。

步骤 2：创建 ASP.NET 应用程序，命名为“综合实验六”。

步骤 3：在网站的根目录上右击，创建 WebUserControl1.ascx 自定义控件，添加相应控件，具体如图 6.23 所示。

图 6.23　导航条页面布局

步骤 4：按如下源文件设置控件的相关属性值。

```
<%@Control Language="C#" AutoEventWireup="true" CodeBehind="WebUserControl1.
ascx.cs" Inherits="WebApplication1.WebUserControl1" %>
<table>
    <tr>
```

```
        <td>
            <asp:HyperLink ID="HyperLink1" runat="server" ImageUrl="~/image/1.
jpg" NavigateUrl="~/首页.aspx">HyperLink</asp:HyperLink>
        </td>
        <td>
            <asp:HyperLink ID="HyperLink2" runat="server" ImageUrl="~/image/2.
jpg"  NavigateUrl="~/分类.aspx">HyperLink</asp:HyperLink>
        </td>
        <td>
            <asp:HyperLink ID="HyperLink3" runat="server" ImageUrl="~/image/3.
jpg"  NavigateUrl="~/购物车.aspx">HyperLink</asp:HyperLink>
        </td>
        <td>
            <asp:HyperLink ID="HyperLink4" runat="server" ImageUrl="~/image/4.
jpg"  NavigateUrl="~/我的.aspx">HyperLink</asp:HyperLink>
        </td>
        <td>
            <asp:HyperLink ID="HyperLink5" runat="server" ImageUrl="~/image/5.
jpg"  NavigateUrl="~/优惠券.aspx">HyperLink</asp:HyperLink>
        </td>
        <td>
            <asp:HyperLink ID="HyperLink6" runat="server" ImageUrl="~/image/6.
jpg"  NavigateUrl="~/会员.aspx">HyperLink</asp:HyperLink>
        </td>
    </tr>
    </table>
```

步骤 5：在网站的根目录上右击，创建 ShoppingIndex.aspx 窗体并将 WebUserControl1.ascx 自定义控件拖曳到该页面，页面源代码如下。

```
<%@Page Language="C#" AutoEventWireup="true" CodeBehind="ShoppingIndex.aspx.
cs" Inherits="WebApplication1.ShoppingIndex" %>
<%@Register src="WebUserControl1.ascx" tagname="WebUserControl1" tagprefix=
"uc1" %>
<! DOCTYPE html>
<html xmlns="http://www.w3.org/1999/xhtml">
    <head runat="server">
        <meta http-equiv="Content-Type" content="text/html; charset=utf-8"/>
        <title></title>
    </head>
    <body>
        <form id="form1" runat="server">
        <div>
            <uc1:WebUserControl1 ID="WebUserControl11" runat="server" />
        </div>
        </form>
    </body>
</html>
```

步骤 6：运行 ShoppingIndex 页面，可以直接使用 WebUserControl1.ascx 用户控件的导航条功能，具体如图 6.24 所示。

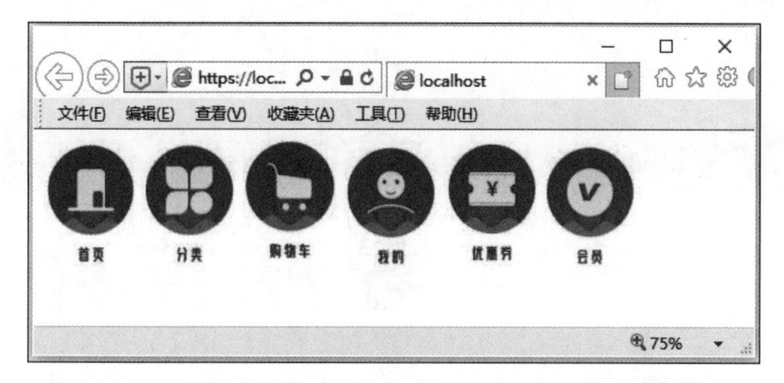

图 6.24　ShoppingIndex 页面运行测试

第7章 导航控件

本章学习目标

- 了解站点地图的配置
- 熟练掌握树状图控件的使用方法
- 熟练掌握菜单控件的使用方法
- 了解站点路径控件的使用方法

本章首先介绍站点地图的作用及基本属性，然后介绍树状图控件、菜单控件，最后介绍站点路径控件的使用方法。

7.1 站点地图

大型网站通常包含成百上千的网页，在不同页面之间实现快捷方便的导航显得尤为重要，传统网站导航需要在页面上通过超链接实现，页面修改或移动时每个页面都需要进行修改，特别麻烦。在 ASP.NET 网站中可以建立站点地图，把所有的链接地址放在一个专门的文件中进行统一管理，通过绑定相应的导航控件提供显示和导航链接，方便进行页面导航管理。

站点地图是以.sitemap 为后缀名的 XML 文件，该文件是一个标准的有固定格式的 XML 文件，所有导航控件都可以使用 Web.sitemap 文件作为数据源。XML 文件中的 <sitemap>和 HTML 文件中的<html>一样，表示整个文件的开始和结束，siteMapNode 节点表示一个链接或目录节点，允许嵌套，默认情况下该节点有 3 个属性，其中 url 表示链接地址，如果是父节点，则 url 可以为空，title 表示在节点上显示的文字，Description 表示鼠标停留时的内容提示。

【示例 7-1】 在 E 盘 ASP.NET 项目代码目录中创建 chapter7 子目录，将其作为网站根目录,添加一个站点地图文件,设置相关属性存储某一大学的组织层次结构。

(1) 在网站根目录“chapter7”上右击,从弹出的快捷菜单中选择“添加”→“添加新项”选项,在已安装的模板中选择“站点地图”,具体如图 7.1 所示。

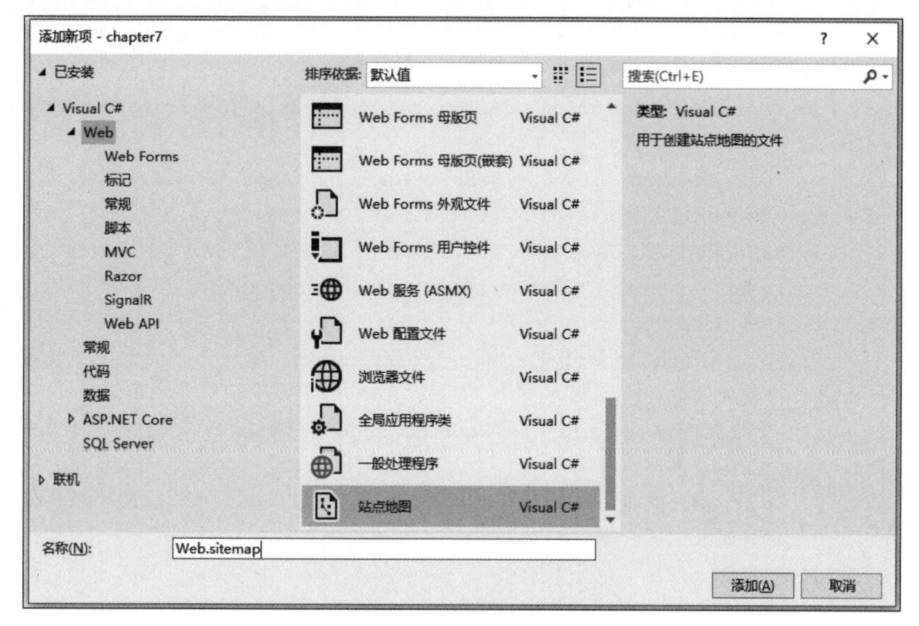

图 7.1 添加“站点地图”模板

(2) 默认站点地图 Web.sitemap 文件中包含如下的 xml 文件。

```
<?xml version="1.0" encoding="utf-8" ?>
<siteMap xmlns="http://schemas.microsoft.com/AspNet/SiteMap-File-1.0" >
<siteMapNode url="" title=""  description="">
    <siteMapNode url="" title=""  description="" />
    <siteMapNode url="" title=""  description="" />
</siteMapNode>
</siteMap>
```

说明:

只有位于根目录下且名称为 Web.sitemap 的站点地图文件才会被自动加载。站点地图的主要属性如下。

- Version:表示 xml 的版本号。
- Encoding:表示编码方式。
- url:表示网页的资源路径。
- title:表示显示的文本。
- description:表示鼠标停留在文本上的内容提示。

(3) 在窗口中编辑该 xml 文件如下。

```
<?xml version="1.0" encoding="utf-8" ?>
<siteMap xmlns="http://schemas.microsoft.com/AspNet/SiteMap-File-1.0" >
    <siteMapNode url="~/Website/Main.aspx"  title="外国语大学"  description="
大学主页">
```

```
            < siteMapNode url ="~/Website/School.aspx"   title ="学院设置"
description="学院介绍" >
              < siteMapNode url ="~/Website/English.aspx"  title ="英语学院"
description="英语学院主页" />
              < siteMapNode url ="~/Website/Japanese.aspx"  title ="日语学院"
description="日语学院主页" />
              < siteMapNode url ="~/Website/Russia.aspx "  title ="俄语学院"
description="俄语学院主页" />
              < siteMapNode url ="~/Website/Germany.aspx "  title ="德语学院"
description="德语学院主页" />
        </siteMapNode>
         < siteMapNode url ="~/Website/Department.aspx "  title ="机构设置"
description="机构介绍" >
              < siteMapNode url ="~/Website/Academic.aspx "  title ="教务处"
description="教务处主页" />
              < siteMapNode url ="~/Website/Student.aspx "  title ="学生处"
description="学生处主页" />
              < siteMapNode url ="~/Website/Research.aspx "  title ="科研处"
description="科研处主页" />
       </siteMapNode>
</siteMapNode>
</siteMap>
```

(4) 该站点地图文件表示了 10 个页面的层次关系,具体结构如图 7.2 所示。

外国语大学
　　学院设置
　　　　英语学院
　　　　日语学院
　　　　俄语学院
　　　　德语学院
　　机构设置
　　　　教务处
　　　　学生处
　　　　科研处

图 7.2　站点地图对应的结构图

说明:

(1) 在 url 属性值中,如果列出了不存在的 URL,将导致请求 Web 应用程序失败。

(2) 在 url 属性值中,如果添加了相关参数,如 url="Second.aspx?id=1",也有可能导致请求 Web 应用程序失败。

7.2 树状图控件 TreeView

7.2.1 TreeView 控件的属性、方法和事件

TreeView 控件又称树状图控件，在工具箱中图标为“TreeView”，封装在 System.Web.UI.Control.WebControl 命名空间的 TreeView 类中。TreeView 控件呈现的外观类似数据结构中的树，可以在用户单击某个节点时作出响应，更改节点的展开或者折叠状态，并提供超链接或者复选框的部分功能，在信息管理中广泛应用。TreeView 控件的属性众多，主要属性如表 7.1 所示，主要方法如表 7.2 所示，主要事件如表 7.3 所示。

表 7.1 TreeView 控件的主要属性

属性名	说明
CollapseImageUrl	设置节点折叠后显示的图像，默认以“+”表示可展开指示图像
ExpandImageUrl	设置节点展开后显示的图像，默认以“-”表示可折叠指示图像
ExpandDepth	获取或设置初次显示时 TreeView 控件的展开层数
ImageSet	获取或设置用于 TreeView 控件的图像组
Nodes	获取或设置 TreeView 控件中的节点集合
SelectedNode	获取当前在 TreeView 控件中选定的节点
ShowLines	获取或设置是否显示连接子节点和父节点之间的连线
ShowCheckBoxes	获取或设置哪些类型节点将在 TreeView 控件中显示复选框
Target	获取或设置节点被选定时页面的打开目标

说明：

(1) ShowCheckBoxes 属性具有如下几个枚举值。

- All：复选框均显示所有节点。
- Leaf：对于所有叶子节点显示复选框。
- None：不显示复选框。
- Parent：复选框均显示所有父节点。
- Root：复选框均显示为所有根节点。

(2) Target 属性具有如下几个枚举值。

- _blank：表示在没有框架的新窗口打开链接。
- _parent：表示直接在框架父级中打开链接。
- _search：表示在搜索窗格打开链接。
- _self：表示在具有焦点的框架打开链接。
- _top：表示在没有框架的完整窗口打开链接。

表 7.2 TreeView 控件的主要方法

方法名	说明
ExpandAll()	打开 TreeView 控件中的每个节点
FindNode()	检索 TreeView 控件中的指定节点

表 7.3 TreeView 控件的主要事件

事件名	说明
SelectedNodeChanged	当单击 TreeView 控件中的节点时触发
TreeNodeCheckChanged	当 TreeView 控件节点所对应的复选框选中状态发生改变时触发
TreeNodeCollapsed	当折叠 TreeView 控件节点时触发
TreeNodeExpanded	当展开 TreeView 控件节点时触发

7.2.2 TreeNodeCollection 类

TreeView 控件中所有的菜单项都是一个 TreeNoe 类的对象，所有的菜单项构成一个 TreeNodeCollection 类型的 Nodes 集合。TreeNodeCollection 类的主要属性和主要方法如表 7.4 和表 7.5 所示。

表 7.4 TreeNodeCollection 类的主要属性

属性名	说明
Count	获取当前 TreeNodeCollection 对象所包含的菜单项数目

表 7.5 TreeNodeCollection 类的主要方法

属性名	说明
Add()	向 TreeNodeCollection 集合中添加一个 TreeNode 对象
AddAt()	向 TreeNodeCollection 集合中指定索引位置添加一个 TreeNode 对象
Clear()	清空 TreeNodeCollection 集合中的所有 TreeNode 对象
IndexOf()	返回 TreeNodeCollection 集合中指定值的 TreeNode 对象索引，若索引不存在，则返回－1
Remove()	从 TreeNodeCollection 集合中移除指定值的 TreeNode 对象
RemoveAt()	从 TreeNodeCollection 集合中移除指定索引的 TreeNode 对象

7.2.3 TreeView 控件的应用

扫一扫

可以采用如下步骤向 TreeView 控件中添加节点。

（1）通过手工方式添加，具体如示例 7-2 所示。

（2）通过数据源控件绑定数据、数据库内容或者站点地图，具体如示例 7-3 所示。

（3）通过程序动态添加菜单项，在节点集合 Nodes 中使用 Add()方法动态添加 Node 节点对象，具体如示例 7-4 所示。

【示例 7-2】 在 E 盘 ASP.NET 项目代码的 chapter7 目录下，创建名为 example7-2 的网页，练习采用图形界面方式添加 TreeView 控件的节点。

(1) 在页面上添加一个 TreeView 控件，单击 TreeView 控件右上角的“>”图标，选择“编辑节点…”，弹出“TreeView 节点编辑器”对话框，如图 7.3 所示。

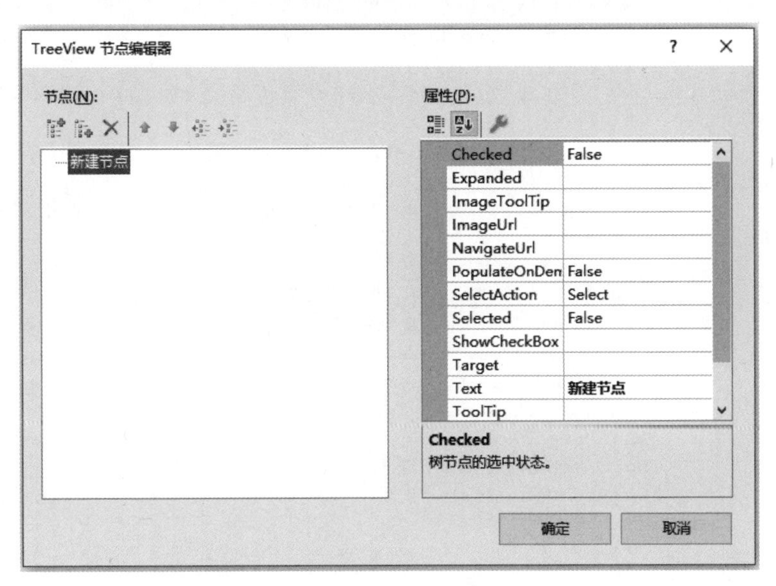

图 7.3 “TreeView 节点编辑器”对话框

(2) 添加相应的根节点和子节点，具体如图 7.4 所示。

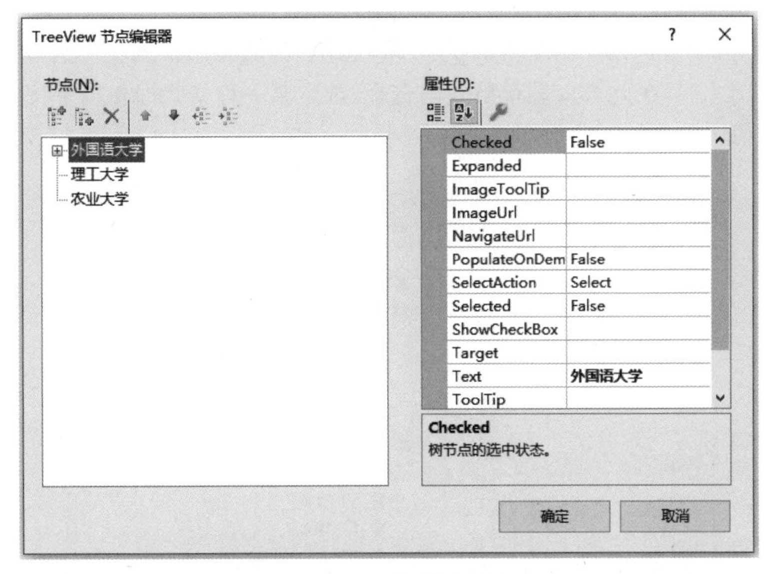

图 7.4 TreeView 节点示意图

(3) 按如下源文件设置控件的相关属性值。

```
<form id="form1" runat="server">
    < asp: TreeView  ID =" TreeView1 "  runat =" server "  ImageSet =" Inbox "
ShowCheckBoxes="All">
            <Nodes>
                  <asp:TreeNode Text="外国语大学" Value="外国语大学">
                      <asp:TreeNode Text="学院设置" Value="学院设置">
                            <asp:TreeNode Text="英语学院" Value="英语学院"></asp:
TreeNode>
                            <asp:TreeNode Text="日语学院" Value="日语学院"></asp:
TreeNode>
                            <asp:TreeNode Text="俄语学院" Value="俄语学院"></asp:
TreeNode>
                            <asp:TreeNode Text="德语学院" Value="德语学院"></asp:
TreeNode>
                      </asp:TreeNode>
                      <asp:TreeNode Text="部门设置" Value="部门设置">
                          <asp:TreeNode Text="教务处" Value="教务处"></asp:TreeNode>
                          <asp:TreeNode Text="学生处" Value="学生处"></asp:TreeNode>
                          <asp:TreeNode Text="科研处" Value="科研处"></asp:TreeNode>
                      </asp:TreeNode>
                  </asp:TreeNode>
                  <asp:TreeNode Text="理工大学" Value="理工大学"></asp:TreeNode>
                  <asp:TreeNode Text="农业大学" Value="农业大学"></asp:TreeNode>
            </Nodes>
        </asp:TreeView>
    <br />
    <asp:Button ID="btnSelect" runat="server" OnClick="btnSelect_Click" Text
="选择" />
    <asp:Label ID="lblSelect" runat="server" Text=""></asp:Label>
    </form>
```

(4) 单击 TreeView 控件右上角的▶图标，选择“自动套用格式…”，弹出“自动套用格式”对话框(见图 7.5)，在此可以进行格式的选择，选定某一种格式即相当于按照该格式设置了相关的属性值。

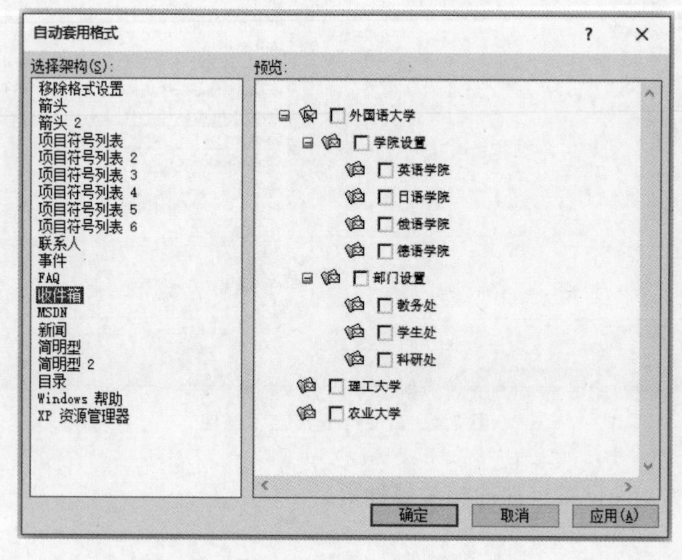

图 7.5　TreeView 控件套用格式

(5) 为控件添加事件,并编辑代码如下。

```
protected void btnSelect_Click(object sender, EventArgs e)
{
    lblSelect.Text="";
    for (int i=0; i<TreeView1.CheckedNodes.Count; i++)
    {
        if (TreeView1.CheckedNodes[i].Checked)
        {
            lblSelect.Text+=TreeView1.CheckedNodes[i].Text+"  ";
        }
    }
}
```

(6) 运行网站,单击某一节点,执行效果如图 7.6 所示;选择若干复选框后单击"选择"按钮,执行效果如图 7.7 所示。

图 7.6 TreeView 控件应用页面演示 1

【示例 7-3】 在 E 盘 ASP.NET 项目代码的 chapter7 目录下,创建名为 example7-3 的网页,在 TreeView 控件中使用站点地图。

(1) 在页面上添加一个 TreeView 控件,单击 TreeView 控件右上角的"[>]"图标,选择"选择数据源"中的"新建数据源…",如图 7.8 所示。在弹出的"数据源配置向导"对话框中选择"站点地图",单击"确定"按钮,如图 7.9 所示。

(2) 页面上新生成一个控件"SiteMapDataSource - SiteMapDataSource1",该控件在网页运行中无外观,可自动与示例 7-1 中创建的 Web.sitemap 进行绑定,在 TreeView 控件中显示站点地图中的层次结构,如图 7.10 所示。

SiteMapDataSource 数据源控件只可以与根目录下名为 Web.sitemap 的文件自动绑定,当网站中需要存在多个站点地图时,则需要在 Web.config 文件中进行相关设定。在 Web.config 文件的＜system.web＞ 节点下添加如下标签。

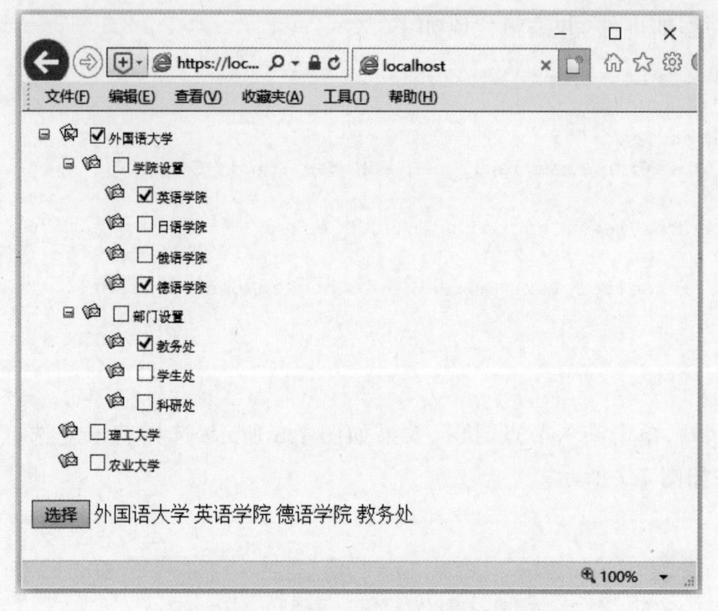

图 7.7　TreeView 控件应用页面演示 2

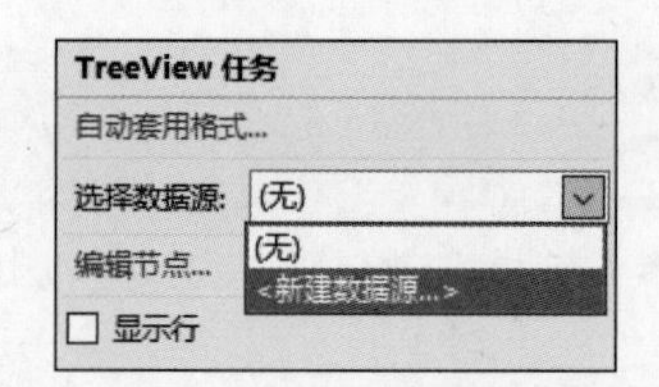

图 7.8　向 TreeView 控件添加数据源

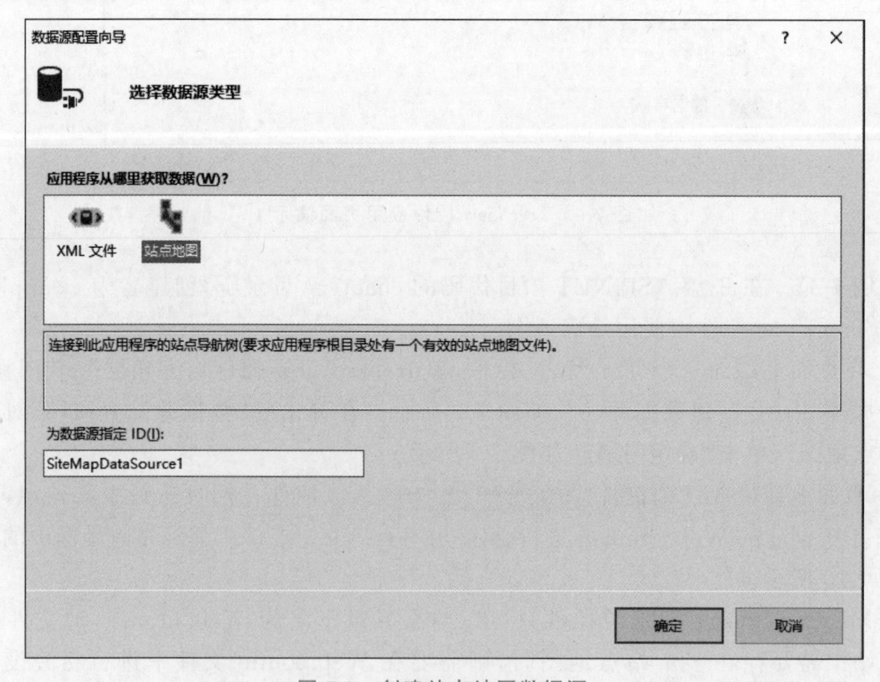

图 7.9　创建站点地图数据源

外国语大学
学院设置
英语学院
日语学院
俄语学院
德语学院
机构设置
教务处
学生处
科研处
SiteMapDataSource - SiteMapDataSource1

图 7.10 站点地图数据源显示

```
<system.web>
    <siteMap defaultProvider="ASPSiteMapProvider">
      <providers >
        <add name="AspSiteMapProvider"  type="System.Web.XmlSiteMapProvider"
siteMapFile="~/Web.sitemap"/>
        <add name="TwoSiteMapProvider"  type="System.Web.XmlSiteMapProvider"
siteMapFile="~/Manager/Web.sitemap"/>
      </providers >
    </siteMap>
    </system.web>
```

说明：

<siteMap>标签具有如下几个属性。

- defaultProvider：默认提供程序。
- Name：提供程序名(自定义名称)。
- Type：提供程序类型 System.Web.XmlSiteMapProvider(固定写法)。
- siteMapFile：站点地图路径(相对路径)。

对于网站中存在的"～/Web.sitemap"和" ～/ Manager/Web.sitemap"两个站点地图文件，可以将它们与导航控件，如 SiteMapPath 、TreeView 和 Menu 等一起使用，方法是将 SiteMapDataSource 数据源控件的 SiteMapProvider 属性设置为 AspSiteMapProvider 或 TwoSiteMapProvider 即可。

【示例 7-4】 在 E 盘 ASP.NET 项目代码的 chapter7 目录下，创建名为 example7-4 的网页，练习在 TreeView 控件中动态添加节点。

(1) 在页面中添加一个 TreeView 控件，其 ID 属性为 TreeView1。

(2) 为页面的 Page_Load 事件添加如下代码。

```
protected void Page_Load(object sender, EventArgs e)
{
    if (! IsPostBack)
    {
        TreeNode node;
        //为 TreeView1 控件添加 1 级节点
        node=new TreeNode("外国语大学");
        TreeView1.Nodes.Add(node);
        node=new TreeNode("理工大学");
```

```
            TreeView1.Nodes.Add(node);
            node=new TreeNode("农业大学");
            TreeView1.Nodes.Add(node);
            //为 TreeView1 控件的第一个子节点添加 2 级节点
            node=new TreeNode("院系设置");
            TreeView1.Nodes[0].ChildNodes.Add(node);
            node=new TreeNode("部门设置");
            TreeView1.Nodes[0].ChildNodes.Add(node);
            //为 TreeView1 控件的第一个子节点中的第一个子节点添加 3 级节点
            node=new TreeNode("英语学院");
            TreeView1.Nodes[0].ChildNodes[0].ChildNodes.Add(node);
            node=new TreeNode("日语学院");
            TreeView1.Nodes[0].ChildNodes[0].ChildNodes.Add(node);
            node=new TreeNode("俄语学院");
            TreeView1.Nodes[0].ChildNodes[0].ChildNodes.Add(node);
            node=new TreeNode("德语学院");
            TreeView1.Nodes[0].ChildNodes[0].ChildNodes.Add(node);
            //为 TreeView1 控件的第一个子节点中的第二个子节点添加 3 级节点
            node=new TreeNode("教务处");
            TreeView1.Nodes[0].ChildNodes[1].ChildNodes.Add(node);
            node=new TreeNode("学生处");
            TreeView1.Nodes[0].ChildNodes[1].ChildNodes.Add(node);
            node=new TreeNode("科研处");
            TreeView1.Nodes[0].ChildNodes[1].ChildNodes.Add(node);
        }
    }
```

（3）页面运行效果如图 7.11 所示。

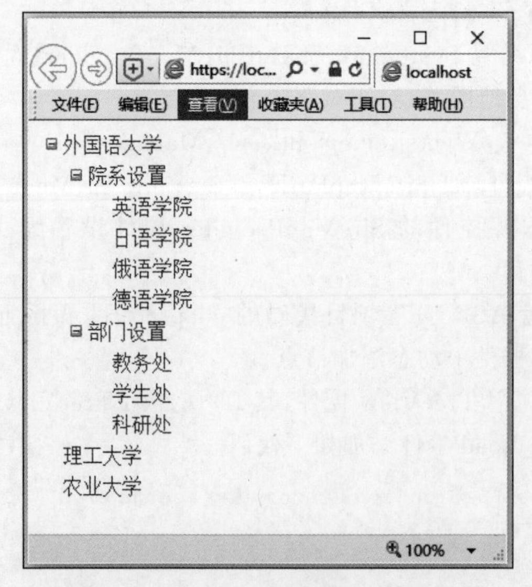

图 7.11　TreeView 控件应用页面演示

7.3 菜单控件 Menu

7.3.1 Menu 控件的属性、方法和事件

Menu 控件又称菜单控件，在工具箱中图标为“Menu”，封装在 System.Web.UI.Control.WebControl 命名空间的 Menu 类中。Menu 控件呈现与 Windows 应用程序类似的菜单，允许用户快速选择不同页面，实现导航功能，在 Web 开发中应用广泛。Menu 控件的属性众多，主要属性如表 7.6 所示，主要事件如表 7.7 所示。

表 7.6　Menu 控件的主要属性

属 性 名	说　　明
DisappearAfter	获取或设置鼠标指针离开节点后动态菜单的持续显示时间
Items	获取或设置 Menu 控件中的菜单集合
IteamsWrap	获取或设置 Menu 控件中的菜单项文本是否换行
Orientation	获取或设置 Menu 控件的呈现方向
SelectedItem	获取当前在 Menu 控件中选定的菜单项
SelectedValue	获取当前在 Menu 控件中选定的菜单项值
StaticDiaplayLevels	获取当前在 TreeView 控件中选定的节点
Target	获取或设置节点被选定时页面的打开目标

说明：

Orientation 属性具有如下几个枚举值。

- Horizontal：水平方向。
- Vertical：竖直方向。

表 7.7　Menu 控件的主要事件

事 件 名	说　　明
MenuItemClick	当单击 Menu 控件中的菜单项时触发

7.3.2 MenuItemCollection 类

Menu 控件所有的菜单项都是一个 Item 类的对象，所有的菜单项构成一个 MenuItemCollection 类型的 Items 集合。MenuItemCollection 类的主要属性和主要方法如表 7.8 和表 7.9 所示。

表 7.8　MenuItemCollection 类的主要属性

属 性 名	说　　明
Count	获取当前 MenuItemCollection 对象所包含的菜单项数目

表 7.9 MenuItemCollection 类的主要方法

方法名	说明
Add()	向 MenuItemCollection 集合中添加一个 MenuItem 对象
AddAt()	向 MenuItemCollection 集合中指定索引位置添加一个 MenuItem 对象
Clear()	清空 MenuItemCollection 集合中的所有 MenuItem 对象
IndexOf()	返回 MenuItemCollection 集合中指定值的 MenuItem 对象索引，若不存在索引，则返回－1
Remove()	从 MenuItemCollection 集合中移除指定值的 MenuItem 对象
RemoveAt()	从 MenuItemCollection 集合中移除指定索引的 MenuItem 对象

7.3.3 Menu 控件的应用

扫一扫

可以采用如下方法向 MenuItem 控件中添加节点。

(1) 通过手工方式添加，与 TreeView 控件在示例 7-2 中的内容基本一致，这里不提供具体示例。

(2) 通过数据源控件绑定数据、数据库内容或者站点地图，与 TreeView 控件在示例 7-3 中采用的方法基本一致，这里不提供具体示例。

(3) 通过程序动态添加菜单项，在菜单集合 Items 中使用 Add()方法动态添加 Item 节点对象，具体如示例 7-5 所示。

【示例 7-5】 在 E 盘 ASP.NET 项目代码的 chapter7 目录下，创建名为 example7-5 的网页，在 Menu 控件中动态添加节点。

(1) 在页面中添加一个 Menu 控件，其 ID 属性为 Menu1。

(2) 设置 Menu1 控件的相关属性，如下列源文件所示。

```
<form id="form1" runat="server">
    <div>
            <asp:Menu ID="Menu1" runat="server" BackColor="#B5C7DE"
DynamicHorizontalOffset="2" Font-Names="Verdana" Font-Size="Larger"
ForeColor="#284E98" Orientation="Horizontal" StaticDisplayLevels="2"
StaticSubMenuIndent="10px">
            <DynamicHoverStyle BackColor="#284E98" ForeColor="White" />
            <DynamicMenuItemStyle HorizontalPadding="5px" VerticalPadding="2px" />
            <DynamicMenuStyle BackColor="#B5C7DE" />
            <DynamicSelectedStyle BackColor="#507CD1" />
            <StaticHoverStyle BackColor="#284E98" ForeColor="White" />
            <StaticMenuItemStyle HorizontalPadding="5px" VerticalPadding="2px" />
            <StaticSelectedStyle BackColor="#507CD1" />
        </asp:Menu>
    </div>
</form>
```

(3) 为页面的 Page_Load 事件添加如下代码。

```
protected void Page_Load(object sender, EventArgs e)
{
    if (!IsPostBack)
```

```
        {
            MenuItem item;
            //为 Menu1 控件添加 1 级节点
            item=new MenuItem("外国语大学");
            Menu1.Items.Add(item);
            item=new MenuItem("理工大学");
            Menu1.Items.Add(item);
            item=new MenuItem("农业大学");
            Menu1.Items.Add(item);
            //为 Menu1 控件的第一个子节点添加 2 级节点
            item=new MenuItem("院系设置");
            Menu1.Items[0].ChildItems.Add(item);
            item=new MenuItem("部门设置");
            Menu1.Items[0].ChildItems.Add(item);
            //为 Menu1 控件的第一个子节点中的第一个子节点添加 3 级节点
            item=new MenuItem("英语学院");
            Menu1.Items[0].ChildItems[0].ChildItems.Add(item);
            item=new MenuItem("日语学院");
            Menu1.Items[0].ChildItems[0].ChildItems.Add(item);
            item=new MenuItem("俄语学院");
            Menu1.Items[0].ChildItems[0].ChildItems.Add(item);
            item=new MenuItem("德语学院");
            Menu1.Items[0].ChildItems[0].ChildItems.Add(item);
            //为 Menu1 控件的第一个子节点中的第二个子节点添加 3 级节点
            item=new MenuItem("教务处");
            Menu1.Items[0].ChildItems[1].ChildItems.Add(item);
            item=new MenuItem("学生处");
            Menu1.Items[0].ChildItems[1].ChildItems.Add(item);
            item=new MenuItem("科研处");
            Menu1.Items[0].ChildItems[1].ChildItems.Add(item);
        }
    }
```

（4）页面运行效果如图 7.12 所示。

图 7.12　Menu 控件应用页面演示

7.4 站点路径控件 SiteMapPath

7.4.1 SiteMapPath 控件的属性、方法和事件

SiteMapPath 控件又称站点路径控件，在工具箱中图标为" SiteMapPath "，封装在 System.Web.UI.Control.WebControl 命名空间的 SiteMapPath 类中。SiteMapPath 控件也被形象地称为"面包条导航"控件，提供一种站点定位的方式，动态显示当前页在站点中的相对位置，并提供了从当前页向上级页面跳转的链接。

SiteMapPath 控件与站点地图密切相关，可以直接使用站点地图中配置的数据，而无须通过 SiteMapDataSource 数据源控件，若页面未在站点地图中标示，则无法显示该控件。SiteMapPath 控件的属性众多，主要属性如表 7.10 所示。

表 7.10 SiteMapPath 控件的主要属性

属 性 名	说 明
CurrentNodeStyle	定义当前节点的样式，包括字体、颜色、样式等
NodeStyle	定义导航上每个节点的样式
ParentLevelsDiaplayed	获取或设置在导航路径上显示的相对当前节点的父节点层数
PathSeparator	指定导航路径中节点之间的分隔符
RootNodeStyle	定义根节点的样式
ShowToolTips	当鼠标指针悬停于导航路径的某个节点时，是否显示相应的工具提示信息
SiteMapProvider	获取或设置用于呈现站点导航控件的站点提供程序的名称

扫一扫

7.4.2 SiteMapPath 控件的应用

【示例 7-6】 在 E 盘 ASP.NET 项目代码目录的 chapter7 子文件中创建 Website 文件夹，按示例 7-1 站点地图中标识的页面结构在文件夹内添加页面，练习使用 SiteMapPath 控件。

（1）在 Website 文件夹内添加母版页 MasterPage.master，母版页内包含一个 SiteMapPath 控件，其 ID 属性为 SiteMapPath1。

（2）设置母版页中 SiteMapPath1 控件的属性值，源文件如下。

```
<form id="form1" runat="server">
        <asp:SiteMapPath ID="SiteMapPath1" runat="server" Font-Names="
Verdana" Font-Size="Larger" PathSeparator="-&gt;">
            <CurrentNodeStyle ForeColor="#333333" />
            <NodeStyle Font-Bold="True" ForeColor="#284E98" />
            <PathSeparatorStyle Font-Bold="True" ForeColor="#507CD1" />
            <RootNodeStyle Font-Bold="True" ForeColor="#507CD1" />
        </asp:SiteMapPath>
```

```
        <asp:ContentPlaceHolder id="ContentPlaceHolder1" runat="server">
        </asp:ContentPlaceHolder>
    </form>
```

(3) 在 Website 文件夹内添加继承自 MasterPage.master 母版页的内容页，按图 7.13 所示命名每个页面，并在页面内添加简单的内容。

(4) 运行 English.aspx 页面，可显示当前页面在站点地图中标示的网站位置，单击路径中的父节点，可实现超链接功能，具体页面效果如图 7.14 所示。

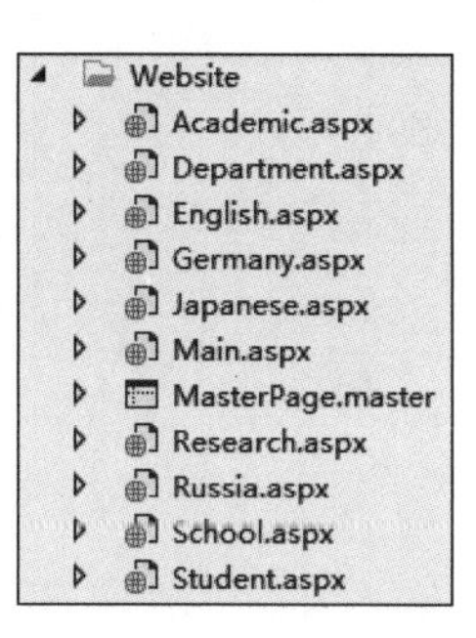

图 7.13　添加的内容页

图 7.14　SiteMapPath 控件应用页面演示

综合实验七　图书商城菜单栏

主要任务

创建 ASP.NET 应用程序，添加 XML 文件制作图书商城菜单栏。

实验步骤

步骤 1：在 VS 2019 菜单上选择“文件”→“新建”→“项目”命令。

步骤 2：创建 ASP.NET 应用程序，命名为“综合实验七”。

步骤 3：在网站的根目录上右击，添加 XML 文件，命名为 Menusite.xml，具体如图 7.15 所示。

步骤 4：按如下文件设置 XML 文件。

```
<xml version="1.0" encoding="utf-8" >
<Menus value="">
  <TopMenu id="100" value="" ImageUrl="image/1.png" Height="100" NavigateUrl="">
      <TopMenuItem id="101" value="纸质图书" NavigateUrl="~/100/book.aspx">
      </TopMenuItem>
      <TopMenuItem id="102" value="电子书" NavigateUrl="~/100/ebook.aspx">
      </TopMenuItem>
      <TopMenuItem id="103" value="连载系列" NavigateUrl="~/100/onlinebook.
aspx">
      </TopMenuItem>
```

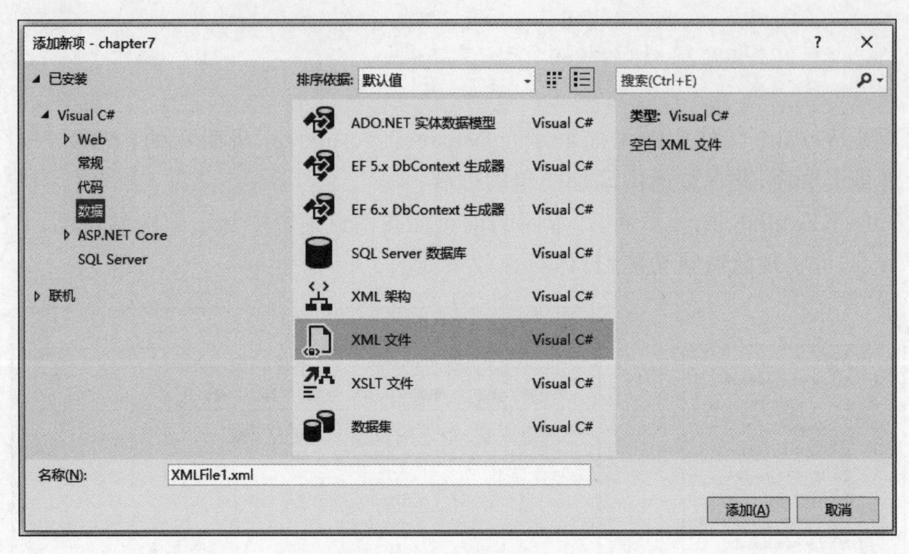

图 7.15　添加 XML 文件

```
    </TopMenu>
    <TopMenu id="200" value="" ImageUrl="image/2.png" NavigateUrl="">
        <TopMenuItem id="202" value="收藏夹" NavigateUrl="~/200/favourite.
aspx">
        </TopMenuItem>
        <TopMenuItem id="203" value="书架" NavigateUrl="~/200/bokstack.aspx">
        </TopMenuItem>
    </TopMenu>
    <TopMenu id="300" value="" ImageUrl="image/3.png" NavigateUrl="">
        <TopMenuItem id="301" value="我的购物车" NavigateUrl="~/300/
shoppingcart.aspx">
        </TopMenuItem>
        <TopMenuItem id="302" value="我的订单" NavigateUrl="~/300/order.aspx">
        </TopMenuItem>
    </TopMenu>
    <TopMenu id="400" value="" ImageUrl="image/4.png" NavigateUrl="~/400/
other.aspx">
    </TopMenu>
</Menus>
```

步骤 5：在网站的根目录上右击，添加 image 文件夹，并添加相关图片，具体如图 7.16 所示。

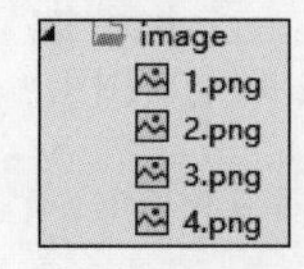

图 7.16　添加文件夹及相关图片

步骤 6：在网站的根目录上右击，创建 BookMallMenu.aspx 窗体，添加 Menu 和 XmlDataSource 控件，并按如下源文件设置控件的相关属性值。

```
<form id="form1" runat="server">
        <asp:XmlDataSource ID="XmlDataSource1" runat="server" DataFile="~/综
合实验七/Menusite.xml"></asp:XmlDataSource>
         <asp:Menu DataSourceID="XmlDataSource1" runat="server" ID="Menu1"
MaximumDynamicDisplayLevels="4"  Orientation="Horizontal" StaticDisplayLevels=
"2" StaticEnableDefaultPopOutImage="False"   DynamicEnableDefaultPopOutImage=
"false" StaticSubMenuIndent="" ItemWrap="True">
            <DataBindings>
             <asp:MenuItemBinding DataMember="TopMenu" ImageUrlField="ImageUrl"
TextField="value"  NavigateUrlField="NavigateUrl" ValueField="value"/>
              <asp:MenuItemBinding DataMember="TopMenuItem" NavigateUrlField=
"NavigateUrl" TextField="value"  ValueField="value"/>
            <asp:MenuItemBinding DataMember="Menus" TextField="value" ValueField
="value"/>
            </DataBindings>
        </asp:Menu>
</form>
```

步骤7：运行 BookMallMenu.aspx 页面，可以直接使用 XML 文件中定义的菜单功能，具体如图 7.17 所示。

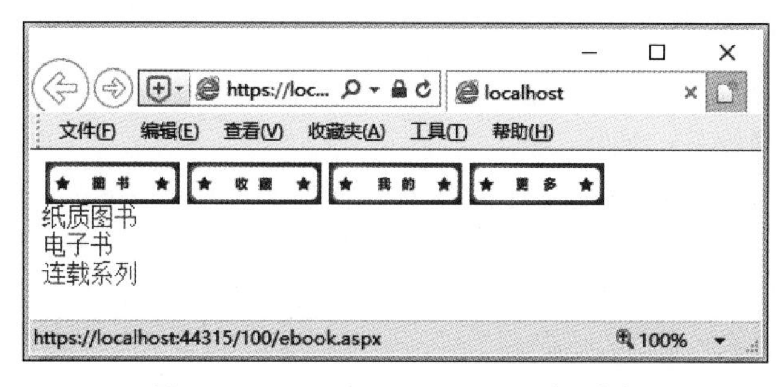

图 7.17　BookMallMenu 页面运行测试

第8章 ASP.NET AJAX控件

本章学习目标

- 了解 AJAX 的特点及作用
- 了解脚本管理控件的使用方法
- 了解脚本管理代理控件的使用方法
- 熟练掌握更新区域控件的使用方法
- 了解更新进度控件的使用方法
- 熟练掌握时钟控件的使用方法

本章首先介绍 AJAX 的特点和基本应用，然后讲述 ASP.NET 中的 AJAX 控件，最后对脚本管理控件、脚本管理代理控件、更新区域控件、更新进度控件、时钟控件的使用方法进行详细讲解。

8.1 ASP.NET AJAX 概述

AJAX(Asynchronous JavaScript and XML)又称为“异步 JavaScript 和 XML”，是一种创建交互式网页应用的网页开发技术，也是当前 Web 开发领域流行的技术。在传统的 Web 开发中，页面操作往往需要进行多次“回发”，通过“回发”才能实现页面的刷新，而使用 AJAX 则无须产生“回发”就可实现刷新的效果，可以提升用户体验，更加方便地进行 Web 应用程序的交互。

8.1.1 AJAX 基础

在 C/S 应用程序开发中，应用程序基本都安装在本地，客户端响应用户事件的时间非常短，且 C/S 应用程序能够及时捕捉和响应用户的操作，用户体验很好。而 B/S 应用程序开发中，在 Web 端由于每次的交互都需要向服务器发送请求，因此服务器接受请求和返回请求的时间取决于服务器的响应时间，感觉上比在本地机运行要慢很多。

为了解决这一问题，在用户浏览器和服务器之间设计一个中间层——AJAX 层，改变了传统 Web 中客户端和服务器的“请求—等待—请求—等待”模式，通过使用 AJAX 层向服务器发送和接收需要的数据，从而不产生页面的刷新，应用模型如图 8.1 所示。

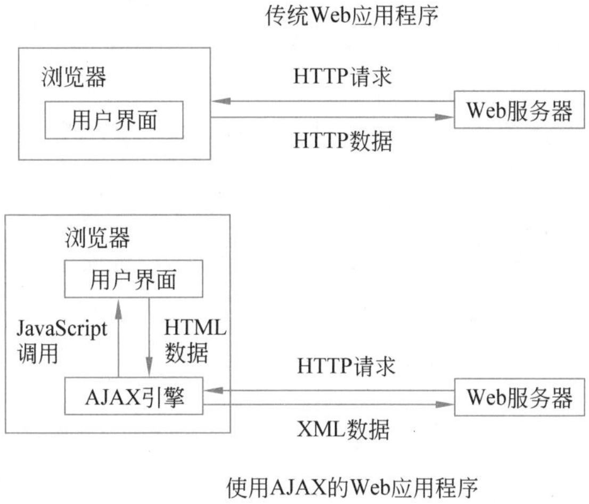

传统Web应用程序

浏览器
用户界面
HTTP请求
HTTP数据
Web服务器

浏览器
用户界面
JavaScript调用
HTML数据
AJAX引擎
HTTP请求
XML数据
Web服务器

使用AJAX的Web应用程序

图 8.1　传统 Web 应用和 AJAX Web 应用模型

AJAX 应用使用 SOAP 和其他一些基于 XML 的 Web Service 接口，在客户端采用 JavaScript 处理来自服务器的响应，减少服务器和浏览器之间的“请求—回发”操作，从而减少了数据传送量，提升了速度，当服务器和客户端之间进行信息通信时，用户会感觉到操作更快了。结合 AJAX 的原理和特点与传统 Web 应用相比，总结如下。

1. 传统的 HTML 整页刷新

传统的 HTML 访问时，客户端浏览器向服务器发送访问请求，服务器接收到请求后，对客户请求进行相应的运算和处理，生成结果后发送回客户端浏览器，客户端浏览器对结果进行处理，实现整页刷新。

2. AJAX 的局部刷新

相对于传统的整页刷新，AJAX 的局部更新更加智慧和人性化，当用户在客户端浏览器页面进行相关操作后，AJAX 将自动访问服务器端，对局部页面进行更新。

3. AJAX 交互

第一次请求发回一个完整的 Web 页面，以后更新数据并不将整个页面重新载入，仅将响应的内容回传。AJAX 是 JavaScript、CSS、DOM、XmlHttpRequest 等技术的集合体，主要应用于异步获取后台数据和局部刷新。

8.1.2　ASP.NET 中的 AJAX

AJAX 技术看似非常复杂，其实 AJAX 并不是新技术，只是一些已有技术的混合体，将这些技术进行一定的修改、整合和发扬就形成了 AJAX 技术。AJAX 包括的主要技术如下。

- XHTML：基于 XHTML 1.0 规范的 XHTML 技术。
- CSS：基于 CSS 2.0 的 CSS 布局的 CSS 编程技术。
- DOM：HTML DOM、XML DOM 等技术。
- JavaScript：JavaScript 编程技术。
- XML：XML DOM、XSLT、XPath 等 XML 编程技术。

使用 AJAX 需要熟练掌握 JavaScript 等技术，具体应用中有一定难度。ASP.NET 将这些技术做了封装，在 2007 年年初推出第一个正式版本，并将早期的 Atlas 更名为 ASP.NET AJAX，在服务器端和客户端分别对应有 ASP.NET 服务器端编程模型和 ASP.NET 客户端编程模型。

ASP.NET AJAX 作为一个完整的开发框架，其编程模型比较简单，很容易与现有的 ASP.NET 程序相结合，只需要在页面中拖几个控件，而不必了解深层次的工作原理。ASP.NET 4.7 中，AJAX 已经成为.NET 框架的原生功能，创建 ASP.NET Web 应用程序就能够直接使用 AJAX 功能。在 ASP.NET 4.7 中，可以直接拖动 AJAX 控件进行 AJAX 开发，并能够同普通控件一同使用，实现页面局部刷新功能。ASP.NET 4.7 中 AJAX 扩展控件如图 8.2 所示。

图 8.2 ASP.NET 4.7 中 AJAX 扩展控件

8.1.3 AJAX 简单应用

【示例 8-1】 在 E 盘 ASP.NET 项目代码目录中创建 chapter8 子目录，将其作为网站根目录，创建名为 example8-1 的网页，练习简单的 AJAX 应用。

(1) 向页面中添加一个脚本管理控件和一个标签控件。

(2) 向页面中添加一个局部更新面板控件。

(3) 向更新面板控件内添加一个标签控件和一个按钮控件。

(4) 修改各控件属性值，如下列源文件所示。

```
<form id="form1" runat="server">
    <div>
     < asp: ScriptManager  ID =" ScriptManager1 "  runat =" server " > </asp:
ScriptManager>
        当前时间:<asp:Label ID="lblTime1" runat="server"></asp:Label>
        <asp:UpdatePanel ID=" UpdatePanel11" runat="server">
            <ContentTemplate>
                当前时间(异步处理):<asp:Label ID="lblTime2" runat="server"></
asp:Label>
            <asp:Button ID="btnRef" runat="server" OnClick="btnRef_Click" Text
="更新" />
                <br />
            </ContentTemplate>
        </asp:UpdatePanel>
    </div>
    </form>
```

（5）在 UpdatePanel1 控件的属性窗口中单击 Trigger 属性的“...”图标，在弹出的“UpdatePanelTrigger 集合编辑器”窗口中添加异步回发触发器，并设置相关属性，如图 8.3 所示。

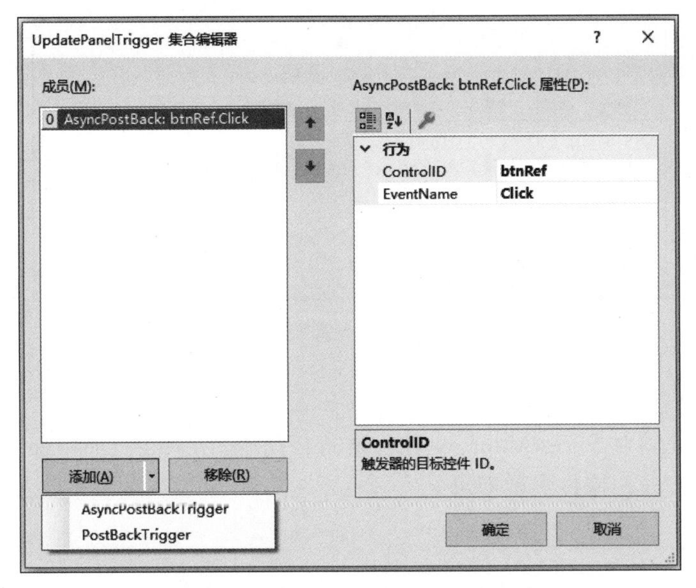

图 8.3　UpdatePanelTrigger 属性设置

（6）为相关事件添加下列事件代码。

```
protected void Page_Load(object sender, EventArgs e)
{
    //赋值控件 1 和控件 2 的文本值为当前时间
    lblTime1.Text=  DateTime.Now.ToString();
    lblTime2.Text=DateTime.Now.ToString();
}
protected void btnRef_Click(object sender, EventArgs e)
{
     //赋值控件 2 的文本值为当前时间
     lblTime2.Text=DateTime.Now.ToString();
}
```

（7）不论页面是首次加载或者是“回发”后呈现到客户端，都能够很明显地感觉到页面整体被刷新，两次显示的时间一致，如图 8.4 所示；使用 UpdatePanel 控件后，单击“更新”按

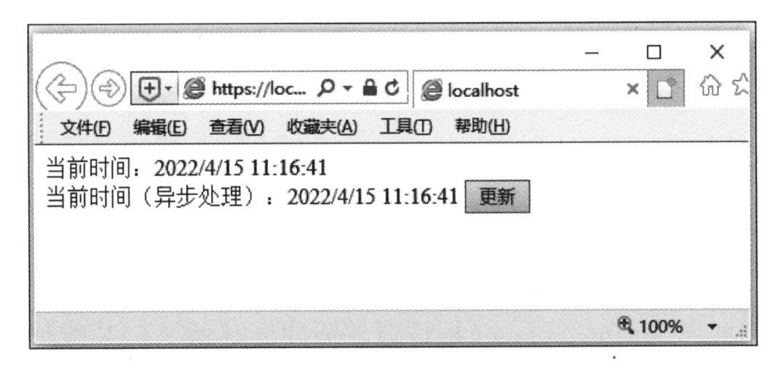

图 8.4　页面整体刷新演示

钮页面局部刷新，只会将 UpdatePanel 控件内的内容进行更新，而不会影响 UpdatePanel 外的内容，运行的页面如图 8.5 所示。

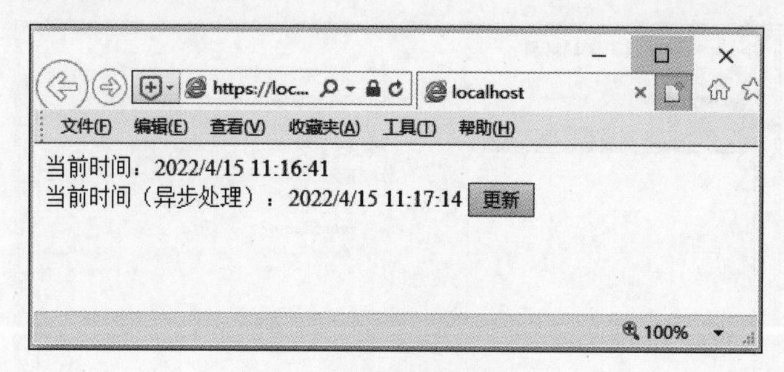

图 8.5　页面局部刷新演示

说明：

页面设计时无须对 ScriptManager 控件进行任何配置，只在页面中保证 ScriptManager 控件位于 UpdatePanel 控件之前即可。

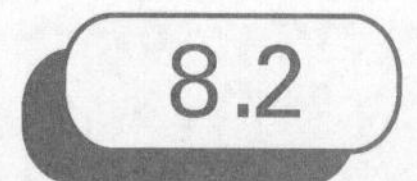

8.2　ASP.NET AJAX 控件

8.2.1　脚本管理控件 ScriptManager

脚本管理控件（ScriptManager）是 ASP.NET AJAX 中非常重要的控件，通过使用 ScriptManager 能够进行页面局部更新的管理。ScriptManager 用来处理页面上的局部更新，同时生成相关的代理脚本，实现通过 JavaScript 访问 Web Service。

ScriptManager 只能在页面中被使用一次，每个页面只能有一个 ScriptManager 控件，用来进行页面的全局管理，以及整个页面的局部更新管理，其常用属性如表 8.1 所示。

表 8.1　ScriptManager 控件的常用属性

属性名	说明
AllowCustomErrorRedirect	获取或设置在异步回发过程中是否进行自定义错误重定向
AsyncPostBackTimeout	获取或设置异步回发的超时事件，默认为 90s
EnablePageMethods	获取或设置是否启用页面方法，默认值为 False
EnablePartialRendering	获取或设置在支持的浏览器上为 UpdatePanel 控件启用异步回发
LoadScriptsBeforeUI	获取或设置在浏览器中呈现 UI 之前是否应加载脚本引用
ScriptMode	获取或设置在多个类型时可加载的脚本类型，默认为 Auto

在 ASP.NET AJAX 应用中，ScriptManager 控件相当于一个总指挥官，只是进行指挥而不进行实际操作，所以 ScriptManager 控件通常需要同其他 AJAX 控件搭配使用。

8.2.2 脚本管理代理控件 ScriptManagerProxy

ScriptManager 控件作为整个页面的管理者，能够实现 AJAX 功能，但是一个页面只能使用一个 ScriptManager 控件，如果有多个，则会出现异常。但如果在母版页中使用了 ScriptManager 控件，那么内容窗体中就不能使用 ScriptManager 控件，否则整合在一起的页面也会出现错误。

为了解决这个问题，就需要使用脚本管理代理控件 ScriptManagerProxy，ScriptManagerProxy 控件和 ScriptManager 控件的用法十分相似，当母版页和内容页都需要进行局部更新时，可以在母版页使用 ScriptManager 控件进行脚本管理，在内容页中使用 ScriptManagerProxy 脚本管理代理控件。

【示例 8-2】 在 chapter8 网站根目录下创建一个名为 MasterPage.master 的母版页，并创建继承自该母版页的内容页 example8-2，练习使用 ScriptManagerProxy 控件。

(1) 向母版页和内容页中分别添加 UpdatePanel 控件及相关控件，具体如图 8.6 和图 8.7 所示。

图 8.6　母版页面设计

图 8.7　内容页面设计

(2) 设置各控件的属性如下。

母版页 MasterPage：

```
<form id="form1" runat="server">
    <div>
        <asp:ScriptManager ID="ScriptManager1" runat="server">
        </asp:ScriptManager>
        <asp:UpdatePanel ID="UpdatePanel1" runat="server">
            <ContentTemplate>
                母版页时间:<asp:Label ID="lblTime1" runat="server" Text="Label"></
asp:Label>
                 <asp:Button ID="btnRef1" runat="server" OnClick="btnRef1_
Click" Text="母版页刷新" />
            </ContentTemplate>
            <Triggers>
                  <asp:AsyncPostBackTrigger ControlID="btnRef1" EventName="
Click" />
            </Triggers>
        </asp:UpdatePanel>
        <asp:ContentPlaceHolder id="ContentPlaceHolder1" runat="server">
        </asp:ContentPlaceHolder>
    </div>
</form>
```

内容页 example8-2：

```
<asp:Content ID="Content1" ContentPlaceHolderID="head" Runat="Server">
</asp:Content>
<asp:Content ID="Content2" ContentPlaceHolderID="ContentPlaceHolder1" Runat
="Server">
    <asp:ScriptManagerProxy ID="ScriptManagerProxy1" runat="server">
    </asp:ScriptManagerProxy>
    <asp:UpdatePanel ID="UpdatePanel1" runat="server">
        <ContentTemplate>
            内容页时间:<asp:Label ID="lblTime2" runat="server" Text="Label"></
asp:Label>
            <asp:Button ID="btnRew2" runat="server" Text="内容页刷新" OnClick="
btnRew2_Click" />
        </ContentTemplate>
        <Triggers>
            <asp:AsyncPostBackTrigger ControlID="btnRew2" EventName="Click" />
        </Triggers>
</asp:UpdatePanel>
</asp:Content>
```

(3) 为控件添加事件,并编辑代码如下。

```
    MasterPage.master.cs:
protected void btnRef1_Click(object sender, EventArgs e)
{
    lblTime1.Text=DateTime.Now.ToString();
}
    example8-2.cs:
protected void btnRew2_Click(object sender, EventArgs e)
{
    lblTime2.Text=DateTime.Now.ToString();
}
```

(4) 母版页和内容都支持 AJAX,均可进行异步处理,运行内容页如图 8.8 所示。

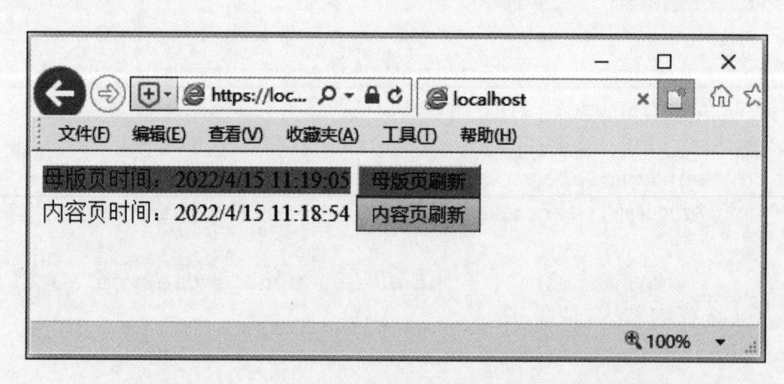

图 8.8 ScriptManagerProxy 控件应用页面演示

8.2.3 更新区域控件 UpdatePanel

更新区域控件(UpdatePanel)是 ASP.NET AJAX 中最常用的控件。UpdatePanel 控件的外观及使用方法与 Panel 控件类似,在 UpdatePanel 控件中放入需要刷新的控件可以实现局部刷新。

UpdatePanel 控件可以用来创建局部更新,开发人员无须编写任何客户端脚本,直接使用 UpdatePanel 控件就能进行局部更新，整个页面中只有 UpdatePanel 控件中的服务器控件会进行刷新操作，页面其他地方都不会被刷新。UpdatePanel 控件的常用属性如表 8.2 所示。

表 8.2 UpdatePanel 控件的常用属性

属 性 名	说 明
RenderMode	指明 UpdatePanel 控件内呈现的标记应为＜div＞或＜span＞
ChildrenAsTriggers	获取或设置在 UpdatePanel 控件的子控件的回发是否导致 UpdatePanel 控件更新，其默认值为 True
EnableViewState	获取或设置控件是否自动保存其往返过程
Triggers	设置可以导致 UpdatePanel 控件更新的触发器的集合
UpdateMode	获取或设置 UpdatePanel 控件回发属性是在每次进行事件时进行更新，还是使用 UpdatePanel 控件的 Update 方法后再进行更新
Visible	获取或设置 UpdatePanel 控件的可见性

说明：

Triggers 属性的枚举值如下。

- AsyncPostBackTrigger：用来指定某个服务器端控件以及将其触发的服务器事件，AsyncPostBackTrigger 属性需要配置控件的 ID 和控件产生的事件名。
- PostBackTrigger：用来指定在 UpdatePanel 中的某个控件，并指定其控件产生的事件将使用传统的“回发”方式进行触发。

UpdatePanel 控件要进行动态更新，必须依赖于 ScriptManager 控件。当 ScriptManager 控件允许局部更新时，它会以异步的方式发送到服务器，服务器接收请求后，执行操作并通过 DOM 对象替换局部代码，其原理如图 8.9 所示。

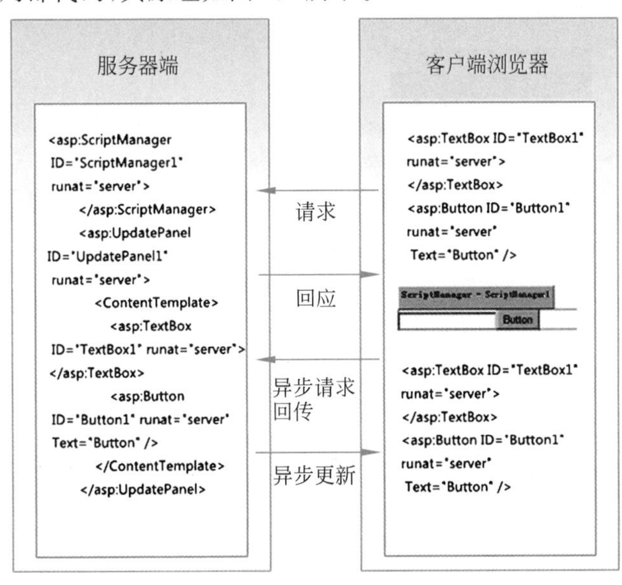

图 8.9 UpdatePanel 控件异步请求示意图

8.2.4 更新进度控件 UpdateProgress

使用 ASP.NET AJAX 时页面并没有整体刷新，只是进行了局部刷新，有时会给用户造成网页提交失败的疑惑，以至于用户可能产生重复操作，甚至非法操作，更新进度控件(UpdateProgress)就可以解决这个问题。

当服务器端与客户端进行异步通信时，可以使用 UpdateProgress 控件告知用户页面正在执行中，在服务器和客户端之间等待时间内，ProgressTemplate 标记会呈现在用户面前，提示用户应用程序正在运行。例如，当用户单击按钮提交表单后，系统应该提示“正在提交中，请稍候…”，提升用户体验友好度。

【示例 8-3】 在 chapter8 网站根目录下创建一个名为 example8-3 的页面，练习使用更新进度控件 UpdateProgress。

(1) 在 chapter8 网站根目录下添加图像文件 progress bar.gif。

(2) 在 example8-3 页面中添加 ScriptManager 和 UpdateProgress 控件，并在 UpdateProgress 控件内添加 Image 控件和相关文字，具体如图 8.10 所示。

图 8.10 UpdateProgress 控件应用页面布局

(3) 设置各控件的属性，具体如下列源文件。

```
<form id="form1" runat="server">
<div>
    <asp:ScriptManager ID="ScriptManager1" runat="server">
    </asp:ScriptManager>
    <asp:UpdatePanel ID="UpdatePanel1" runat="server">
        <ContentTemplate>
            <asp:UpdateProgress ID="UpdateProgress1" runat="server">
                <ProgressTemplate>
                    努力加载中,请稍候 ...<br />
                    <asp:Image ID="Image1" runat="server" ImageUrl="~/progress
bar.gif" />
                </ProgressTemplate>
            </asp:UpdateProgress>
            更新时间:<asp:Label ID="lblTime" runat="server" Text=""></asp:
Label>
        <asp:Button ID="btnRef" runat="server" Text="更新" OnClick="btnRef_
Click" />
        </ContentTemplate>
        <Triggers>
            <asp:AsyncPostBackTrigger ControlID="btnRef" EventName="Click" />
        </Triggers>
    </asp:UpdatePanel>
</div>
</form>
```

(4) 为控件添加事件,并编辑代码如下。

```
protected void btnRef_Click(object sender, EventArgs e)
{
    //设置进程挂起 3000ms
    System.Threading.Thread.Sleep(3000);
    lblTime.Text=DateTime.Now.ToString();
}
```

(5) 运行页面,当用户单击"更新"按钮时触发事件,在执行的方法中调用 System.Threading.Thread.Sleep()方法,将系统线程挂起的时间设置为 3000ms,在这 3000ms 的时间内会呈现"努力加载中,请稍候..."的文字及进度条图片,具体如图 8.11 所示;当 3000ms 过后,执行后面的 C#语句,效果如图 8.12 所示。

图 8.11　UpdateProgress 控件页面挂起演示

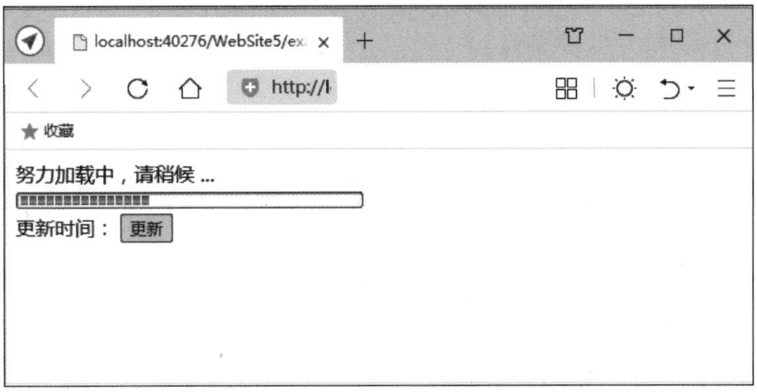

图 8.12　UpdateProgress 控件页面运行演示

用户提交表单后,如果服务器和客户端之间的通信需要较长时间,则会较长时间显示 UpdateProgress 控件;如果服务器和客户端之间交互的时间很短,基本上看不到 UpdateProgress 控件的显示。UpdateProgress 控件在大量的数据访问和数据操作中能够提高用户友好度,避免错误操作发生。

8.2.5 时钟控件 Timer

在 C/S 应用程序开发中，Timer 控件是最常用的控件，使用 Timer 控件能够进行时间控制，周期性触发某个事件。但是，在 Web 应用中，由于 Web 应用的无状态性，要以复杂的编程和大量的性能要求为代价，通过 JavaScript 实现该时间事件。

ASP.NET AJAX 中提供了一个 Web 版的 Timer 控件，用于执行局部更新，能够控制应用程序周期性进行事件刷新。Timer 控件的常用属性如表 8.3 所示，常用事件如图 8.4 所示。

表 8.3 Timer 控件的常用属性

属 性 名	说 明
Enabled	获取或设置是否启用 Tick 事件触发
Interval	获取或设置 Tick 事件之间的间隔时间，单位为 ms

表 8.4 Timer 控件的常用事件

事 件 名	说 明
Tick	每当经过指定的时间间隔触发

【示例 8-4】 在 chapter8 网站根目录下创建名为 example8-4 的页面，练习使用时钟控件 Timer。

(1) 在 example8-4 页面中添加 ScriptManager 控件和 UpdatePanel 控件，并在 UpdatePanel 控件内添加 Timer 控件及相关控件，具体如图 8.13 所示。

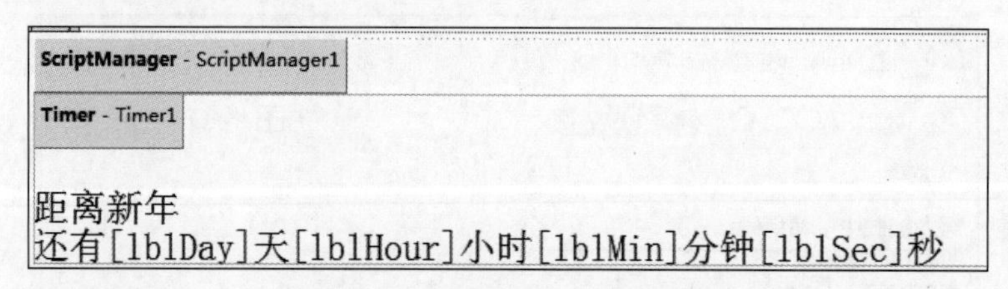

图 8.13 Timer 控件应用页面布局

(2) 设置各控件的属性，具体如下列的源文件。

```
<form id="form1" runat="server">
<div>
    <asp:ScriptManager ID="ScriptManager1" runat="server">
    </asp:ScriptManager>
    <asp:UpdatePanel ID="UpdatePanel1" runat="server">
        <ContentTemplate>
            <asp:Timer ID="Timer1" runat="server" Interval="1000" OnTick="
Timer1_Tick">
            </asp:Timer>
            <br/>
            <span">距离新年</span><br/><span class="auto-style1">还有</span
><asp:Label ID="lblDay" runat="server" CssClass="auto-style1"></asp:Label>
```

```
            <span>天</span><asp:Label ID="lblHour" runat="server" CssClass="
auto-style1"></asp:Label>
            <span>小时</span><asp:Label ID="lblMin" runat="server" CssClass="
auto-style1"></asp:Label>
            <span>分钟</span><asp:Label ID="lblSec" runat="server" CssClass="
auto-style1"></asp:Label>
            <span>秒 </span>
        </ContentTemplate>
    </asp:UpdatePanel>
    </div>
</form>
```

(3) 为控件添加事件,并编辑代码如下。

```
protected void Timer1_Tick(object sender, EventArgs e)
{
    //设置 targetDate 为目标时间
    DateTime targetDate =new DateTime(DateTime.Now.Year, 12, 31, 23, 59, 59);
    //计算剩余的天数、小时数、分钟数和秒数
    int days =targetDate.DayOfYear -DateTime.Now.DayOfYear;
    int hours =targetDate.Hour -DateTime.Now.Hour;
    int mins =targetDate.Minute -DateTime.Now.Minute;
    int secs =targetDate.Second -DateTime.Now.Second;
    //将所有剩余时间转换为秒数
    int allSecs =days * 24 * 3600 +hours * 3600 +mins * 60 +secs +1;
    //将剩余时间折算为天数、小时数、分钟数和秒数
    int lastDays =allSecs / 24 / 3600;
    int lastHours =allSecs %(24 * 3600) / 3600;
    int lastMins =allSecs %3600 / 60;
    int lastSecs =allSecs %60;
    if (allSecs >=0)
    {
        lblDay.Text =lastDays.ToString();
        lblHour.Text =lastHours.ToString();
        lblMin.Text =lastMins.ToString();
        lblSec.Text =lastSecs.ToString();
    }
    else
    {
        Timer1.Enabled =false;
        Response.Write("<script>alert('新年快乐!');</script>");
    }
}
```

(4) 运行页面,每间隔一秒刷新一次,计算剩余时间并显示在倒计时牌中,执行效果如图 8.14 所示。

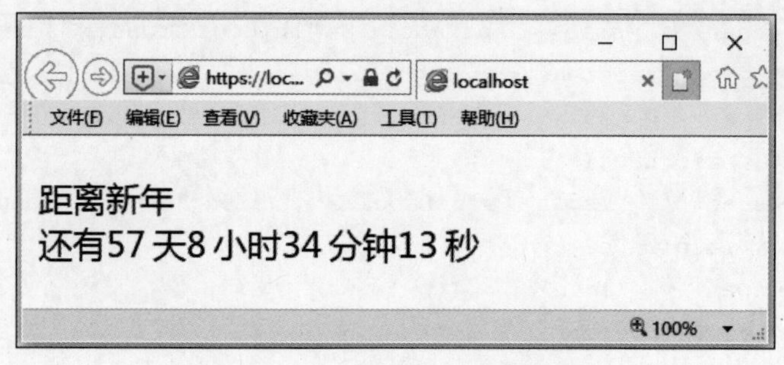

图 8.14　Timer 控件应用页面演示

Timer 控件无须复杂的 JavaScript 就可以直接实现时间控制，同时 Timer 控件也会占用大量的服务器资源，如果不停地进行客户端服务器的信息通信操作，将造成服务器死机。

综合实验八　基于 AJAX 的简易聊天室

主要任务

创建 ASP.NET 应用程序，使用 AJAX 扩展控件制作简易聊天室。

实验步骤

步骤 1：在 VS 2019 菜单上选择"文件"→"新建"→"项目"命令。

步骤 2：创建 ASP.NET 应用程序，命名为"综合实验八"。

步骤 3：在网站的根目录上右击，创建 Login.aspx 页面，添加相应控件，具体如图 8.15 所示。

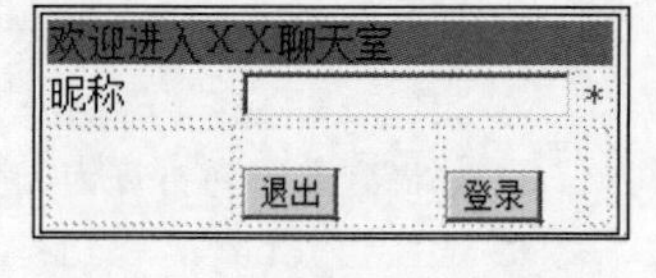

图 8.15　聊天登录页面布局

步骤 4：按如下源文件设置控件的相关属性值。

```
<form id="form1" runat="server">
        <div>
             <table border="0">
            <tr>
                <td colspan="4">
                    欢迎进入 XX 聊天室</td>
            </tr>
            <tr>
                <td>
                    昵称</td>
                <td colspan="2">
                     <asp:TextBox ID="txtName" runat="server" Width="129px">
</asp:TextBox></td>
                <td>
                     <asp:RequiredFieldValidator ID="RequiredFieldValidator1"
runat="server" ControlToValidate="TextBox1"
                          ErrorMessage=" *" SetFocusOnError="True" ForeColor=
"Red"></asp:RequiredFieldValidator></td>
```

```
            </tr>
            <tr>
                <td>
                </td>
                <td> <br />
                        <asp:Button ID="btnExist" runat="server" Text="退出"
CausesValidation="False" OnClick="btnExist_Click" />
                      </td>
                <td>
                    <br />
                        <asp:Button ID="btnLogin" runat="server" Text="登录"
OnClick="btnLogin_Click" />
                </td>
                <td></td>
            </tr>
        </table>
    </div>
</form>
```

步骤 5：编辑 Login.aspx.cs 代码如下。

```
protected void btnLogin_Click(object sender, EventArgs e)
{
    Session["id"]=TextBox1.Text;
    Application["id"]=Application["id"].ToString()+"@"+Session["id"].ToString();
    Response.Redirect("main.aspx");
}
protected void btnExist_Click(object sender, EventArgs e)
{
    Response.Write("<script>close();</script>");
}
```

步骤 6：在网站的根目录上右击，创建 Main.aspx 页面，添加相应控件，具体如图 8.16 所示。

步骤 7：按如下源文件设置控件的相关属性值。

```
<form id="form1" runat="server">
      <asp:ScriptManager ID="ScriptManager1" runat="server" />
      <div>
            <table>
                <tr>
                    <td>
                        聊天室</td>
                </tr>
                <tr>
                    <td>
                        欢迎</td>
                    <td>
                        <asp:Label ID="lblName" runat="server"></asp:Label></td>
                    <td>
                        在线会员</td>
                </tr>
                <tr>
```

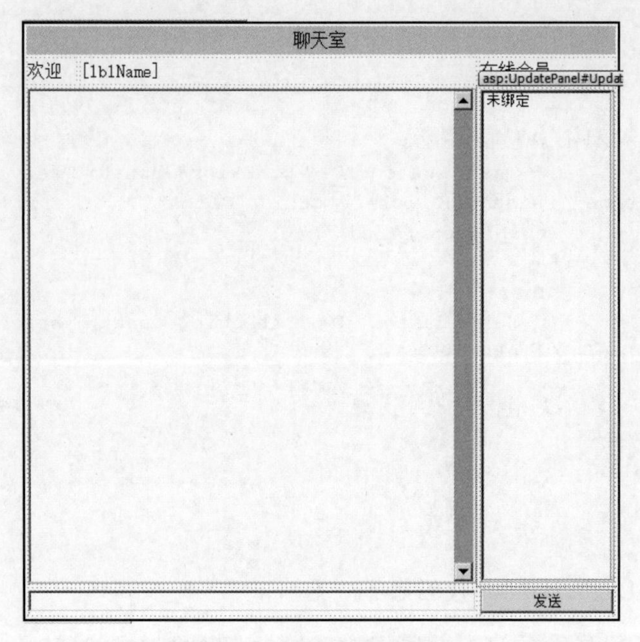

图 8.16 聊天主页面布局

```
                <td colspan="2">
                    <asp:UpdatePanel ID="UpdatePanel1" runat="server">
                        <ContentTemplate>
                            <asp:TextBox ID="txtChart" runat="server" Height
="431px" ReadOnly="True" TextMode="MultiLine"></asp:TextBox>
                        </ContentTemplate>
                        <Triggers>
                            <asp:AsyncPostBackTrigger ControlID="Timer1"
EventName="Tick" />
                            <asp:AsyncPostBackTrigger ControlID="btnSend"
EventName="Click" />
                        </Triggers>
                    </asp:UpdatePanel>
                </td>
                <td>
                    <asp:UpdatePanel ID="UpdatePanel2" runat="server">
                        <ContentTemplate>
                            <asp:ListBox ID="lbUsers" runat="server" Height
="435px" SelectionMode="Multiple"  Width="120px"></asp:ListBox>
                        </ContentTemplate>
                        <Triggers>
                            <asp:AsyncPostBackTrigger ControlID="Timer1"
EventName="Tick" />
                        </Triggers>
                    </asp:UpdatePanel>
                </td>
            </tr>
            <tr>
                <td colspan="2">
```

```
                        <asp:TextBox ID="txtMessage" runat="server" Width="
399px"></asp:TextBox></td>
                    <td>
                        <asp:Button ID="btnSend" runat="server" OnClick="
Button1_Click" Text="发送" Width="120px" /></td>
                </tr>
            </table>
        </div>
        <asp:Timer ID="Timer1" runat="server" Interval="1000" OnTick="Timer1_
Tick">
        </asp:Timer>
        <asp:Timer ID="Timer2" runat="server" OnTick="Timer2_Tick">
        </asp:Timer>
</form>
```

步骤 8：编辑 Main.aspx.cs 代码如下。

```
protected void Page_Load(object sender, EventArgs e)
{
    if (Session["id"].ToString()=="")
    {
        Response.Redirect("~/login.aspx");
    }
    else
    {
        lblName.Text=Session["id"].ToString();
    }
}
protected void Button1_Click(object sender, EventArgs e)
{
     Application["chat"]=Application["chat"].ToString() +
     Session["id"].ToString()+"说:" +
     txtMessage.Text+"("+DateTime.Now.ToString()+")\n";
     txtMessage.Text="";
}
protected void Timer1_Tick(object sender, EventArgs e)
{
     txtChart.Text=Application["chat"].ToString();
     lbUsers.Items.Clear();
     string[] str=Application["id"].ToString().Split('@');
     foreach (string t in str)
     {
         lbUsers.Items.Add(t);
     }
     lbUsers.Items.Remove("");
}
protected void Timer2_Tick(object sender, EventArgs e)
{
}
```

步骤 9：在网站的根目录上右击，创建 Global.asax 文件，在 Application_Start()中添加如下代码。

```
protected void Application_Start(object sender, EventArgs e)
{
        Application["chat"]="";
        Application["id"]="";
}
```

步骤 10：运行 Login 页面，可以使用昵称登录 Main 页面使用聊天功能，具体如图 8.17 和图 8.18 所示。

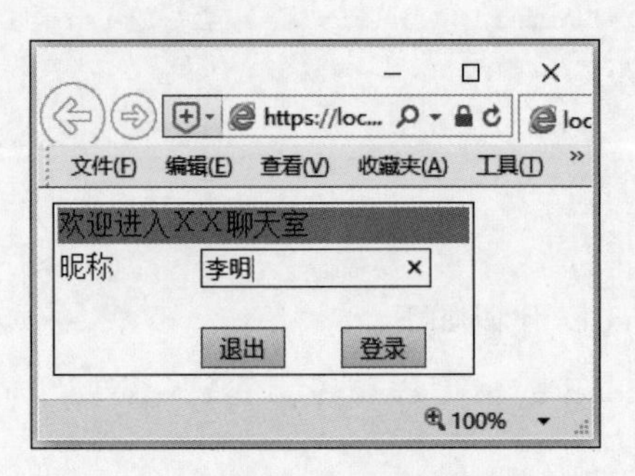

图 8.17　聊天登录页面运行测试

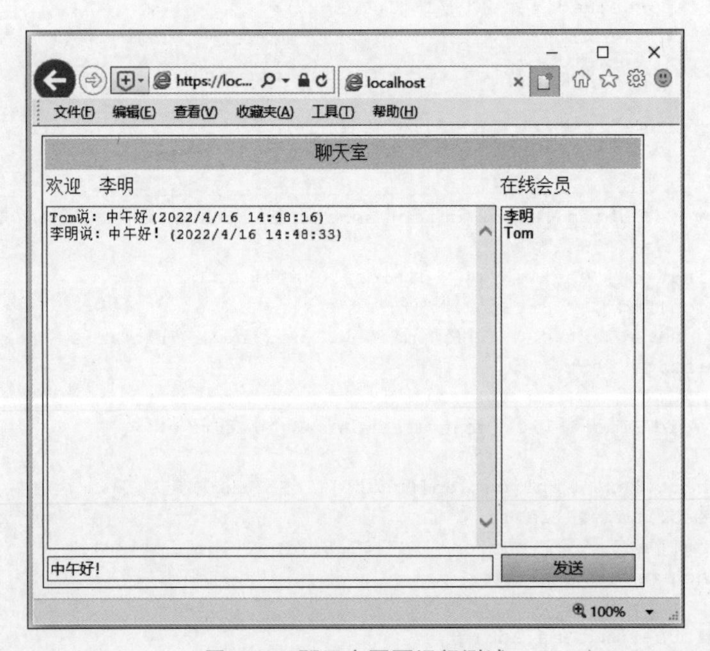

图 8.18　聊天主页面运行测试

第9章 ADO.NET数据库访问

本章学习目标

- 了解 ADO.NET 原理及特点
- 熟练掌握 SqlConnection 类的使用方法
- 熟练掌握 SqlCommand 类的使用方法
- 熟练掌握 SqlDataReader 类的使用方法
- 熟练掌握 DataSet 类的使用方法
- 熟练掌握 SqlDataAdapter 类的使用方法

本章首先介绍 ADO.NET 的基本原理及特点，然后讲述 SqlConnection 类、SqlCommand 类、SqlDataReader 类、DataSet 类以及 SqlDataAdapter 类的属性和方法，最后在 SQL Server 2012 数据库中进行了相关操作应用。

9.1 ADO.NET 基础

ADO.NET 是.NET Framework 中的一个类库，能够让开发人员在应用程序中更加方便地访问和操作数据。与 C#.NET、VB.NET 不同的是，ADO.NET 并不是一种语言，而是对象的集合。在 ADO.NET 中封装了大量复杂的数据操作代码，使得在 ASP.NET 应用程序开发中只编写少量代码即可完成大量数据操作。

9.1.1 ADO.NET 介绍

扫一扫

ADO.NET 可以处理多样的数据源，既可以处理存储在内存中的数据，也可以处理存储在存储区域的数据，如文本文件、XML、关系数据库等。作为.NET 框架的重要组成部分，ADO.NET 封装在 System.Data.dll 中，并与 System.Xml.dll 中的 XML 类集成。ADO.NET

中的各类既分工明确，又相互协作，提供了表格数据的访问服务。ADO.NET 中主要包含两组重要的类：一组负责处理软件内部的实际数据(DataSet)；另一组负责与外部数据系统通信(Data Provider)。ADO.NET 架构如图 9.1 所示。

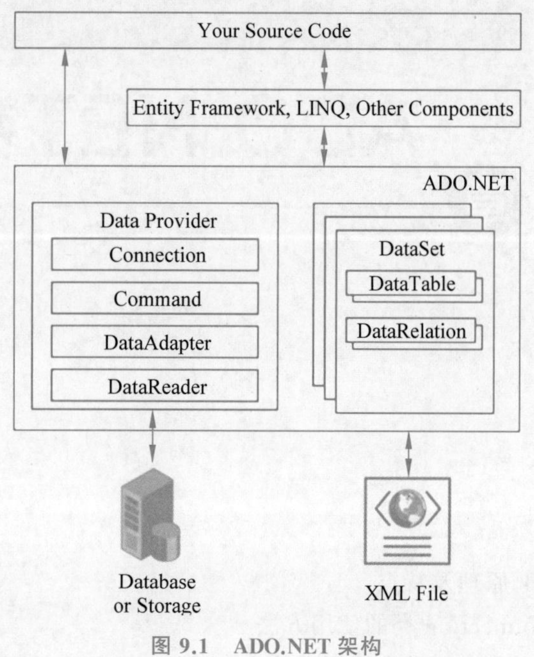

图 9.1　ADO.NET 架构

9.1.2　ADO.NET 与 ADO

ADO.NET 的名称起源于 ADO，全称为 ActiveX Data Objects，是.NET 编程环境和 Windows 环境中优先使用的数据访问接口。但 ADO.NET 并不只是 ADO 的简单升级版本，ADO.NET 和 ADO 是两种截然不同的数据访问方式。ADO 使用 OLE DB 接口基于微软的.NET 体系架构，是.NET 未实施之前访问数据的组件，ADO.NET 拥有自己的 ADO.NET 接口，随着.NET 的发展，ADO.NET 以其显著的优越性逐步取代 ADO。

ADO 技术和 ADO.NET 技术的主要对比如下。

1. 数据在内存中的存在形式

ADO 中的数据以 RecordSet 记录集的形式存放在内存中，而 ADO.NET 中的数据以 DataSet 数据集的形式存放在内存中。

2. 数据的表现形式

在 ADO 中，记录集的表现形式像一个表。如果需要包含来自多个数据库的表的数据，必须使用 JOIN 语句将各个数据库中表的数据组合到单个记录集中。而在 ADO.NET 中，数据集是一个表或多个表的集合，数据集可以保存多个独立的表并维护有关表之间关系的信息。

3. 数据的连接和断开

ADO 是为连接式的访问而设计的，而 ADO.NET 仅在操作数据时连接数据库，可以通过数据集进行读入，当需要将数据集中的资源更新到数据库时，ADO.NET 再与数据库连接并更新。

4. 构架设计

在构架设计上 ADO.NET 与 ADO 也是不同的，ADO.NET 相对于 ADO 更加方便、简洁，在设计角度 ADO.NET 的设计更加完善。

9.1.3 ADO.NET 中的常用对象

在 ADO.NET 中可以创建对象来进行数据库的操作，从而简化开发。ADO.NET 的常用对象如下。

- Connection 对象：连接对象，提供与数据库的连接。
- Command 对象：命令对象，表示要执行的数据库命令。
- Parameter 对象；参数对象，表示数据库命令中的参数。
- DataReader 对象：数据流对象，表示从数据源中提供的快速的、只向前的、只读数据流。
- DataAdapter 对象：数据适配器对象，提供连接 DataSet 对象和数据源的桥梁。DataAdapter 使用 Command 对象在数据源中执行 SQL 命令，将数据加载到 DataSet 中，实现对 DataSet 中数据的更改与数据源保持一致。
- DataSet 对象：数据集对象，表示命令的结果数据集。DataSet 是包含一个或多个 DataTable 对象的集合，由数据行、数据列以及主键、外键、约束数据的关系等信息组成。

9.1.4 ADO.NET 数据库操作过程

使用 ADO.NET 中的对象，不仅能够通过控件绑定数据源，也可以通过程序实现数据源的访问。在 ADO.NET 中对数据库的操作基本上需要三个步骤：创建一个连接；执行命令对象并显示；释放连接。ADO.NET 的使用过程如图 9.2 所示。

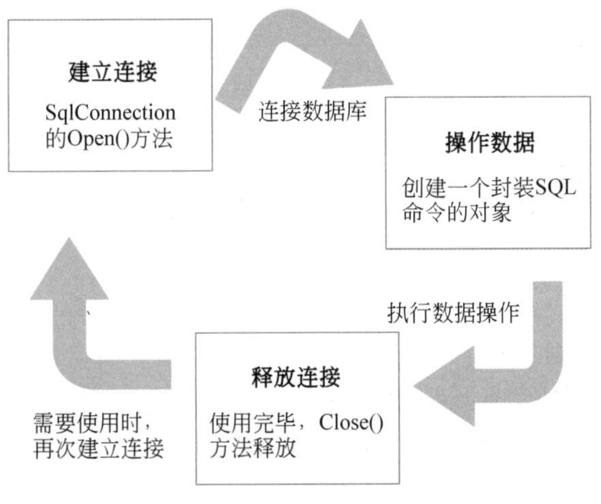

图 9.2　ADO.NET 的使用过程

从图 9.2 中可以归纳出使用 ADO.NET 访问数据库的规范步骤，具体如下。

(1) 创建一个连接对象。

(2) 使用对象的 Open()方法打开连接。

(3) 创建一个封装 SQL 命令的对象。

(4) 调用执行命令对象。

(5) 执行数据库操作。

(6) 执行完毕，释放连接。

本书以 SQL Server 2012 数据库为例，将具体讲解数据库的连接建立、操作数据以及释放连接等操作，掌握了这些基本知识，就能够使用 ADO.NET 进行基本的数据库开发。

9.2 SqlConnection 连接对象

扫一扫

9.2.1 SqlConnection 对象的属性与方法

在 ASP.NET 开发中，连接 SQL Server 数据库需要使用 SqlConnection 对象，对应的类包含在 System. Data. SqlClient 命名空间中，使用该类首先需要添加命名引用。SqlConnection 类可以用来联系.NET Framework 和 SQL Server 类型数据库通信会话，SqlConnection 类与后续章节的 SqlDataAdapter 类以及 SqlCommand 类一起实现了.NET Framework 与 SQL Server 类型数据库的交互。

SqlConnection 类的主要属性、方法及构造函数如表 9.1～表 9.3 所示。

表 9.1　SqlConnection 类的主要属性

属　性　名	说　　明
ConnectionString	获取或设置用于打开 SQL Server 数据库的字符串
ConnectionTimeout	获取终止尝试并生成错误之前在尝试建立连接时所等待的时间
Database	获取当前数据库的名称或打开连接后要使用的数据库的名称
DataSource	获取要连接的 SQL Server 的实例的名称
State	最近在连接上执行网络操作时 SqlConnection 的状态
ServerVersion	获取一个字符串，该字符串包含客户端所连接到的 SQL Server 实例的版本

说明：

State 属性具有如下几个枚举值。

- Broken：表示数据连接中断，只有连接曾经打开过才会出现此状态。
- Open：表示连接处于打开状态。
- Connecting：表示连接对象正与数据源连接。
- Executing：表示连接对象正在执行命令。
- Fetching：表示连接对象正在检索数据。
- Closed：表示连接对象处于关闭状态。

表 9.2　SqlConnection 类的主要方法

方 法 名	说　明
Close()	关闭与数据库之间的连接，此方法是关闭已打开连接的首选方法
Dispose()	释放使用的所有资源
Open()	使用由 ConnectionString 指定的属性设置打开一个数据库连接

表 9.3　SqlConnection 类的构造函数

构 造 函 数	说　明
SqlConnection()	初始化 SqlConnection 类的新实例
SqlConnection(String)	使用连接字符串初始化 SqlConnection 类的新实例

9.2.2　创建连接字符串 ConnectionString

扫一扫

在连接数据库前，需要为连接设置连接字符串。连接字符串明确了数据库管理系统位置、数据库系统登录方式、数据库连接对象等信息，从而正确地与 SQL Server 数据库建立连接。连接字串示例代码如下。

```
server='服务器地址';database='数据库名称';uid='数据库用户名';pwd='数据库密码';
```

上述代码是数据库连接字串的基本格式，如果需要连接到本地的 demo 数据库，则编写的 SQL 连接字串如下。

```
server='(local) /sqlexpress ';database='demo';uid='sa';pwd='sa';
```

其中，server 是 SQL Server 服务器的地址，如果数据库服务器是本地服务器，则配置为(loacal)/sqlexpress 或者./sqlexpress 即可；如果是远程服务器，则需要填写具体的 IP 地址。uid 是数据库登录时的用户名，pwd 是数据库登录时使用的密码。此外，数据库服务器也可以使用 Windows 身份认证，对应属性设置为 trusted_connection＝true。ConnectionString 对象的属性如表 9.4 所示。

表 9.4　ConnectionString 对象的属性

属 性 名	说　明
Server 或 Data Source	访问的 SQL Server 服务器的地址
Database 或 Initial Catalog	访问的数据库名
Uid 或 User Id	数据库登录时的用户名
Pwd 或 Password	数据库登录时使用的密码
trusted_connection 或 Integrated Security	属性值赋值为 True，表示使用信任模式，即以 Windows 身份认证登录数据库

说明：

(1) 连接字符串“server＝'服务器地址';database＝'数据库名称';uid＝'数据库用户名';pwd='数据库密码'”中的单引号可以缺省，即该连接字符串也可写为“server＝服务器地址；

database=数据库名称;uid=数据库用户名;pwd=数据库密码”。

(2) 连接字符串中的属性值与数据库中的 SQL 语句对应,不区分大小写。

扫一扫

9.2.3 Web.config 文件中的连接字符串

对于应用程序而言,可能需要在多个页面的程序代码中使用数据连接字符串。当数据库连接字符串发生改变时,要修改所有连接字符串,此时可以在＜connectionStrings＞配置字节中定义应用程序的数据库连接字符串,所有程序从该配置节读取字符串,当需要改变连接时,只在配置节中重新设置即可。

1. 在 Web.config 文件中配置数据库连接字符串

在 Web.config 文件中添加如下代码,即可将应用程序的数据库连接字符串存储在＜connectionStrings＞配置节中。

```
<connectionStrings>
    <add name="ConnectionName"connectionString="Server=.\SQLEXPRESS;Database
=Demo;UserID=sa;Password=abc123"  providerName="System.Data.SqlClient"/>
</connectionStrings>
```

2. 获取 Web.config 文件中的数据库连接字符串

在页面的.cs 代码内通过一段代码获取＜connectionStrings＞ 标签里的数据库连接的字符串,代码如下。

引用命名空间:

```
Using System.Configuration;
```

定义变量并赋值:

```
string conStr = ConfigurationManager. ConnectionStrings [ " ConnectionName "].
ToString();
```

conStr 变量值即 Web.config 中保存的连接字符串。

9.2.4 SqlConnection 对象的应用

Microsoft SQL Server 2012 是微软发布的新一代数据库管理平台,与 VS 2019 兼容性好,本书中示例选用 SQL Server 2012 为后台数据库,首先在平台中创建 Demo 数据库,然后新建 Student 数据表并添加部分测试数据。Student 表的结构及表中内容如图 9.3 和图 9.4 所示。

YY-PC\SQLEXPRES...o - dbo.student ×

列名	数据类型	允许 Null 值
sno	char(8)	☐
sname	char(20)	☑
sex	char(3)	☑
age	int	☑
dept	varchar(15)	☑

图 9.3 Student 表的结构

	sno	sname	sex	age	dept
1	05880101	周一	男	20	计算机系
2	05880102	吴二	女	20	信息系
3	05880103	张三	女	19	计算机系
4	05880104	李四	男	22	信息系
5	05880105	王五	男	22	数学系
6	05880106	赵六	男	19	数学系
7	05880107	陈七	女	23	日语系
8	05880108	刘八	男	21	日语系

图 9.4 Student 表内容

【示例 9-1】 在 E 盘 ASP.NET 项目代码目录中创建 chapter9 子目录，将其作为网站根目录，创建名为 example9-1 的网页，使用 SqlConnection 对象进行数据的连接。

(1) 双击打开网站根目录的 Web.Config 文件，在<connectionStrings>标签内添加连接字符串如下。

```
<connectionStrings>
<add name="DemoConnection" connectionString="Server=.\SQLEXPRESS;Database=
Demo;trusted_connection=true;"  providerName="System.Data.SqlClient" />
</connectionStrings>
```

(2) 在 example9-1 的网页中添加一个按钮控件，设置源文件如下。

```
<form id="form1" runat="server">
    <div>
        <asp:Button ID="btnConnect" runat="server" OnClick="btnConnect_Click"
Text="连接数据库"/>
    </div>
</form>
```

(3) 在 example9-1.cs 内添加对命名空间的引用。

```
using System.Data.SqlClient;
using System.Configuration;
```

(4) 为按钮添加并编辑代码如下。

```
protected void btnConnect_Click(object sender, EventArgs e)
{
    //从 web.config 配置文件取出数据库连接串
    string sqlConStr = ConfigurationManager.ConnectionStrings["DemoConnection"].
    ConnectionString;
    //实例化数据库连接对象
    SqlConnection sqlCon=new SqlConnection(sqlConStr);
    //打开连接
    sqlCon.Open();
    if (sqlCon.State==System.Data.ConnectionState.Open)
    {
        Response.Write("<script>alert('数据库连接成功!')</script>");
    }
    //关闭连接
    sqlCon.Close();
    //释放连接对象
    sqlCon.Dispose();
}
```

(5) 运行页面，单击“连接数据库”按钮测试数据库连接是否成功，结果如图 9.5 所示。

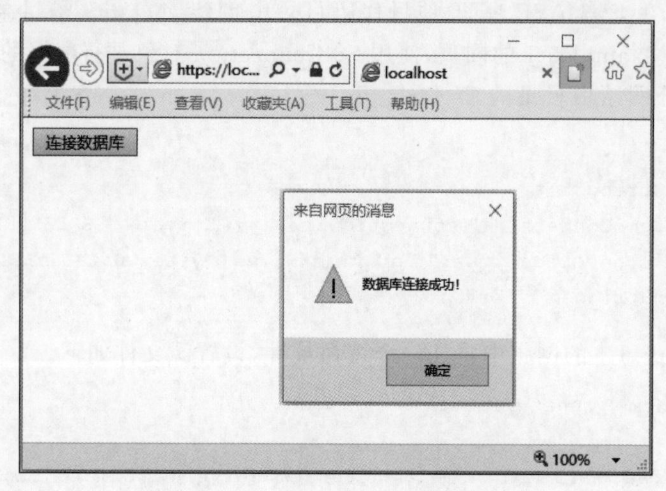

图 9.5　SqlConnection 对象应用页面演示

9.3 SqlCommand 命令对象

扫一扫

9.3.1 SqlCommand 对象的属性与方法

SqlCommand 对象可以使用连接命令直接与数据源进行通信。当需要向数据库内插入或删除数据时，就可以使用 SqlCommand 对象，对应的类包含在 System.Data.SqlClient 命名空间中。SqlCommand 类中包括了数据库在执行命令时的全部信息，当 SqlCommand 对象的属性设置好之后，可以调用 SqlCommand 对象的方法对数据库中的数据进行处理。SqlCommand 对象的主要属性、方法及构造函数如表 9.5～表 9.7 所示。

表 9.5　SqlCommand 类的主要属性

属 性 名	说 明
Name	获取或设置 Command 的程序化名称
Connection	获取或设置对 Connection 对象的引用
CommandType	获取或设置指定使用 SQL 语句或存储过程，默认情况下是 SQL 语句
CommandText	获取或设置命令对象包含的 SQL 语句或存储过程名
Parameters	命令对象的参数集合

说明：

(1) CommandType 属性具有如下 3 种枚举值。

- Text：表示 Command 对象用于执行 SQL 文本，该属性的默认值为 Text。
- StoredProcedure：表示 Command 对象用于执行存储过程。
- TableDirect：表示 Command 对象用于直接处理某张表。

(2) CommandText 属性根据 CommandType 属性的取值决定 CommandText 属性表

示的意义，具体分为下列 3 种情况。

- 如果 CommandType 属性取值为 Text，则 CommandText 属性表示 SQL 语句的内容。
- 如果 CommandType 属性取值为 StoredProcedure，则 CommandText 属性表示存储过程的名称。
- 如果 CommandType 属性取值为 TableDirect，则 CommandText 属性表示表的名称。

表 9.6　SqlCommand 对象的主要方法

方 法 名	说　　明
ExecuteReader()	执行查询操作，返回一个具有多行多列数据的数据流
ExecuteScalar()	执行查询操作，返回单个值
ExecuteNonQuery()	执行插入、修改或删除操作，返回本次操作受影响的行数

表 9.7　SqlCommand 对象的构造函数

构 造 函 数	说　　明
SqlCommand ()	初始化 SqlCommand 类的新实例
SqlCommand (cmdText))	初始化包含命令文本的命令对象，cmdText 为命令文本
SqlCommand (cmdText, connection)	初始化包含命令文本和连接对象的命令对象，cmdText 为命令文本，connection 为连接对象

9.3.2　ExecuteNonQuery()方法

当指定一个 SQL 语句后，可以通过 ExecuteNonQuery()方法执行该语句的操作。ExecuteNonQuery()不仅可以执行 SQL 语句，也可以执行存储过程。同时，可以通过 Command 对象和 Parameters 进行参数传递，示例代码如下所示。

```
string str=" server=. /sqlexpress ;database=demo; trusted_connection=true; ";
SqlConnection con=new SqlConnection(str);
con.Open();
SqlCommand cmd=new SqlCommand("insert into student values ('05880110')",con);
cmd.ExecuteNonQuery();
```

运行上述代码，执行“insert into news values ('05880110')”这条 SQL 语句向数据库中插入数据。执行数据库的 insert、update 以及 delete 等不返回任何行的语句或存储过程时，可以使用 ExecuteNonQuery()方法，该方法返回一个整数，表示执行 SQL 语句或存储过程时表中受影响的行数。

【示例 9-2】 在 chapter9 网站根目录下创建名为 example9-2 的网页，页面内包含若干控件，练习使用 SqlCommand 对象的 ExecuteNonQuery()方法实现数据的增加、删除、修改操作。

(1) 按图 9.6 所示添加相应控件。

(2) 按如下源文件设置控件的相关属性值。

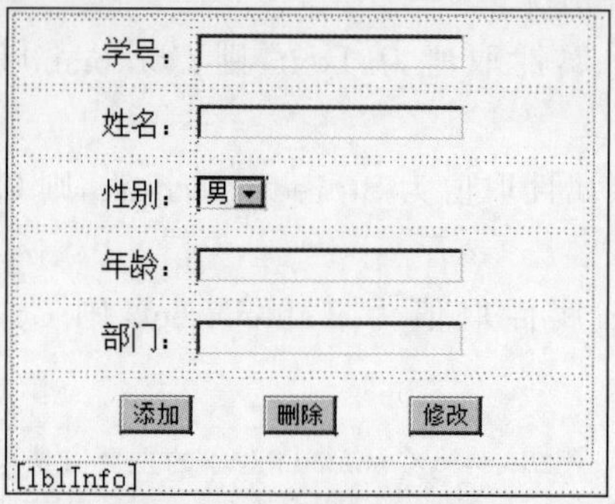

图 9.6 SqlCommand 对象应用页面设计

```
<form id="form2" runat="server">
        <div>
            <table>
                <tr>
                    <td>学号:</td>
                    <td>
                        <asp:TextBox ID="txtNum" runat="server"></asp:TextBox>
                    </td>
                </tr>
                <tr>
                    <td>姓名:</td>
                    <td>
                        <asp:TextBox ID="txtName" runat="server"></asp:TextBox>
                    </td>
                </tr>
                <tr>
                    <td>性别:</td>
                    <td>
                        <asp:DropDownList ID="ddlSex" runat="server">
                            <asp:ListItem Selected="True">男</asp:ListItem>
                            <asp:ListItem>女</asp:ListItem>
                        </asp:DropDownList>
                    </td>
                </tr>
                <tr>
                    <td>年龄:</td>
                    <td>
                        <asp:TextBox ID="txtAge" runat="server"></asp:TextBox>
                    </td>
                </tr>
                <tr>
                    <td>部门:</td>
                    <td>
                        <asp:TextBox ID="txtDept" runat="server"></asp:TextBox>
                    </td>
                </tr>
                <tr>
                    <td colspan="2">
```

```
                        <asp:Button ID="btnInsert" runat="server"  Text="添加"
OnClick="btnInsert_Click"/>
                        <asp:Button ID="btnDelete" runat="server"  Text="删除"
OnClick="btnDelete_Click"/>
                      
                        <asp:Button ID="btnUpdate" runat="server"  Text="修改"
  OnClick="btnUpdate_Click"/>
                    </td>
                </tr>
            </table>
        </div>
        <asp:Label ID="lblInfo" runat="server" Text=""></asp:Label>
</form>
```

(3) 添加命名空间的引用。

```
using System.Data.SqlClient;
using System.Configuration;
```

(4) 为控件添加事件,并编辑代码如下。

```
SqlConnection con;
SqlCommand cmd;
protected void Page_Load(object sender, EventArgs e)
{
    string sqlconnstr=ConfigurationManager.ConnectionStrings["DemoConnection"].
ConnectionString;
    con=new SqlConnection(sqlconnstr);
    cmd=new SqlCommand();
    cmd.Connection=con;
}
protected void btnInsert_Click(object sender, EventArgs e)
{
    //设置 cmd 对象的 CommandText 属性值
    cmd.CommandText=string.Format("insert into student values ('{0}','{1}',
'{2}','{3},'{4}')", txtNum.Text, txtName.Text, ddlSex.SelectedValue, txtAge.
Text, txtAge.Text, txtDept.Text);
    con.Open();
    int n=cmd.ExecuteNonQuery();
    con.Close();
    if (n==1)
    {
        lblInfo.Text="添加新学生信息成功";
    }
    else
    {
        lblInfo.Text="添加新学生信息失败";
    }
}
protected void btnDelete_Click(object sender, EventArgs e)
{
    cmd.CommandText=string.Format("delete from student where sno='{0}'",
txtNum.Text);
```

```
        con.Open();
        int n=cmd.ExecuteNonQuery();
        con.Close();
        if (n==1)
        {
            lblInfo.Text="删除学生信息成功";
        }
        else
        {
            lblInfo.Text="删除学生信息失败";
        }
    }
    protected void btnUpdate_Click(object sender, EventArgs e)
    {
        cmd.CommandText=string.Format("update student set sname='{0}',sex='{1}',
    age={2},dept='{3}' where sno='{4}'", txtName.Text, ddlSex.SelectedValue,
    txtAge.Text, txtDept.Text, txtNum.Text);
        con.Open();
        int n=cmd.ExecuteNonQuery();
        con.Close();
        if (n==1)
        {
            lblInfo.Text="修改学生信息成功";
        }
        else
        {
            lblInfo.Text="修改学生信息失败";
        }
    }
```

说明：

String.Format（String，Object[]）方法可以将String中的格式项替换为指定数组中相应Object实例的值。如String.Format（"a{0}b{1}c{2}d{0}efg"，'A',"12",3);等价于字符串“aAb12c3dAefg”。

(5) 运行网站，执行效果如图9.7所示。

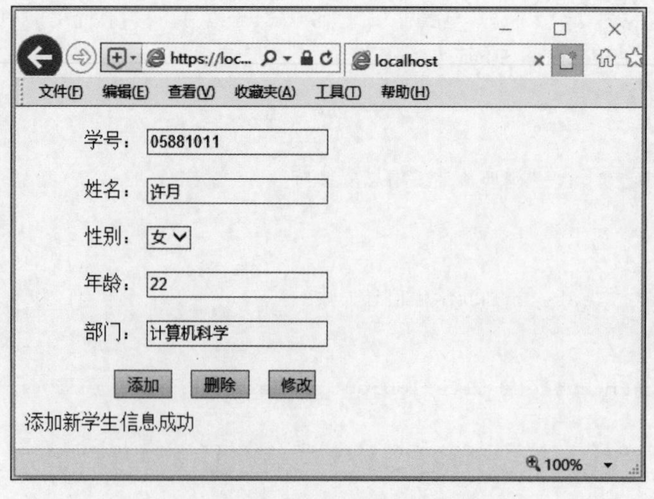

图9.7 SqlCommand对象应用页面演示

ExecuteNonQuery()方法不仅能够执行 SQL 语句，同样可以执行存储过程和数据定义语言来对数据库或目录进行构架操作，如 CREATE TABLE 等。但在执行存储过程之前，必须预先创建该存储过程。

9.3.3 ExecuteScalar()方法

SqlCommand 类中的 ExecuteScalar()方法提供了返回单个值的功能，如果需要获取数据值或者 Count(*)、Sum(Money)等聚合函数的结果，可以使用 ExecuteScalar()方法。示例代码如下。

```
string str=" server=. /sqlexpress ;database=demo; trusted_connection=true; ";
SqlConnection con=new SqlConnection(str);
con.Open();
SqlCommand cmd=new SqlCommand("select count( * ) from student", con);
object  obj=cmd.ExecuteScalar();
```

上述代码创建了连接，实例化 SqlCommand 对象，执行 ExecuteScalar()方法返回单个值。

【示例 9-3】 在 chapter9 网站根目录下创建名为 example9-3 的网页，页面内包含若干控件，练习使用 SqlCommand 对象的 ExecuteScalar ()方法实现数据的查询。

(1) 按图 9.8 所示添加相应控件。

学号：	
姓名：	[lblName]
性别：	[lblSex]
年龄：	[lblAge]
部门：	[lblDept]
查询	

图 9.8 SqlCommand 对象查询应用页面设计

(2) 设置控件的相关属性值如下。

```
<form id="form1" runat="server">
    <div>
            <table>
                <tr>
                    <td>学号:</td>
                    <td>
                        <asp:TextBox ID="txtNum" runat="server"></asp:TextBox>
                    </td>
                </tr>
                <tr>
                    <td>姓名:</td>
                    <td>
                        <asp:Label ID="lblName" runat="server" Text =""></asp:
Label>
```

```
                    </td>
                </tr>
                <tr>
                    <td>性别:</td>
                    <td>
                            <asp:Label ID="lblSex" runat="server" Text=""></asp:
Label>
                    </td>
                </tr>
                <tr>
                    <td>年龄:</td>
                    <td>
                            <asp:Label ID="lblAge" runat="server" Text=""></asp:
Label>
                    </td>
                </tr>
                <tr>
                    <td>部门:</td>
                    <td>
                        <asp:Label ID="lblDept" runat="server"></asp:Label>
                    </td>
                </tr>
                <tr>
                    <td colspan="2">
                        <asp:Button ID="btnSelect" runat="server"  Text="查询"
OnClick="btnSelect_Click"/>
                        </td>
                </tr>
            </table>
    </div>
    </form>
```

(3) 添加命名空间的引用。

```
using System.Data.SqlClient;
using System.Configuration;
```

(4) 为控件添加事件,并编辑代码如下。

```
SqlConnection con;
SqlCommand cmd;
protected void Page_Load(object sender, EventArgs e)
{
     string sqlconnstr=ConfigurationManager.ConnectionStrings["DemoConnection"].
ConnectionString;
    con=new SqlConnection(sqlconnstr);
    cmd=new SqlCommand();
    cmd.Connection=con;
}
protected void btnSelect_Click(object sender, EventArgs e)
{
    object obj;
    cmd.CommandText="select sname  from student where sno='"+txtNum.Text+"'";
    con.Open();
```

```
        obj=cmd.ExecuteScalar();
        con.Close();
        lblName.Text=obj.ToString();
        cmd.CommandText="select sex  from student where sno='"+txtNum.Text+"'";
        con.Open();
        obj=cmd.ExecuteScalar();
        con.Close();
        lblSex.Text=obj.ToString();
        cmd.CommandText="select age  from student where sno='"+txtNum.Text+"'";
        con.Open();
        obj=cmd.ExecuteScalar();
        con.Close();
        lblAge.Text=obj.ToString();
        cmd.CommandText="select dept  from student where sno='"+txtNum.Text+"'";
        con.Open();
        obj=cmd.ExecuteScalar();
        con.Close();
        lblDept.Text=obj.ToString();
    }
```

(5) 运行网站,执行效果如图 9.9 所示。

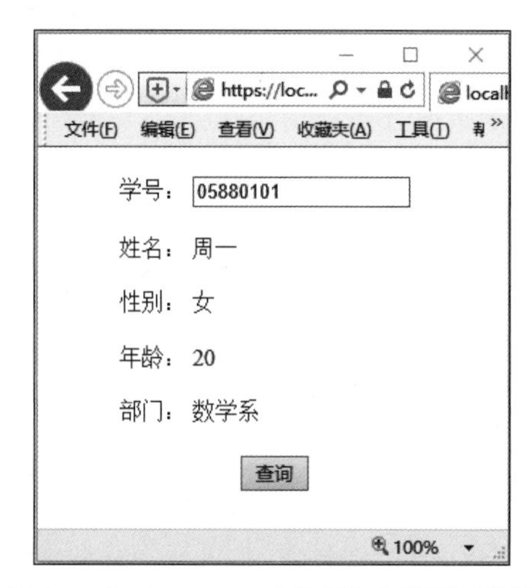

图 9.9　SqlCommand 对象查询应用页面演示

9.3.4　SqlParameter 参数对象

如果 SqlCommand 对象执行 SQL 语句或者存储过程时,在 SQL 语句中或者存储过程中含有参数,此时可以使用 SqlParameter 对象方便地设置 SQL 语句或存储过程的参数。SqlParameter 对象的主要属性如表 9.8 所示。

表 9.8　SqlParameter 对象的主要属性

属性名	说明
ParameterName	获取或设置参数的名称
SqlDbType	获取或设置指定参数的数据类型,如整型、字符型等

续表

属 性 名	说 明
Direction	获取或设置指定参数的方向
Value	获取或设置指定输入参数的值

说明：

Direction 属性具有如下几个枚举值。

- ParameterDirection.Input：表示输入参数，也是 Direction 属性的默认值。
- ParameterDirection.Output：表示输出参数。
- ParameterDirection.InputOutput：表示输入参数和输出参数。
- ParameterDirection.ReturnValue：表示返回值。

使用 SqlCommand 对象执行带参数的 SQL 语句时，可以创建一个 SqlParameter 对象，直接添加到 SqlCommand 对象的参数集合中来指定 SQL 语句中参数的值。

下面将示例 9-2 中的插入命令：

```
cmd.CommandText=string.Format("insert into student values ('{0}','{1}','{2}',
{3}, '{4}')", txtNum.Text, txtName.Text, ddlSex.SelectedValue, txtAge.Text,
txtDept.Text);
```

改写为用 SqlCommand 对象来实现，代码如下。

```
cmd.CommandText="insert into student values (@Sno, @Sname, @Sex, @Age, @Dept)";
cmd.AddWithValue ("@Sno", txtNum.Text);
cmd.AddWithValue ("@Sname", txtName.Text);
cmd.AddWithValue ("@Sex", ddlSex.SelectedValue);
cmd.AddWithValue ("@Age", txtAge.Text);
cmd.AddWithValue ("@Dept", txtDept.Text);
```

在指定“参数值”时，示例 9-2 中的数据类型必须与后台数据库中对应列的数据类型一致，否则会产生错误。而使用了 SqlParameter 对象会自动转换匹配，不需要再明确声明 Age 列的数据类型为 int 型，Sex 列的“参数值”为可变字符型等。

9.4 SqlDataReader 数据访问对象

扫一扫

9.4.1 SqlDataReader 对象的属性与方法

SqlDataReader 对象可以从数据库中得到只读、只向前的数据流，对应的类包含在 System.Data.SqlClient 命名空间中。SqlDataReader 每次的访问或操作只有一个记录保存在服务器的内存中，具有较快的访问能力，占用较少的服务器资源。SqlDataReader 对象的主要属性、方法如表 9.9 和表 9.10 所示。

表 9.9 SqlDataReader 对象的主要属性

属 性 名	说 明
FieldCount	获取当前行中的列数
IsClosed	获取一个布尔值，指示 SqlDataReader 对象是否关闭
RecordAffect	获取执行 SQL 语句时修改的行数

表 9.10 SqlDataReader 对象的主要方法

属 性 名	说 明
Read()	获取当前行中的列数
Close()	关闭 SqlDataReader 对象
IsDBNull	返回布尔值，表示列是否包含 NULL 值
GetName()	返回指定索引的字段名
GetBoolean()	获取指定列的值，类型为布尔值
GetString()	获取指定列的值，类型为字符串
GetByte()	获取指定列的值，类型为字节
GetInt32()	获取指定列的值，类型为整型值
GetDouble()	获取指定列的值，类型为双精度值
GetDateTime()	获取指定列的值，类型为日期时间值
GetValue(i)	获取索引为 i 指定列的值，类型为对象

SqlDataReader 类没有构造函数，如果要创建 SqlDataReader 类的对象，只能通过 SqlCommand 类的 ExcuteReader()方法得到。

9.4.2 使用 SqlDataReader 对象读取数据

SqlDataReader 对象的 Read()方法可以判断 SqlDataReader 对象中的数据是否有下一行数据，并将游标下移到该行。通过 Read()方法可以判断 SqlDataReader 对象中的数据是否读完，示例代码如下。

```
while (dr.Read())
```

同样，通过 Read()方法可以遍历读取数据库中行的信息，在读取到一行时，如果需要获取某列的值，只使用索引器即可，即使用“[”和“]”运算符确定某一列的值，示例代码如下。

```
while (dr.Read())
{
    Response.Write(dr["sname"].ToString()+"<hr/>");
}
```

上述代码通过 dr["sname"]获取数据库中 sname 列的值，同样也可以通过列索引获取某一列的值，示例代码如下。

```
while (dr.Read())
{
    Response.Write(dr[1].ToString()+"<hr/>");
}
```

【示例 9-4】 在 chapter9 网站根目录下创建名为 example9-4 的网页，页面内包含若干控件，练习使用 SqlDataReader 对象读取数据库中的数据。

（1）按图 9.10 所示添加相应控件。

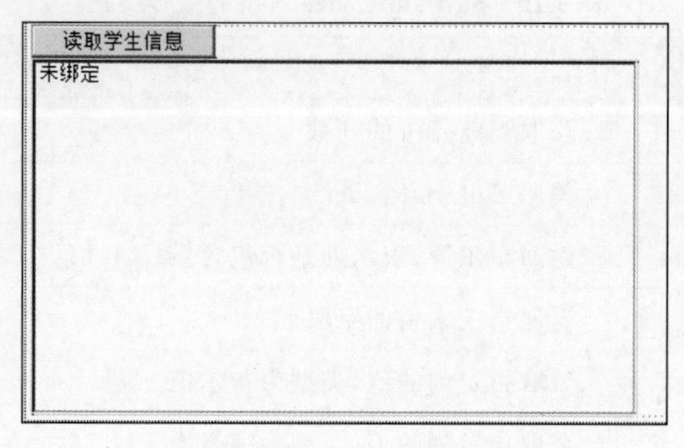

图 9.10 SqlDataReader 对象应用页面设计

（2）按如下源文件设置控件的相关属性值。

```
<form id="form1" runat="server">
    <div>
        <asp:Button ID="btnRead" runat="server" OnClick="btnRead_Click" Text="
读取学生信息"/>
        <br/>
        <asp:ListBox ID="ListBox1" runat="server"></asp:ListBox>
    </div>
</form>
```

（3）添加命名空间的引用。

```
using System.Data.SqlClient;
using System.Configuration;
```

（4）为控件添加事件，并编辑代码如下。

```
protected void btnRead_Click(object sender, EventArgs e)
{
    string sqlconnstr=ConfigurationManager.ConnectionStrings["DemoConnection"].
ConnectionString;
    SqlConnection con=new SqlConnection(sqlconnstr);
    SqlCommand cmd=new SqlCommand();
    cmd.Connection=con;
    cmd.CommandText="select * from student";
    con.Open();
    SqlDataReader sdr=cmd.ExecuteReader();
    int n=sdr.FieldCount;
    string strFileName="";
    for (int i=0; i<n; i++)
```

```
        {
            strFileName+=sdr.GetName(i)+"   ";
        }
        ListBox1.Items.Add(strFileName);
        while (sdr.Read())
        {
            string strContent="";
            for (int i=0; i<n; i++)
            {
                strContent+=sdr.GetValue(i)+"   ";
            }
            ListBox1.Items.Add(strContent);
        }
        sdr.Close();
        con.Close();
    }
```

(5) 运行网站,执行效果如图 9.11 所示。

图 9.11 SqlDataReader 对象应用页面演示

在 SqlDataReader 对象没有关闭之前,数据库连接会一直保持 Open 状态,一个连接只能被一个 SqlDataReader 对象使用,故在使用 SqlDataReader 时,使用完毕后应立即调用 SqlDataReader.Close()方法将其关闭。

9.5 DataSet 数据集对象

DataSet 是 ADO.NET 中用来访问数据库的对象,对应的类包含在 System.Data 命名空间中。DataSet 可以用来存储从数据库查询到的数据结果,也可简单地把 DataSet 想象成若干张虚拟的表,但是这些表不是简单地只存储数据,而是一个具有数据结构的表,并且存放在内存中。在获得数据或更新数据后,DataSet 应立即与数据库断开,这样可以高效地访问和操作数据库。

DataSet 对象具有离线访问数据库的特性，能用来接收海量的数据信息。DataSet 对象本身不与数据库发生关系，而是通过 DataAdapter 对象从数据库里获取数据并把修改后的数据更新到数据库。关于 SqlDataAdapter，将在 9.6 节讲述。

扫一扫

9.5.1 DataSet 数据集对象介绍

DataSet 能够支持多表、表间关系、数据库约束等，可以用来模拟简单的数据库模型。DataSet 对象模型如图 9.12 所示。

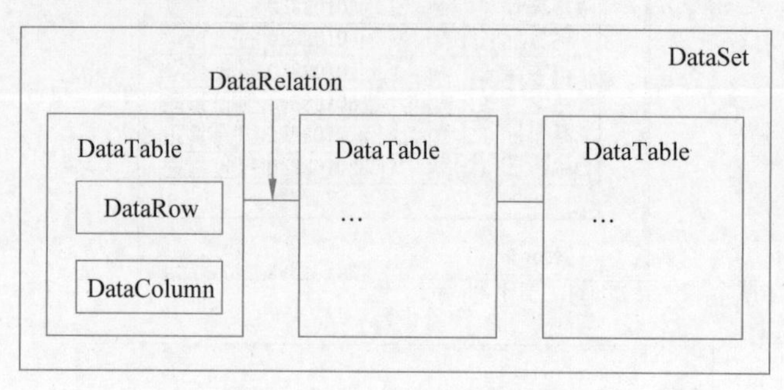

图 9.12　DataSet 对象模型

DataSet 对象的主要属性、方法及构造函数如表 9.11～表 9.13 所示。

表 9.11　DataSet 对象的主要属性

属 性 名	说　明
CaseSensitive	获取或设置表中的字符串比较是否区分大小写。如果区分大小写，则为 False，默认值为 False
DataSetName	获取或设置当前 DataSet 对象名
Tables	获取包含在 DateSet 对象中表的集合
Relations	获取数据集中表的关系集合
Namespace	获取或设置的命名空间 DataSet

表 9.12　DataSet 对象的主要方法

方 法 名	说　明
Clear()	清除 DataSet 的所有表
Copy()	将 DataSet 中的结构和数据进行复制
Dispose()	释放使用的所有资源
Merge()	合并指定 DataSet 到当前及其架构 DataSet
Reset()	清除所有表并从 DataSet 中删除所有关系、外部约束和表

表 9.13　DataSet 对象的构造函数

构造函数	说　明
DataSet()	初始化 DataSet 类的新实例
DataSet(string)	初始化 DataSet 并指定其名称

9.5.2 DataTable 数据表对象

DataSet 的 Tables 属性表示表的集合,每个表都是一个 DataTable 对象。应用中每个 DataTable 对象对应数据库中的一张表或者是多表查询得到的一个结果,可以通过 Tables 集合的索引器访问。索引器的参数可以是字符串类型的表名,也可以是索引值。

1. Tables 集合的属性和方法

Tables 集合的主要属性和主要方法如表 9.14 和表 9.15 所示。

表 9.14 Tables 集合的主要属性

属 性 名	说 明
Counts	获取 Tables 集合中表的个数

表 9.15 Tables 集合的主要方法

方 法 名	说 明
Add()	向 Tables 集合中添加一张表
AddRange()	向 Rows 集合中添加一个表的数组
Clear()	移除 Tables 集合中所有的表
Contains()	判断指定表是否在 Tables 集合中
Insert()	向 Tables 集合指定位置插入一张表
IndexOf()	检索指定表在 Tables 集合中的索引
Remove()	从 Tables 集合中移除指定的表
RemoveAt()	从 Tables 集合中移除指定索引位置的表

2. DataTable 对象

DataTable 表示内存中关系数据的表,可以独立创建和使用,也可以由.NET Framework 对象使用。DataTable 中的集合形成了二维表的数据结构,具有 Rows 集合和 Columns 集合等属性,主要属性、方法及构造函数如表 9.16～表 9.18 所示。

表 9.16 DataTable 对象的主要属性

属 性 名	说 明
Columns	获取属于此表的列的集合
Constraints	获取此表中所有约束的集合
DataSet	获取该表对象所属的 DataSet
HasErrors	获取一个值,该值指示该表所属的 DataSet 中所有表的行中是否有错误
MinimumCapacity	获取或设置此表的初始起始大小
PrimaryKey	获取或设置列数组为数据表主键
TableName	获取或设置 DataTable 的名称
Rows	获取属于此表的行集合

表 9.17　DataTable 对象的主要方法

方 法 名	说　明
Clear()	清除 DataTable 的所有数据
Copy()	将结构和数据复制到此 DataTable
GetErrors()	获取一个数组 DataRow 包含错误的对象
Merge()	合并指定 DataTable 与当前 DataTable
NewRow()	创建一个新 DataRow，使其具有与表相同的架构
Reset()	重置 DataTable 到其原始状态。重置中删除所有数据、索引、关系和表的列。如果数据集包含一个数据表，该表重置之后仍可是数据集的一部分

表 9.18　DataTable 对象的构造函数

构造函数	说　明
DataTable()	初始化 DataTable 类的新实例
DataTable(string)	初始化 DataTable 类的对象并指定其名称

9.5.3　DataColumn 数据列对象

DataTable 的 Columns 属性表示表的列集合，每个 DataColumn 对象可以表示数据库中的一个字段或者由聚合函数等查询得到的新属性，可以通过 Columns 集合的索引器进行访问，索引器的参数可以为字符串类型的列名，也可以是索引值。

1. Columns 集合的属性和方法

Columns 集合的主要属性和主要方法如表 9.19 和表 9.20 所示。

表 9.19　Columns 集合的主要属性

属 性 名	说　明
Counts	获取 Columns 集合中列的个数

表 9.20　Columns 集合的主要方法

方 法 名	说　明
Add()	向 Columns 集合中添加一列
AddRange()	向 Rows 集合中添加一个列的数组
Clear()	移除 Columns 集合中所有的列
Contains()	判断指定列是否在 Columns 集合中
Insert()	向 Columns 集合指定位置插入一列
IndexOf()	检索指定列在 Columns 集合中的索引
Remove()	从 Columns 集合中移除指定的列
RemoveAt()	从 Columns 集合中移除指定索引位置的列

2. DataColumn 对象

DataColumn 用来模拟物理数据库中的列，DataColum 的组合组成了 DataTable 中列的架构。DataColumn 对象的主要属性、方法及构造函数如表 9.21～表 9.23 所示。

表 9.21　DataColumn 对象的主要属性

属 性 名	说　　明
AllowDBNull	获取或设置一个值，该值表示是否允许空值在本专栏中属于表的行
AutoIncrement	获取或设置一个值，表示添加新行时该列是否以递增值自动添加
Caption	获取或设置列标题
ColumnName	获取或设置列名称
DefaultValue	获取或设置列的默认值，在创建新行时使用
MaxLength	获取或设置文本列的最大长度
ReadOnly	获取或设置一个值，表示列是否允许更改

表 9.22　DataColumn 对象的主要方法

方 法 名	说　　明
Dispose()	释放使用的所有资源

表 9.23　DataColumn 对象的构造函数

构 造 函 数	说　　明
DataColumn()	初始化 DataColumn 类的对象
DataColumn(string)	初始化 DataColumn 类的对象并指定其列名称

9.5.4　DataRow 数据行对象

DataTable 的 Rows 属性表示表的行集合，每个 DataRow 对象可以表示数据库中的一条记录或者是多表查询得到的一条数据，可以通过 Rows 集合的索引器进行访问，索引器的参数为行的索引值。

1. Rows 集合的属性和方法

Rows 集合的主要属性和主要方法如表 9.24 和表 9.25 所示。

表 9.24　Rows 集合的主要属性

属 性 名	说　　明
Counts	获取 Rows 集合中行的个数

表 9.25　Rows 集合的主要方法

方 法 名	说　　明
Add()	向 Rows 集合中添加一行
AddRange()	向 Rows 集合中添加一个行的数组

续表

方 法 名	说 明
Clear()	移除 Rows 集合中所有的行
Contains()	判断指定行是否在 Rows 集合中
Insert()	向 Rows 集合指定位置插入一行
IndexOf()	检索指定行在 Rows 集合中的索引
Remove()	从 Rows 集合中移除指定的行
RemoveAt()	从 Rows 集合中移除指定索引位置的行

2. DataRow 对象

在创建了表和表中列的集合，并使用约束定义表的结构后，可以使用 DataRow 对象向表中添加新的数据库行。插入一个新行，首先要声明一个 DataRow 类型的变量，DataRow 对象没有构造函数，只能使用 DataTable 对象的 NewRow 方法返回一个新的 DataRow 对象。

DataTable 会根据 DataColumnCollection 定义的表的结构创建 DataRow 对象。DataRow 对象的主要属性及主要方法如表 9.26 和表 9.27 所示。

表 9.26 DataRow 对象的主要属性

属 性 名	说 明
CaseSensitive	表示在表中的字符串比较是否区分大小写
Columns	获取属于此表列的集合
Constraints	获取此表是有约束的集合
DataSet	获取该对象所属 DataSet
MinimumCapacity	获取或设置此表的初始起始大小
PrimaryKey	获取或设置数据表的主键字段数组
TableName	获取或设置 DataTable 的名称
Rows	获取属于此表的行的集合

表 9.27 DataRow 对象的主要方法

方 法 名	说 明
Clear()	清除 DataTable 中的所有数据
Copy()	将结构和数据复制到 DataTable
GetErrors()	获取一个数组 DataRow，使其包含错误的对象
Merge()	合并指定 DataTable 与当前 DataTable
NewRow()	创建一个新 DataRow，使其具有与表相同的架构

9.5.5 DataSet 数据集的应用

通过本章前几节的讲述，可知 DataSet 数据集中包含若干 DataTable，而数据表 DataTable 又由若干数据行 DataRow 和数据列 DataColumn 构成。构建数据集通常有如下 4 个步骤。

（1）通过 DataSet 类的构造函数创建一个 DataSet 对象。

（2）通过 DataTable 类的构造函数创建一个 DataTable 对象，并将其加入 DataSet 对象中。

（3）通过 DataColumn 类的构造函数创建数据列对象，并依次添加到 DataTable 的 Columns 属性中，从而得到数据表的结构，然后将其加入 DataTable 对象中。

（4）通过 DataRow 类的构造函数创建符合当前表结构的 DataRow 对象，并为该对象设置对应字段的数据，然后将该 DataRow 对象添加到 DataTable 类的 Rows 属性中。

【示例 9-5】 在 chapter9 网站根目录下创建名为 example9-5 的网页，创建数据集、数据表、数据列以及数据行对象，构建一个 DataSet 数据集的应用。

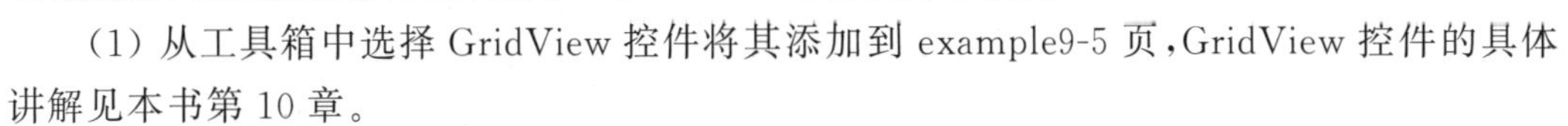

（1）从工具箱中选择 GridView 控件将其添加到 example9-5 页，GridView 控件的具体讲解见本书第 10 章。

（2）添加命名空间的引用。

```
using System.Data;
```

（3）为页面添加事件，并编辑代码如下。

```
protected void Page_Load(object sender, EventArgs e)
{
    //实例化一个 DataSet 对象 ds
    DataSet ds=new DataSet();
    //实例化一个 DataTable 对象 table
    DataTable table=new DataTable("teacher");
    //将 table 对象添加到 ds 对象的 Tables 集合中
    ds.Tables.Add(table);
    //声明一个 DataColumn 类型的成员 column
    DataColumn column;
    //实例化 DataColumn 对象 column,列名为教职工号,字段类型为 String
    column=new DataColumn("教职工号", System.Type.GetType("System.String"));
    //将 column 对象添加到 table 对象的 Columns 集合中
    table.Columns.Add(column);
    column=new DataColumn("姓名", System.Type.GetType("System.String"));
    table.Columns.Add(column);
    column=new DataColumn("性别", System.Type.GetType("System.String"));
    table.Columns.Add(column);
    column=new DataColumn("出生年月", System.Type.GetType("System.DateTime"));
    table.Columns.Add(column);
    //声明一个 DataRow 类型的成员 row
    DataRow row;
    //实例化 DataRow 对象 row,每一行中包含教职工号、姓名、性别、出生年月 4 个字段
    row=table.NewRow();
    //设置行内"教职工号"字段的值为"T001"
```

```
    row["教职工号"]="T001";
    row["姓名"]="张老师";
    row["性别"]="男";
    row["出生年月"]="1981-05-25";
    //将 row 对象添加到 table 对象的 Rows 集合中
    ds.Tables["teacher"].Rows.Add(row);
    row=table.NewRow();
    row["教职工号"]="T002";
    row["姓名"]="王老师";
    row["性别"]="女";
    row["出生年月"]="1980-11-11";
    ds.Tables["teacher"].Rows.Add(row);
    //设置 GridView1 控件的 DataSource 属性为 ds 数据集对象
    GridView1.DataSource=ds;
    //为 GridView1 控件进行数据绑定
    GridView1.DataBind();
}
```

(4) 运行网站,执行效果如图 9.13 所示。

图 9.13　DataSet 数据集应用页面演示

9.6　SqlDataAdapter 数据适配器对象

扫一扫

SqlDataAdapter 是 DataSet 和 SQL Server 之间的桥梁,对应的类包含在 System.Data.SqlClient 命名空间中。SqlDataAdapter 可以将数据源中的数据填充到 DataSet 中,也可以更改数据源中的数据以匹配 DataSet 中的数据。

9.6.1　SqlDataAdapter 类的属性与方法

DataSet 与数据源进行交互时,SqlDataAdapter 就提供 DataSet 对象和数据源之间的连接。SqlDataAdapter 类的主要属性、主要方法及构造函数如表 9.28～表 9.30 所示。

表 9.28　SqlDataAdapter 类的主要属性

属　性　名	说　　明
DeleteCommand	获取或设置 SQL 语句或存储过程,从数据集中删除记录
InsertCommand	获取或设置 SQL 语句或存储过程,将记录插入数据源
UpdateCommand	获取或设置 SQL 语句或存储过程,用于更新数据源中的记录

表 9.29　SqlDataAdapter 类的主要方法

方 法 名	说　明
Fill(DataSet)	添加或刷新 DataSet
Fill(DataSet, String)	添加或刷新 DataSet 以匹配使用数据源的 DataSet 和 DataTable 名称
Fill(DataTable)	添加或刷新指定范围中的 DataSet 以匹配数据源的 DataTable 名称
Update(DataSet)	执行相应的 INSERT、UPDATE 或 DELETE 语句来更新数据库中的 DataSet
Update(DataTable)	执行相应的 INSERT、UPDATE 或 DELETE 语句来更新数据库中的 DataTable

表 9.30　SqlDataAdapter 类的构造函数

构造函数	说　明
SqlDataAdapter()	初始化 SqlDataAdapter 类新实例
SqlDataAdapter (SqlCommand)	使用指定 SqlCommand 作为 SelectCommand 属性初始化 SqlDataAdapter 类实例
SqlDataAdapter (String, SqlConnection)	通过 SelectCommand 和 SqlConnection 对象初始化 SqlDataAdapter 类实例

9.6.2　使用 SqlDataAdapter 对象获取数据

扫一扫

SqlDataAdapter 通常的作用是填充 DataSet 和更新 SQL Server 数据库，SqlDataAdapter 和 DataSet 之间本身没有直接连接，当执行 SqlDataAdpater.Fill(DataSet)方法后，两个对象之间就有了连接。SqlDataAdapter 的 Fill()方法不需要调用活动的 SqlConnection 对象，SqlDataAdapter 会自动打开连接对象，并在获取查询结果后关闭与数据库的连接。

【示例 9-6】　在 chapter9 网站根目录下创建名为 example9-6 的网页，使用数据适配器对象 SqlDataAdapter 填充数据集，获取数据库内数据表中的数据，并将其显示在页面中。

(1) 向 example9-6 页面添加一个 Label 控件，并设置其 ID 属性值为 lblInfo。

(2) 添加命名空间的引用。

```
using System.Data;
using System.Data.SqlClient;
using System.Configuration;
```

(3) 为页面添加事件，并编辑代码如下。

```
protected void Page_Load(object sender, EventArgs e)
{
    string sqlConnStr=ConfigurationManager.ConnectionStrings["DemoConnection"].
ConnectionString;
    SqlConnection con=new SqlConnection(sqlConnStr);
    //建立 DataSet 对象 ds
    DataSet ds=new DataSet();
    //建立 SqlDataAdapter 对象 sda
    SqlDataAdapter sda=new SqlDataAdapter("select * from student", con);
    //用 Fill()方法返回的数据填充 DataSet,数据表取名为 tb_student
```

```
        sda.Fill(ds, "tb_student");
        //声明 DataRow 对象 row
        DataRow row;
        //逐行遍历,取出各行的数据
        for (int i=0; i<ds.Tables["tb_student"].Rows.Count; i++)
        {
            row=ds.Tables["tb_student"].Rows[i];
            lblInfo.Text+=" 学号:"+row[0];
            lblInfo.Text+=" 姓名:"+row[1];
            lblInfo.Text+=" 性别:"+row[2];
            lblInfo.Text+=" 年龄:"+row[3];
            lblInfo.Text+=" 部门:"+row[4]+"<br/>";
        }
        con.Dispose();
        sda.Dispose();
    }
```

(4) 运行网站,执行效果如图 9.14 所示。

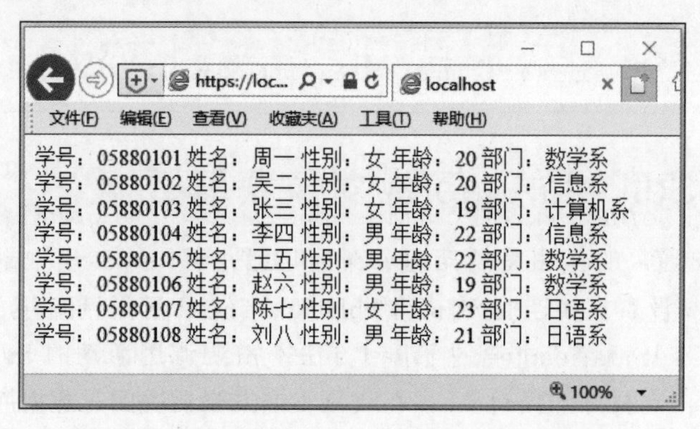

图 9.14 SqlDataAdapter 对象获取数据页面演示

说明:

(1) SqlDataReader 对象和 SqlDataAdapter 对象相似,也需要数据库连接对象。DataAdapter 对象能够自动打开和关闭连接,而 SqlDataReader 对象需要手动管理连接。

(2) DataSet 最大的好处在于能够提供无连接的数据库副本。

扫一扫

9.6.3 使用 SqlDataAdapter 对象更新数据

SqlDataAdapter 对象的 Fill()方法可以将得到的结果填充到数据集的一个数据表 DataTable 对象中,当表内的数据发生改变(如增加、修改、删除等)后,也可以使用 SqlDataAdapter 对象的 Update()方法。执行调用预编译好的 Insert、Delete 和 Update 等 SQL 命令更新数据库,使得内存 DataTable 和数据库实际的表格同步。

【示例 9-7】 在 chapter9 网站根目录下创建名为 example9-7 的网页,页面内包含若干控件,使用数据适配器对象 SqlDataAdapter 填充数据集,使用 SqlDataAdapter 对象的 Update()方法实现数据的增加、删除、修改操作。

(1) 按图 9.15 所示添加相应控件。

(2) 按如下源文件设置控件的相关属性值。

图 9.15　SqlDataAdapter 对象修改数据页面设计

```
<form id="form1" runat="server">
    <div>
        <div>
            <table>
                <tr>
                    <td>学号:</td>
                    <td>
                             <asp:TextBox ID="txtNum" runat="server"></asp:
TextBox>
                    </td>
                </tr>
                <tr>
                    <td>姓名:</td>
                    <td>
                             <asp:TextBox ID="txtName" runat="server"></asp:
TextBox>
                    </td>
                </tr>
                <tr>
                    <td colspan="2">
                          <asp:Button ID="btnInsert" runat="server" OnClick="
btnInsert_Click" Text="添加"/>
                          <asp:Button ID="btnDelete" runat="server" OnClick="
btnDelete_Click" Text="删除"/>
                          <asp:Button ID="btnUpdate" runat="server" OnClick="
btnUpdate_Click" style="width: 40px" Text="修改"/>
                    </td>
                </tr>
            </table>
        </div>
        <asp:Label ID="lblInfo" runat="server" Text=""></asp:Label>
    </div>
</form>
```

(3) 添加命名空间的引用。

```
using System.Data;
using System.Data.SqlClient;
using System.Configuration;
```

（4）为控件添加事件，并编辑代码如下。

```
SqlConnection con;
SqlCommand cmd;
SqlDataAdapter sda;
DataSet ds;
protected void Page_Load(object sender, EventArgs e)
{
    string sqlconnstr=ConfigurationManager.ConnectionStrings["DemoConnection"].
ConnectionString;
    con=new SqlConnection(sqlconnstr);
    cmd=new SqlCommand();
    cmd.Connection=con;
    sda=new SqlDataAdapter("select * from student", con);
    ds=new DataSet();
    sda.Fill(ds, "tb_student");
}
protected void btnInsert_Click(object sender, EventArgs e)
{
    //定义 sda 对象的 InsertCommand 命令
    sda.InsertCommand=new SqlCommand("insert into student (sno,sname) values
(@Sno,@Sname)", con);
    //为 InsertCommand 命令添加参数,@Sname 是长度为 50 的可变字符类型,对应数据表中的
    //sname 字段
    sda.InsertCommand.Parameters.Add("@Sname", SqlDbType.VarChar, 50, "sname");
    sda.InsertCommand.Parameters.Add("@Sno", SqlDbType.VarChar, 50, "sno");
    //构建插入的新行,并赋值
    DataRow row;
    row=ds.Tables["tb_student"].NewRow();
    row["sno"]=txtNum.Text;
    row["sname"]=txtName.Text;
    ds.Tables["tb_student"].Rows.Add(row);
    //如果插入的行数大于 0
    if (sda.Update(ds, "tb_student") >0)
    {
        lblInfo.Text="添加学生信息成功。";
    }
    else
    {
        lblInfo.Text="添加学生信息失败。";
    }
}
protected void btnDelete_Click(object sender, EventArgs e)
{
    //定义 sda 对象的 DeleteCommand 命令
    sda.DeleteCommand=new SqlCommand("delete from student where sno=@Sno",
con);
    //为 DeleteCommand 命令添加参数,@Sno 是长度为 50 的可变字符类型,对应数据表中的
    //sno 字段
    sda.DeleteCommand.Parameters.Add("@Sno", SqlDbType.VarChar, 50, "sno");
    DataRow row;
    //逐行遍历,取出各行数据,删除指定学号的数据
    for (int i=0; i<ds.Tables["tb_student"].Rows.Count; i++)
```

```
        {
            row=ds.Tables["tb_student"].Rows[i];
            //获取要删除的行,并删除
            if (row["sno"].ToString()==txtNum.Text)
            {
                row.Delete();
            }
        }
        //如果修改的行数大于 0
        if (sda.Update(ds, "tb_student") >0)
        {
            lblInfo.Text="删除学生信息成功。";
        }
        else
        {
            lblInfo.Text="删除学生信息失败。";
        }
    }
    protected void btnUpdate_Click(object sender, EventArgs e)
    {
        //定义 sda 对象的 UpdateCommand 命令
        sda.UpdateCommand=new SqlCommand("update student set sname=@Sname where sno=@
Sno", con);
        //为 UpdateCommand 命令添加参数,@Sname 是长度为 50 的可变字符类型,对应数据表中的
        //sname 字段
        sda.UpdateCommand.Parameters.Add("@Sname", SqlDbType.VarChar, 50, "sname");
        sda.UpdateCommand.Parameters.Add("@Sno", SqlDbType.VarChar, 50, "sno");
        DataRow row;
        //逐行遍历,取出各行数据,修改指定学号的数据
        for (int i=0; i<ds.Tables["tb_student"].Rows.Count; i++)
        {
            row=ds.Tables["tb_student"].Rows[i];
            //获取到要修改的行,并修改字段值
            if (row["sno"].ToString()==txtNum.Text)
            {
                row["sname"]=txtName.Text;
            }
        }
        //如果修改的行数大于 0
        if (sda.Update(ds, "tb_student") >0)
        {
            lblInfo.Text="修改学生信息成功。";
        }
        else
        {
            lblInfo.Text="修改学生信息失败。";
        }
    }
```

(5) 运行网站,执行效果如图 9.16 所示。

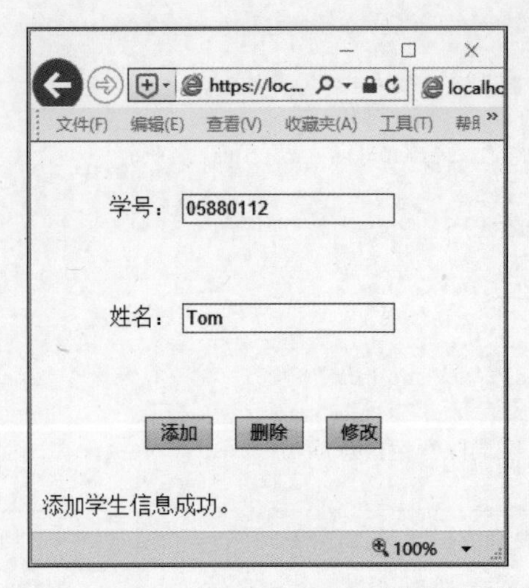

图 9.16 SqlDataAdapter 对象获取数据页面演示

9.6.4 SqlCommandBuilder 类的应用

SqlDataAdapter 对象只能按提供的 SQL 语句从数据源读取数据到 DataSet，在 DataTable 对象中，如果想对修改后的数据使用 Update()方法更新到数据源，必须设置好相关的 InsertCommand、DeleteCommand、UpdateCommand 命令，否则 SqlDataAdapter 对象无法自动完成该操作，针对这一问题，微软创建了构建猜想类 SqlCommandBuilder。

SqlCommandBuilder 对象可以生成用于更新数据库的 SQL 语句，开发者不必自己创建这些语句，只构建一个 SqlCommandBuilder 对象为 SqlDataAdapter 对象服务即可，相当于帮助 SqlDataAdapter 对象做了猜想，构建了 InsertCommand、DeleteCommand、UpdateCommand 命令代码。SqlCommandBuilder 对象的实例化代码如下。

```
SqlCommandBuilder scb=new SqlCommandBuilder(sda);
```

其中 sda 为 SqlDataAdapter 对象。

【示例 9-8】 在 chapter9 网站根目录下创建名为 example9-8 的网页，使用 SqlCommandBuilder 对象生成用于更新数据库的 SQL 语句，测试 SqlCommandBuilder 对象为数据适配器对象创建 SQL 命令。

（1）添加命名空间的引用。

```
using System.Data;
using System.Data.SqlClient;
using System.Configuration;
```

（2）为页面添加事件，并编辑代码如下。

```
SqlConnection con;
SqlCommand cmd;
SqlDataAdapter sda;
DataSet ds;
protected void Page_Load(object sender, EventArgs e)
```

```
{
    string sqlconnstr=ConfigurationManager.ConnectionStrings["DemoConnection"].
ConnectionString;
    con=new SqlConnection(sqlconnstr);
    cmd=new SqlCommand();
    cmd.Connection=con;
    sda=new SqlDataAdapter("select * from student", con);
    ds=new DataSet();
    sda.Fill(ds, "tb_student");
    //建立 SqlCommandBuilder 类的对象 scb 自动生成 SqlDataAdapter 类的对象 sda 的
    //Command 命令
    SqlCommandBuilder scb=new SqlCommandBuilder(sda);
    Response.Write("<br>SqlCommandBuilder 实例的 Insert 命令:" + scb.
GetInsertCommand().CommandText);
    Response.Write("<br>SqlCommandBuilder 实例的 Update 命令:" + scb.
GetUpdateCommand().CommandText);
    Response.Write("<br>SqlCommandBuilder 实例的 Delete 命令:" + scb.
GetDeleteCommand().CommandText);
}
```

(3) 运行网站,可显示 SqlCommandBuilder 对象为数据适配器对象创建的包含数据库中所有关系的 SQL 命令,具体如图 9.17 所示。

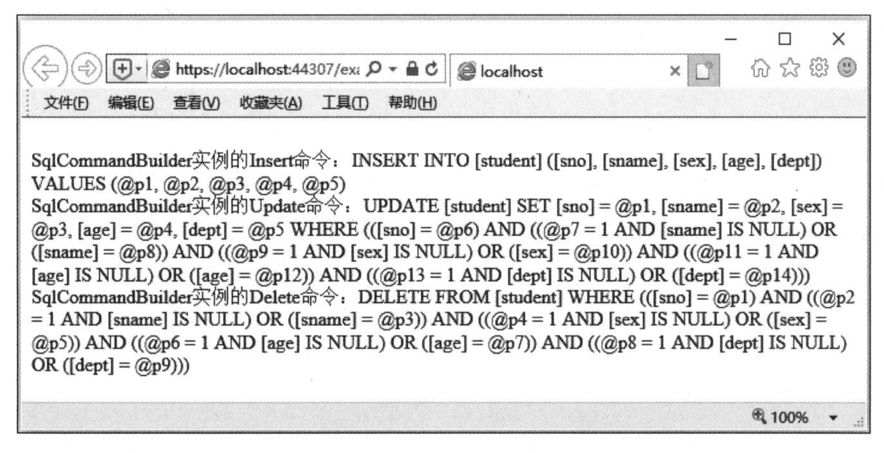

图 9.17 SqlCommandBuilder 对象应用页面演示

SqlCommandBuilder 对象能够使数据库与操作同步,在调用 SqlDataAdapter 对象的 Update()方法时,SqlCommandBuilder 对象会遍历所有修改或增加行,根据行 RowState 标志位构建相应的 SQL 命令。

综合实验九 数据控件绑定

主要任务

创建 ASP.NET 应用程序,使用数据源控件直接绑定 GridView。

实验步骤：

步骤 1：在 VS 2019 菜单上选择“文件”→“新建”→“项目”命令。

步骤 2：创建 ASP.NET 应用程序，命名为“综合实验九”。

步骤 3：在网站的根目录上右击，创建“综合实验九.aspx”页面，添加相应控件，具体如图 9.18 所示。

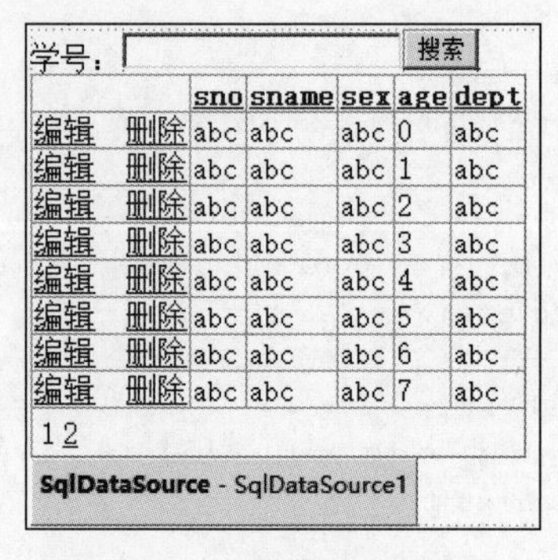

图 9.18　数据管理页面布局

步骤 4：按如下源文件设置控件的相关属性值。

```
<form id="form1" runat="server">
        <div>
            学号:<asp:TextBox ID="txtSno" runat="server"></asp:TextBox>
            <asp:Button ID="btnSearch" runat="server" OnClick="btnSearch_
Click" Text="搜索"/>
            <asp:GridView ID="GridView1" runat="server" AllowPaging="True"
AllowSorting =" True " AutoGenerateColumns =" False " DataKeyNames =" sno "
DataSourceID="SqlDataSource1" PageSize="8">
                <Columns>
                    <asp:CommandField ShowDeleteButton="True" ShowEditButton="
True"/>
                    <asp:BoundField DataField="sno" HeaderText="sno" ReadOnly="
True" SortExpression="sno"/>
                    <asp:BoundField DataField=" sname " HeaderText =" sname "
SortExpression="sname"/>
                    <asp:BoundField DataField =" sex " HeaderText =" sex "
SortExpression="sex"/>
                    <asp:BoundField DataField =" age " HeaderText =" age "
SortExpression="age"/>
                    <asp:BoundField DataField =" dept " HeaderText =" dept "
SortExpression="dept"/>
                </Columns>
            </asp:GridView>
        </div>
        <asp:SqlDataSource ID="SqlDataSource1" runat="server" ConflictDetection="
CompareAllValues"
            ConnectionString="<%$ ConnectionStrings:demoConnectionString %>"
            DeleteCommand="DELETE FROM [student] WHERE [sno]=@original_sno"
            InsertCommand="INSERT INTO [student] ([sno], [sname], [sex], [age],
[dept]) VALUES (@sno, @sname, @sex, @age, @dept)"
```

```
            OldValuesParameterFormatString="original_{0}"
            SelectCommand="SELECT * FROM [student] WHERE ([sno] LIKE '%'+@sno+'%')"
            UpdateCommand="UPDATE [student] SET [sname]=@sname, [sex]=@sex, [age]
=@age, [dept]=@dept WHERE [sno]=@original_sno">
            <DeleteParameters>
                <asp:Parameter Name="original_sno" Type="String"/>
            </DeleteParameters>
            <InsertParameters>
                <asp:Parameter Name="sno" Type="String"/>
                <asp:Parameter Name="sname" Type="String"/>
                <asp:Parameter Name="sex" Type="String"/>
                <asp:Parameter Name="age" Type="Int32"/>
                <asp:Parameter Name="dept" Type="String"/>
            </InsertParameters>
            <SelectParameters>
                <asp:ControlParameter ControlID="txtSno" DefaultValue="%" Name="
sno" PropertyName="Text" Type="String"/>
            </SelectParameters>
            <UpdateParameters>
                <asp:Parameter Name="sname" Type="String"/>
                <asp:Parameter Name="sex" Type="String"/>
                <asp:Parameter Name="age" Type="Int32"/>
                <asp:Parameter Name="dept" Type="String"/>
                <asp:Parameter Name="original_sno" Type="String"/>
            </UpdateParameters>
        </asp:SqlDataSource>
    </form>
```

步骤5：编辑“综合实验九.aspx.cs”，代码如下。

```
public partial class 综合实验九 : System.Web.UI.Page
{
    protected void Page_Load(object sender, EventArgs e)
    {

    }
    protected void btnSearch_Click(object sender, EventArgs e)
    {
         GridView1.DataSourceID=null;
         GridView1.DataSource=SqlDataSource1;
          GridView1.DataBind();
    }
}
```

步骤6：在Web.config中添加连接字符串属性如下。

```
<connectionStrings>
    <add name="demoConnectionString" connectionString="Data Source=.\
sqlexpress;Initial Catalog=demo;Integrated Security=True"
      providerName="System.Data.SqlClient"/>
</connectionStrings>
```

步骤7：运行“综合实验九”页面，初始显示图书信息如图9.19所示。单击“搜索”按钮，可按学号进行模糊查询，如图9.20所示。单击任一行中的“删除”按钮，可删除该条记录，单击任一行中的“编辑”按钮，可编辑保存除学号外的其他字段值，如图 9.21 所示。

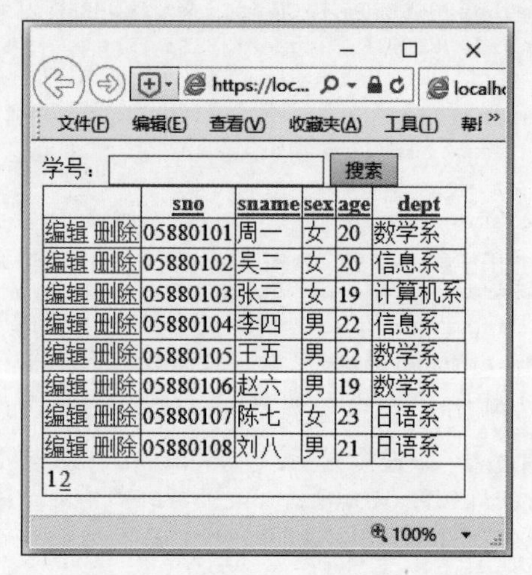

学号：

	sno	sname	sex	age	dept
编辑 删除	05880101	周一	女	20	数学系
编辑 删除	05880102	吴二	女	20	信息系
编辑 删除	05880103	张三	女	19	计算机系
编辑 删除	05880104	李四	男	22	信息系
编辑 删除	05880105	王五	男	22	数学系
编辑 删除	05880106	赵六	男	19	数学系
编辑 删除	05880107	陈七	女	23	日语系
编辑 删除	05880108	刘八	男	21	日语系
12					

图 9.19　显示学生信息运行测试

学号：2

	sno	sname	sex	age	dept
编辑 删除	05880102	吴二	女	20	信息系
编辑 删除	05880120	李明明	男	20	数学系

图 9.20　搜索学生信息运行测试

学号：

	sno	sname	sex	age	dept
更新 取消	05880101	周一	女	18	数学系
编辑 删除	05880102	吴二	女	20	信息系
编辑 删除	05880103	张三	女	19	计算机系
编辑 删除	05880104	李四	男	22	信息系
编辑 删除	05880105	王五	男	22	数学系
编辑 删除	05880106	赵六	男	19	数学系
编辑 删除	05880107	陈七	女	23	日语系
编辑 删除	05880108	刘八	男	21	日语系
12					

图 9.21　编辑学生信息运行测试

第10章 ASP.NET中的数据绑定

本章学习目标

- 了解数据绑定表达式
- 熟练掌握数据源的创建方法
- 熟练掌握 List 控件的数据绑定方法
- 了解 Repeater 控件的使用方法
- 熟练掌握 DataList 控件的使用方法
- 熟练掌握 GridView 控件的使用方法

本章首先介绍数据绑定表达式，然后讲述数据源的创建方法并对 List 数据控件的绑定方法进行讲解，最后对数据控件中的模板进行详细讲解。

10.1 简单数据绑定

在了解 ADO.NET 基础后，就可以使用 ADO.NET 进行数据库开发和操作。ADO.NET 提供了数据库的智能连接配置，可以使用数据绑定技术更加方便、直接地将数据记录在 Web 控件中显示。

数据绑定是把数据集中字段绑定到控件特定属性上的一种技术，完成绑定后，显示的内容将随着数据记录同步变化。ASP.NET 使用的数据绑定表达式，语法如下。

```
<%#属性名称%>
```

表达式是在 ASP.NET 页面进行数据绑定的基础，表达式中的属性名称有以下几种类型。

- 公有或受保护的变量绑定：<%#，strName %>　　strName 为字段名

- 方法结果绑定：<%＃，getName()%>　　getName 为方法名
- 表达式绑定：<%＃，表达式 %>
- 集合绑定：<%＃，myArray %>　　myArray 为集合(数组)名

说明：

控件绑定时必须由控件或其父控件调用 DataBind()方法，如 Page.DataBind()或 Control.DataBind()等调用，数据源才可以绑定到服务器控件。

【示例 10-1】 在 E 盘 ASP.NET 项目代码目录中创建 chapter10 子目录，将其作为网站根目录，创建名为 example10-1 的网页，页面内包含一个 TextBox 控件、一个 CheckBoxList1 控件，使用<%＃...%>表达式实现控件的数据绑定。

(1) 设置控件的相关属性值如下。

```
<form id="form1" runat="server">
     <div>
        <%#getStr()+"表达式 2" %>
       <br/>
        <asp:TextBox ID="TextBox1" runat="server" Text="<%#singleValue %>"></asp:
TextBox>
        <br/>
        <asp:CheckBoxList ID="CheckBoxList1" runat="server" DataSource="<%#arr %>">
        </asp:CheckBoxList>
     </div>
</form>
```

(2) 为控件添加事件，并编辑代码如下。

```
public partial class example10_1 : System.Web.UI.Page
{
  //定义页面类数据成员 singleValue 和数组 arr,在源文件中将通过绑定表达式直接引用这两
个成员
    public string singleValue="数值 1";
    public string[] arr=new string[] { "1", "2", "3", "4", "5" };
    protected void Page_Load(object sender, EventArgs e)
    {
        Page.DataBind();
    }
    public string getStr()
    {
        return "数值 2";
    }
}
```

(3) 在页面内分别使用变量绑定、表达式绑定、方法结果绑定和集合绑定进行数据绑定，运行网站，执行效果如图 10.1 所示。

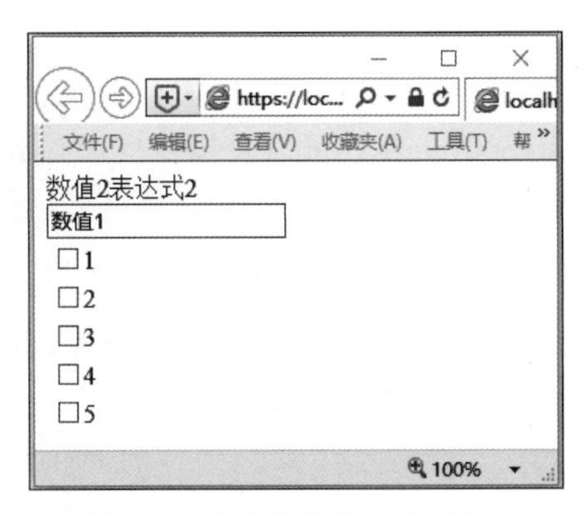

图 10.1 简单数据绑定页面演示

10.2 数据源的创建

顾名思义,数据源(Data Source)是指数据的来源,是提供某种数据的器件或原始媒体,数据源中存储了所有建立数据库连接的信息。数据源中并无真正的数据,数据源定义的是连接到实际数据库的一条路径,它仅记录连接到哪个数据库以及如何连接。一个数据库可以有多个数据源连接,每个数据控件都具有数据源属性,设置控件要显示的数据的来源,是数据控件的核心属性。在 ADO.NET 中获取数据源可以使用 C# 代码创建填充数据集,也可以使用数据源控件直接获取数据源。

10.2.1 使用语句建立数据源

第 9 章讲解了 DataSet 对象和 SqlDataReader 对象的知识点,DataSet 对象可以用来存储从数据库查询到的数据结果,具有离线访问数据库的特性,通过 SqlDataAdapter 对象可以将修改后的数据更新到数据库。SqlDataReader 对象是从数据库中得到的只读、只向前的数据流,每次访问或操作只有一个记录保存在服务器内存中,可以通过 SqlCommand 对象的 ExecuteReader()方法得到。

DataSet 对象和 SqlDataReader 对象都可以作为数据控件的数据源,差别在于 DataSet 对象作为断开式存在的独立虚拟数据库可以同时作为多个控件的数据源,而 SqlDataReader 对象只能作为单一控件的数据源。

10.2.2 使用数据源控件 SqlDataSource 建立数据源

SqlDataSource 控件又称 SQL 数据源控件,在工具箱中图标为"SQL SqlDataSource",封装在 System.Web.UI.Control.WebControl 命名空间的 SqlDataSource 类中。SqlDataSource 控件是一个通过 ADO.NET 连接到 SQL 数据库提供者的数据源,可直接用来配置数据源,绑定数据源控件后,能够通过数据源控件获取数据库中的数据并显示,而无须编写代码。

SqlDataSource 控件支持数据的检索、插入、更新、删除、排序等操作,当 SqlDataSource 控件所属的页面被打开时,SqlDataSource 控件能够自动打开数据库,执行 SQL 语句或存储过程，返回选定的数据，最后关闭连接。SqlDataSource 控件极大地简化了开发流程，缩减了开发代码量，但在性能上不太适应大型开发，通常用于中小型开发。

ASP.NET 提供的 SqlDataSource 控件可以直接拖曳添加到页面，对应的页面生成的 ASP.NET 标签如下。

```
<asp:SqlDataSource ID="SqlDataSource1" runat="server"></asp:SqlDataSource>
```

切换到视图模式，单击 SqlDataSource 控件会显示"配置数据源……"，单击"配置数据源……"连接，会显示智能 SqlDataSource 控件配置向导，如图 10.2 所示。

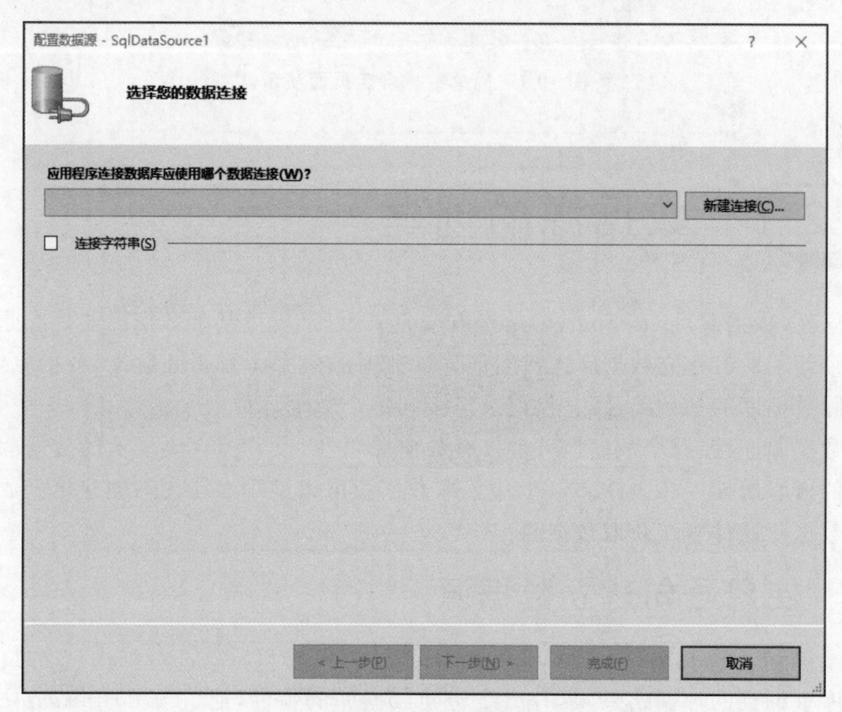

图 10.2　配置 SqlDataSource 控件

单击"新建连接..."，弹出"添加连接"窗口，在此可以填写或选择"服务器名"，选择登录服务器方式，在下拉列表中可选择待连接的数据库，如图 10.3 所示。配置好连接后，单击"测试连接"按钮可测试连接是否成功，如图 10.4 所示。

单击"添加连接"窗口中的"确定"按钮，再次回到"配置数据源"窗口，此时可以显示新创建的连接名(也可在下拉列表中选择其他的连接字符串)，单击下方"连接字符串"前的"+"可展开显示新构建的连接字符串，如图 10.5 所示。

单击"下一步"按钮，选择是否将连接保存的配置文件，并可设置连接字符串名，如图 10.6 所示。勾选"是，将此连接别存为"选项，并在下方文本框中填写新文件名，之后单击"下一步"按钮，在 web.config 配置文件中生成如下标签。

```
<connectionStrings>
    <add name="DemoConnectionString" connectionString="Data Source=.\sqlexpress;
```

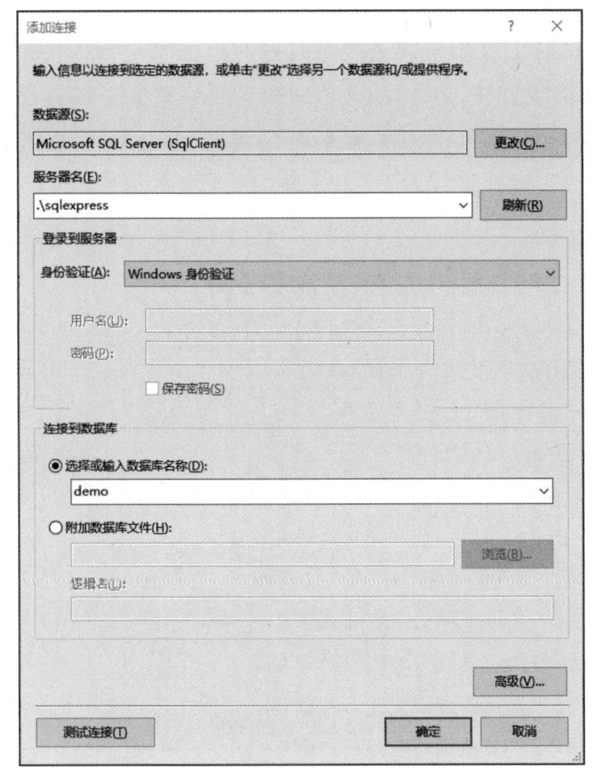

图 10.3　SqlDataSource 控件添加连接

图 10.4　SqlDataSource 控件测试连接

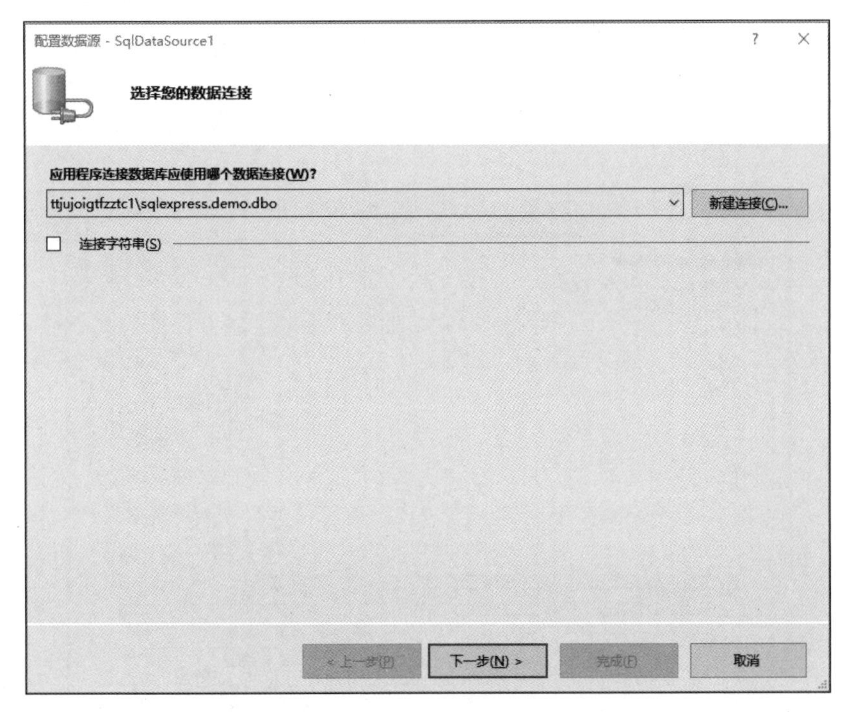

图 10.5　新建连接后的"配置数据源"窗口

```
Initial Catalog = Demo; Integrated Security = True " providerName =" System. Data.
SqlClient"/>
</connectionStrings>
```

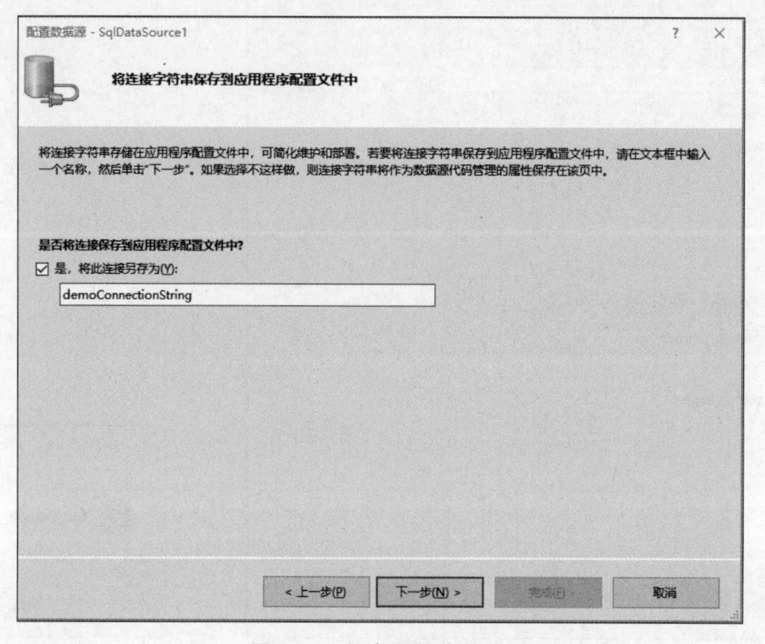

图 10.6　保存连接字符串

数据源控件可以进行 Select 语句的配置和生成。手动编写 Select 语句或其他语句，单击"指定来自表或视图的列"按钮可进行自定义配置。Select 语句的配置和生成如图 10.7 所示。

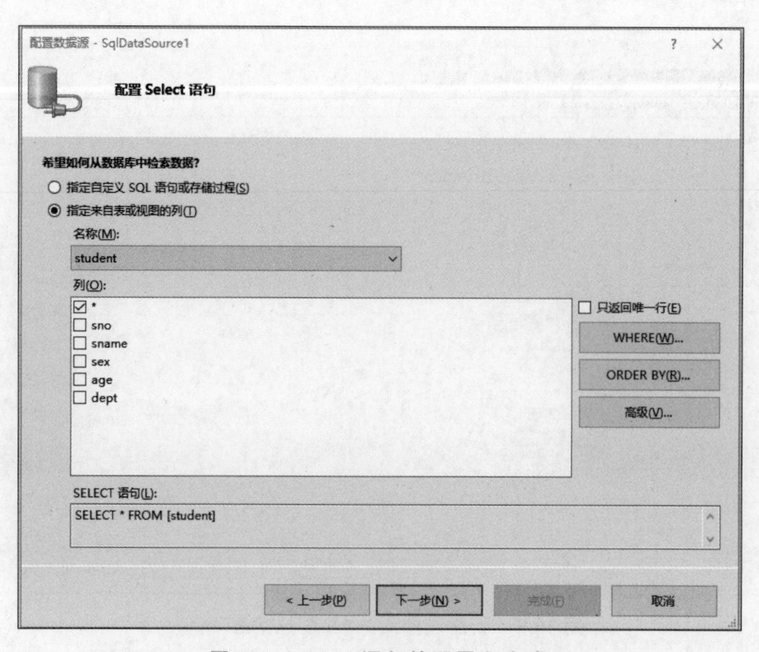

图 10.7　Select 语句的配置和生成

单击 WHERE 按钮，可以对字段进行配置，如同 ADO.NET 中参数化查询一样通过@表示参数变量。添加 WHERE 子句时，SQL 命令中的值可以选择 None、Control、Cookie 或者 Session 等来源，配置完成后可以测试查询，如图 10.8 所示。

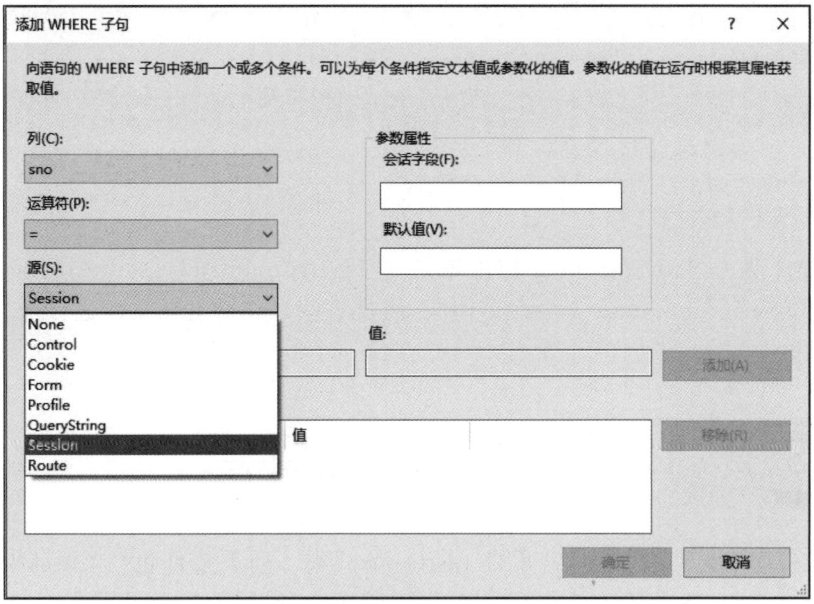

图 10.8 配置添加的 WHERE 条件

单击“高级…”按钮，可以生成更新数据源的 INSERT、UPDATE、DELETE 方法，如图 10.9 所示。

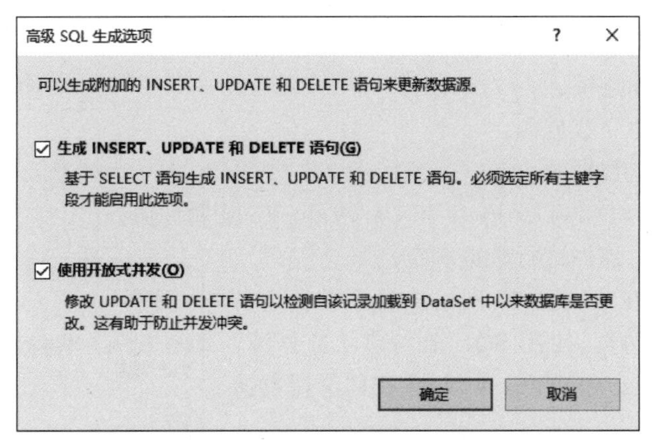

图 10.9 配置添加增加、删除、修改命令

配置相应的查询语句后，SqlDataSource 控件新增的 SQL 代码如下。

```
    <asp:SqlDataSource ID="SqlDataSource1"
DeleteCommand="DELETE FROM [student] WHERE [sno]=@original_sno AND (([sname]=
@original_sname) OR ([sname] IS NULL AND @original_sname IS NULL)) AND (([sex]=@
original_sex) OR ([sex] IS NULL AND @original_sex IS NULL)) AND (([age]=@original
```

```
_age) OR ([age] IS NULL AND @original_age IS NULL)) AND (([dept]=@original_dept)
OR ([dept] IS NULL AND @original_dept IS NULL))"
InsertCommand="INSERT INTO [student] ([sno], [sname], [sex], [age], [dept]) VALUES
(@sno, @sname, @sex, @age, @dept)" OldValuesParameterFormatString="original_{0}"
SelectCommand="SELECT * FROM [student]"
UpdateCommand="UPDATE [student] SET [sname]=@sname, [sex]=@sex, [age]=@age,
[dept]=@dept WHERE [sno]=@original_sno AND (([sname]=@original_sname) OR
([sname] IS NULL AND @original_sname IS NULL)) AND (([sex]=@original_sex) OR
([sex] IS NULL AND @original_sex IS NULL)) AND (([age]=@original_age) OR ([age]
IS NULL AND @original_age IS NULL)) AND (([dept]=@original_dept) OR ([dept] IS
NULL AND @original_dept IS NULL))">
</asp:SqlDataSource>
```

代码中自动为 SqlDataSource 控件增加了 InsertCommand、DelectCommad、UpdateCommand 和 SelectCommand 4 个命令语句,通过这些语句可对数据源进行增加、删除、修改、查询操作。

10.3 List 控件的数据绑定

List 控件简称为列表控件,继承自 ListControl 类。列表控件包括 CheckBoxList、RadioButtonList、DropDownList、ListBox、BulletedList 5 种,ListControl 子类对象具有 DataSource 属性,用于指定要绑定的数据源。如果数据源包含多张表,可以使用 DataMember 属性指定具体的表名,通过设置 DataTextField 和 DataValueField 属性,可将字段绑定到列表控件项的 ListItem.Text 和 ListItem.Value 上显示。通过设置 DataTextFormatString 属性,可以设定列表控件显示文本的格式。

List 控件进行数据绑定时,具体步骤如下。

(1) 创建数据库连接。

(2) 读取数据建立数据源。

(3) 为控件指定数据源。

(4) 设置控件的 DataTextField 和 DataValueField 属性值。

(5) 调用方法对控件进行数据绑定。

【示例 10-2】 在 chapter10 网站根目录下创建名为 example10-2 的网页,使用 SQL 语句建立数据源,并对 RadioButtonList 和 CheckBoxList 控件进行数据绑定。

(1) 按图 10.10 所示添加相应控件。

(2) 按如下源文件设置控件的相关属性值。

请从下列部门中选择一个主办院系:
○未绑定
请选择参加竞赛的学生:
□未绑定
确定
[lblInfo]

图 10.10 列表控件绑定数据页面设计

```
<form id="form1" runat="server">
    <div>
        请从下列部门中选择一个主办院系:<asp:RadioButtonList ID="rbtnlDept" runat
="server" RepeatDirection="Horizontal">
```

```
        </asp:RadioButtonList>
        <br/>
        请选择参加竞赛的学生:<br/>
        <asp:CheckBoxList ID="chklSname" runat="server" RepeatDirection="Hor-
izontal">
        </asp:CheckBoxList>
        <br/>
         <asp:Button ID="btnOk" runat="server" Text="确定" OnClick="btnOk_
Click"/>
        <br/>
        <asp:Label ID="lblInfo" runat="server" Text=""></asp:Label>
    </div>
</form>
```

(3) 添加命名空间的引用。

```
using System.Data;
using System.Data.SqlClient;
```

(4) 为控件添加事件,并编辑代码如下。

```
public partial class example10_2 : System.Web.UI.Page
{
    SqlConnection con;
    SqlCommand cmd;
    SqlDataAdapter sda;
    DataSet ds;
    SqlDataReader sdr;
    protected void Page_Load(object sender, EventArgs e)
    {
        if (! IsPostBack)
        {
            //创建数据连接对象 con
            con=new SqlConnection(@"server=.\sqlexpress;database=demo;trusted
_connection=true;");
            //以数据集作为数据源
            sda=new SqlDataAdapter("select distinct dept from student", con);
            ds=new DataSet();
            sda.Fill(ds, "tb_dept");
            rbtnlDept.DataSource=ds;
            rbtnlDept.DataTextField="dept";
            rbtnlDept.DataValueField="dept";
            rbtnlDept.DataBind();
            //以数据流作为数据源
            cmd=new SqlCommand("select * from student", con);
            con.Open();
            sdr=cmd.ExecuteReader();
            chklSname.DataSource=sdr;
            chklSname.DataTextField="sname";
            chklSname.DataValueField="sno";
            chklSname.DataBind();
            con.Close();
        }
    }
```

```
protected void btnOk_Click(object sender, EventArgs e)
{
    lblInfo.Text ="主办院系:"+rbtnlDept.SelectedValue;
    lblInfo.Text+="<br>参加学生:<br>";
    for (int i=0; i<chklSname.Items.Count; i++)
    {
        if (chklSname.Items[i].Selected)
        {
            lblInfo.Text+="姓名:"+chklSname.Items[i].Text+",学号:"+chkl-
Sname.Items[i].Value+"<br>";
        }
    }
}
}
```

(5) 运行网站,用 DataSet 对象和 SqlDataReader 对象分别对列表控件进行绑定,选择主办院系和参加学生后,执行效果如图 10.11 所示。

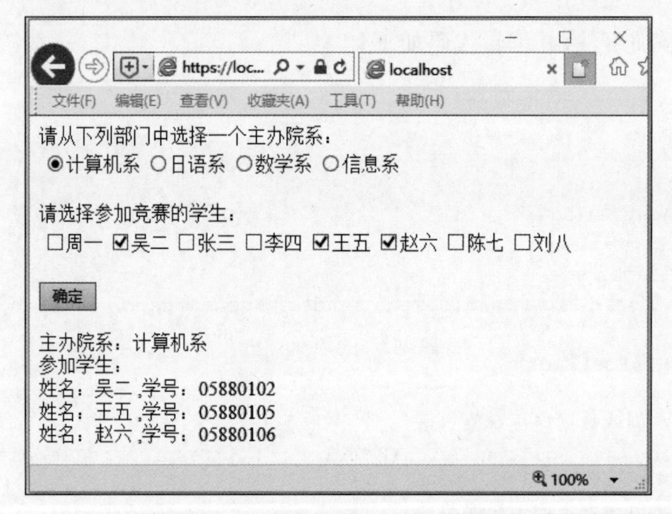

图 10.11 列表控件绑定数据页面演示

10.4 数据控件的数据绑定

10.4.1 数据控件的绑定方法

10.1 节讲解了数据绑定表达式<% #…%>的基本应用,在数据绑定表达式 <% #和%> 分隔符之内,除几种基本形式,还可以使用 Eval()和 Bind()方法进行绑定,其中<% #Eval("数据字段名称")%>为单向绑定,只具有读取功能;<% #Bind("数据字段名称")%>为双向绑定,具有读取和写入的功能。

- Eval()方法

Eval()方法可用于数据控件绑定,如在 GridView、DetailsView 和 FormView 控件的模

板中绑定数据表达式。Eval()方法以数据字段的名称作为参数,从数据源的当前记录返回一个包含该字段值的字符串,并可以使用参数指定返回字符串的格式,字符串格式类似于String 类的 Format()方法的使用。

- Bind()方法

Bind()方法与 Eval()方法相似,但也存在一定的差异。使用 Bind()方法可检索数据绑定字段的值,也可实现数据的修改操作。如果数据源控件定义了 Select、Insert、Delete 和 Update 等 SQL 命令,则通过使用 GridView、DataList 控件模板中的 Bind()方法,就可以将从模板中提取的值传递给数据源控件进行更新。

10.4.2 重复列表控件 Repeater

Repeater 控件又称重复列表控件,在工具箱中图标为"Repeater",封装在 System.Web.UI.Control.WebControl 命名空间的 Repeater 类中。重复列表控件是一个可重复操作的控件,它能够通过模板显示数据源的内容,可以通过配置模板设置标题和页脚等属性。Repeater 控件的常用属性如表 10.1 所示。

表 10.1 Repeater 控件的常用属性

属性名	说明
Adapter	获取控件的浏览器特定适配器
AlternatingItemTemplate	获取或设置对象实现 System.Web.UI.ITemplate,定义交替项的显示
ClientID	获取由 ASP.NET 生成的 HTML 标记的控件 ID
Controls	获取 System.Web.UI.ControlCollection 中包含的子控件
DataMember	获取或设置特定的表中 DataSource 要绑定的控件
DataSource	获取或设置用于填充该列表提供数据的数据源
DataSourceID	获取或设置数据源的 ID 属性
FooterTemplate	获取或设置 System.Web.UI.ITemplate 定义页脚模板
HeaderTemplate	获取或设置 System.Web.UI.ITemplate 定义头部模板
ItemTemplate	获取或设置 System.Web.UI.ITemplate 定义项模板
SeparatorTemplate	获取或设置 System.Web.UI.ITemplate 定义项之间的分隔符模板
Visible	获取或设置一个值,该值指示服务器控件是否呈现

重复列表控件支持 AlternatingItemTemplate、ItemTemplate 、HeaderTemplate 、FooterTemplate 和 SeparatorTemplate 5 种模板,用来显示相应的设计信息,在这 5 种模板中,ItemTemplate 模板是必须设置的。ItemTemplate 源代码如下。

```
<asp:Repeater ID="Repeater1" runat="server" DataSourceID="SqlDataSource1">
    <ItemTemplate>
        <%#Eval("sname")%>
    </ItemTemplate>
</asp:Repeater>
```

在ItemTemplate模板中可以使用HTML制作样式，直接使用“<%#%>”绑定数据库中的列，如在Repeater控件中直接使用“<%#Eval(“sname”)%>”方式显示sname字段的值。

重复列表控件常用的事件为ItemCommand，当重复列表控件中某个按钮被单击时，会触发ItemCommand事件，并可以通过RepeaterCommandEventArgs参数获取Command-Argument、CommandName和CommandSource 3个属性的值。

【示例10-3】 在chapter10网站根目录下创建名为example10-3的网页，使用SqlData-Source控件建立数据源，并对Repeater控件进行数据绑定，编辑其模板项。

(1) 在example10-3页面添加SqlDataSource控件，并配置其数据源，在“数据连接”中选择DemoConnectionString，如图10.12所示。

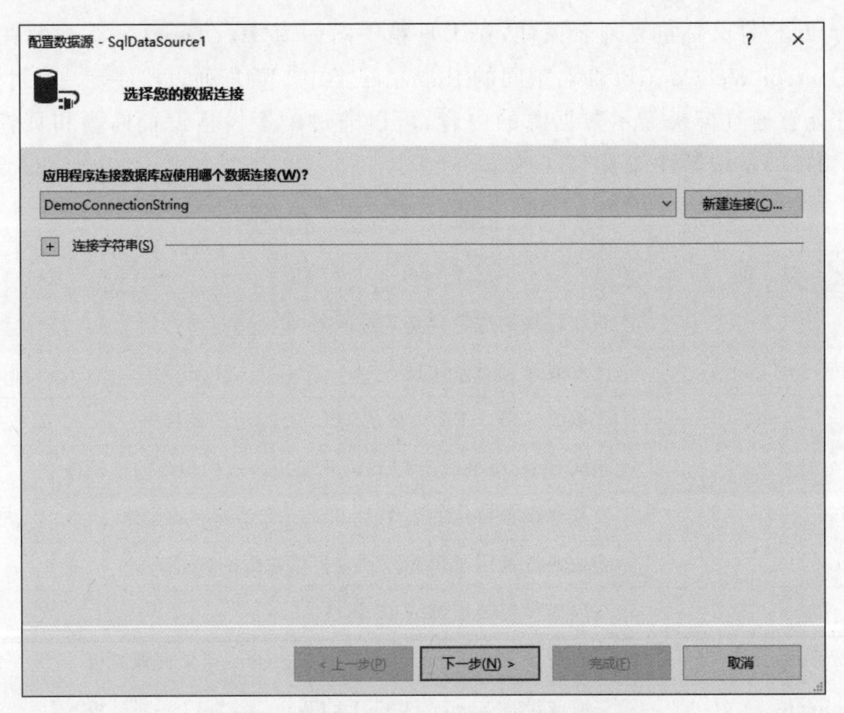

图10.12 应用Repeater控件创建数据连接

(2) 选择student表中的所有数据，并测试查询，如图10.13所示。

(3) 按如下源文件设置控件的相关属性值。

```
<form id="form1" runat="server">
    <div>
        <asp: Repeater  ID =" Repeater1 "  runat =" server "  DataSourceID ="
SqlDataSource1" OnItemCommand="Repeater1_ItemCommand">
            <HeaderTemplate><h2>学生信息选择表</h2></H></HeaderTemplate>
             <ItemTemplate>
            <div style="border-bottom:1px dashed #ccc; padding:5px 5px 5px 5px;">
                <%#Eval("sname")%>   <%#Eval("dept")%>
                <asp:Button ID="btn" runat="server" Text="选择"
                CommandArgument='<%#Eval("sno")%>'/>
```

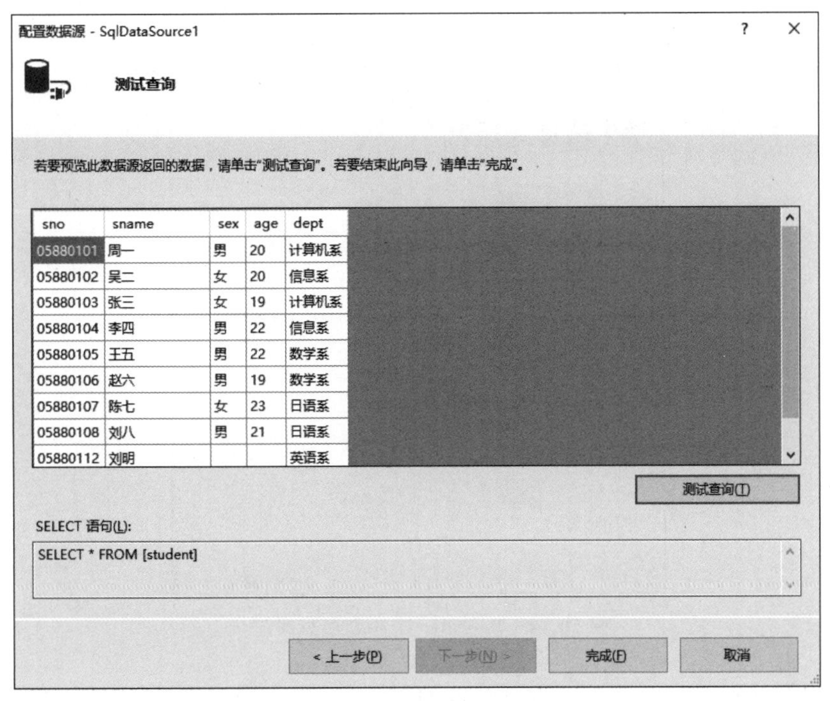

图 10.13　应用 Repeater 控件数据测试查询

```
            </div>
        </ItemTemplate>
            <FooterTemplate>   试用版 v1.0</FooterTemplate>
        </asp:Repeater>
        <asp:SqlDataSource ID="SqlDataSource1" runat="server" ConnectionString="
<%$ ConnectionStrings: DemoConnectionString %>" SelectCommand=" SELECT * FROM
[student]"></asp:SqlDataSource>
        <br/>
    </div>
</form>
```

(4) 为控件添加事件，并编辑代码如下。

```
protected void Repeater1_ItemCommand(object source, RepeaterCommandEventArgs e)
{
    string sno=e.CommandArgument.ToString();
    Response.Write("<script>alert('选择的学生学号为:"+sno+"');</script>");
}
```

(5) 运行网站，单击"选择"按钮，会触发 ItemCommand 事件，执行效果如图 10.14 所示。

使用 Repeater 控件需要具有一定的 HTML 知识，页面设计虽然有一定的复杂度，但是可增加灵活性，能够按照用户的想法显示不同的样式，让数据显示更加丰富。

10.4.3　数据列表控件 DataList

DataList 控件又称数据列表控件，在工具箱中图标为" DataList "，封装在 System.Web.UI.Control.WebControl 命名空间的 DataList 类中。DataList 控件支持各种不同的模

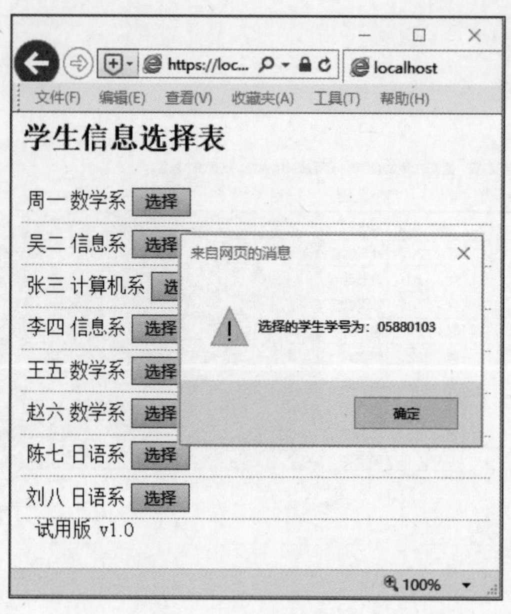

图 10.14　Repeater 控件应用页面演示

板样式，可以创建项、交替项、选定项和编辑项模板，也可以使用标题、脚注和分隔符模板。与 Repeater 控件相同，DataList 控件也支持 HTML，但 DataList 控件的样式属性更丰富。DataList 控件的常用外观样式如表 10.2 所示。DataList 控件的常用模板如表 10.3 所示。

表 10.2　DataList 控件的常用外观样式

样　式　名	说　　明
AltermatingItemStyle	编写交替项样式
EditItemStyle	正在编辑项的样式
FooterStyle	列表结尾处脚注的样式
HeaderStyle	列表头部标头的样式
ItemStyle	单一项的样式
SelectedItemStyle	选定项的样式
SeparatorStyle	各项之间分隔符的样式

通过设置 DataList 控件的属性，能够实现复杂的 HTML 样式，而且 DataList 控件能够套用自定义格式，如图 10.15 所示。DataList 控件属性生成器，如图 10.16 所示。

表 10.3　DataList 控件的常用模板

模板属性名	说　　明
ItemTemplate	内容项模板，数据源中的每一行呈现一次
AlternatingItemTemplate	交替项模板，数据源中的每两行呈现一次
SelectedItemTemplate	选中项模板，当用户选择 DataList 控件中的某一项时呈现

续表

模板属性名	说　　明
EditItemTemplate	编辑项模板，指定当某项处于编辑模式中时呈现
HeaderTemplate	表头模板，在列表的开始处呈现
FooterTemplate	表尾模板，在列表的结束处呈现
SeparatorTemplate	分隔符模板，在每项之间呈现

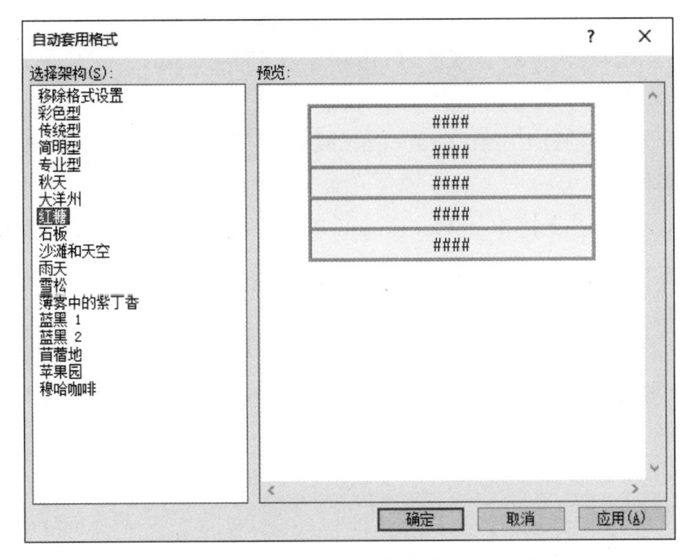

图 10.15　DataList 控件自动套用格式

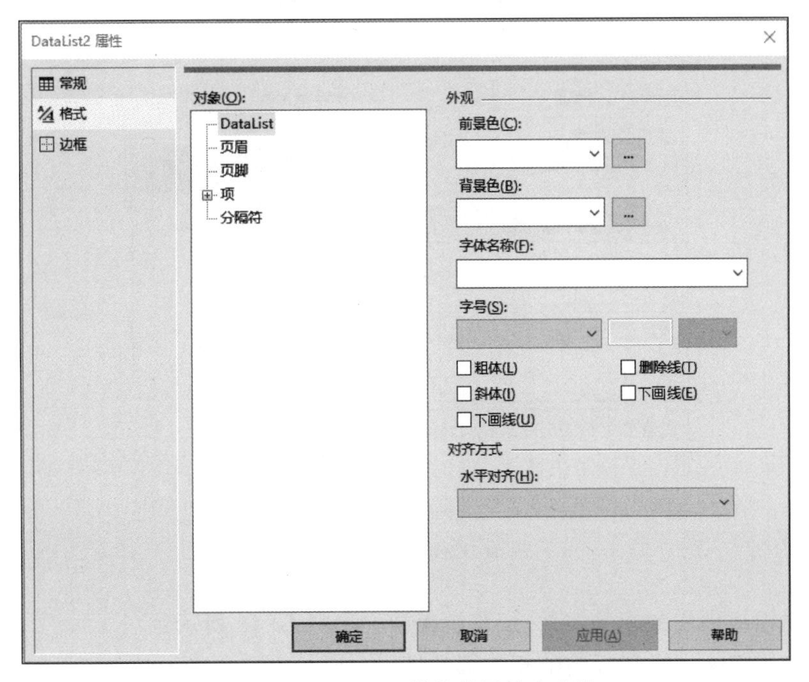

图 10.16　DataList 控件的属性生成器

DataList 控件不仅支持 ItemCommand、ItemCreated、ItemDataBound 事件，还支持其他的服务器事件，如响应列表中的按钮单击而引发的四个事件。

- EditCommand：编辑命令
- DeleteCommand：删除命令
- UpdateCommand：修改命令
- CancelCommand：取消命令

若触发这些事件，需要将 Button、LinkButton 或 ImageButton 等控件添加到 DataList 控件的模板中，并将其 CommandName 属性设置为指定关键字，如 edit、delete、update 或 cancel 等。当用户单击项中的某个按钮时，由单击按钮的 CommandName 属性值确定触发的事件。例如，若某按钮的 CommandName 属性设置为 edit，则单击该按钮时将引发 EditCommand 事件；若某按钮的 CommandName 属性设置为 delete，则单击该按钮时将引发 DeleteCommand 事件。

扫一扫

同时，DataList 控件还支持 ItemCommand 事件，单击某个按钮在触发预定义命令的同时也将同步触发 ItemCommand 事件，可以获取事件传递的 CommandArgument 参数值，编写相应的方法进行应用。

【示例 10-4】 在 chapter10 网站根目录下创建名为 example10-4 的网页，使用 SqlDataSource 控件建立数据源，并对 DataList 控件进行数据绑定，编辑模板项，以列表的形式显示数据。

(1) 在 example10-4 页面添加 SqlDataSource 控件，设置其 ID 属性为 SqlDataSource1，按示例 10-3 的步骤配置数据源。

(2) 在页面添加 DataList 控件，并设置 DataSource 属性为 SqlDataSource1 控件。

(3) 单击 DataList 控件右上角的"[>]"图标，选择"编辑模板"，并显示"项模板"，如图 10.17 所示。

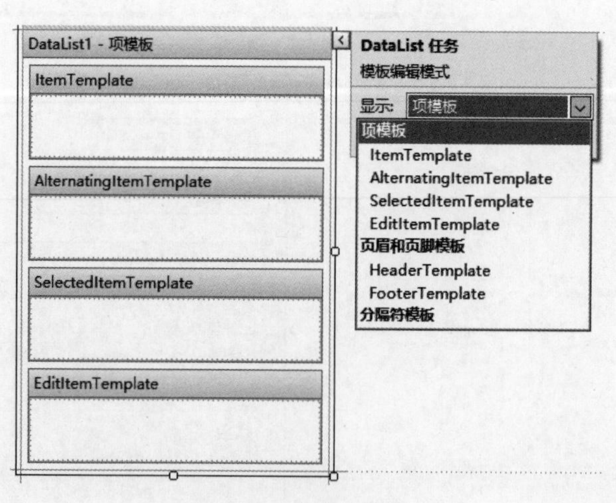

图 10.17 模板编辑界面

(4) 在项模板内添加相关控件，编辑模板样式如图 10.18 所示。

(5) 按如下源文件设置控件的相关属性值。

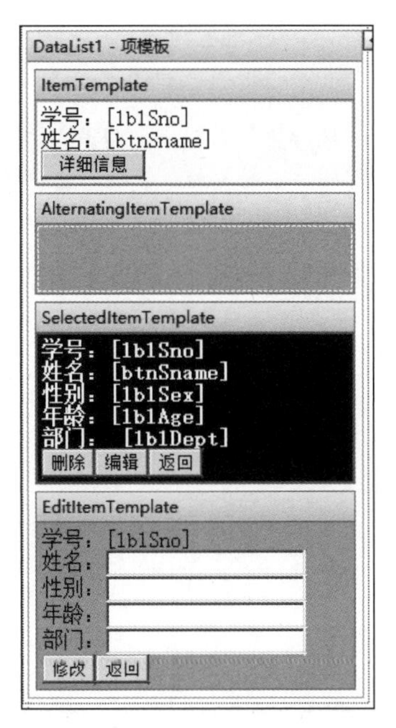

图 10.18　DataList 控件项模板编辑

```
<form id="form1" runat="server">
    <div>
        <asp:DataList ID="DataList1" runat="server" BackColor="#CCCCCC" Borde-
rColor="#999999" BorderStyle="Solid" BorderWidth="3px" CellPadding="4" Cell-
Spacing="2" DataSourceID="SqlDataSource1" ForeColor="Black" GridLines="Both"
OnCancelCommand="DataList1_CancelCommand" OnDeleteCommand="DataList1_Delete-
Command" OnEditCommand="DataList1_EditCommand" OnItemCommand="DataList1_Item-
Command" OnUpdateCommand="DataList1_UpdateCommand" RepeatColumns="4" DataKey-
Field="sno">
            <EditItemTemplate>
                学号:<asp:Label ID="lblSno" runat="server" Text='<%#Eval("
sno")%>'></asp:Label>
                <br/>
                姓名:<asp:TextBox ID="txtSname" runat="server" Text='<%#Eval("
sname")%>'></asp:TextBox>
                <br/>
                性别:<asp:TextBox ID="txtSex" runat="server" Text='<%#Eval("
sex")%>'></asp:TextBox>
                <br/>
                年龄:<asp:TextBox ID="txtAge" runat="server" Text='<%#Eval("
age")%>'></asp:TextBox>
                <br/>
                部门:<asp:TextBox ID="txtDept" runat="server" Text='<%#Eval("
dept")%>'></asp:TextBox>
                <br/>
                <asp:Button ID="btnUpdate" runat="server" Text="修改" Command-
Name="update"/>
               <asp:Button ID="btnBack" runat="server" Text="返回" CommandName="
cancel"/>
```

```
                 
            </EditItemTemplate>
            <FooterStyle BackColor="#CCCCCC"/>
             < HeaderStyle BackColor = " Black " Font - Bold = " True " ForeColor = "
White"/>
            <ItemStyle BackColor="White"/>
            <ItemTemplate>
                学号:< asp: Label ID= "lblSno" runat= "server" Text= '<% # Eval ("
sno") %>'></asp:Label>
                <br/>
                姓名:< asp: Label ID= "btnSname" runat= "server" Text= '<% # Eval ("
sname") %>'></asp:Label>
                <br/>
                <asp:Button ID="btnAllInfo" runat="server" Text="详细信息" Com-
mandName="select"/>
                <br/>
            </ItemTemplate>
            <SelectedItemStyle BackColor="#000099" Font-Bold="True" ForeColor
="White"/>
            <SelectedItemTemplate>
                学号:< asp: Label ID= "lblSno" runat= "server" Text= '<% # Eval ("
sno") %>'></asp:Label>
                <br/>
                姓名:< asp: Label ID= "btnSname" runat= "server" Text= '<% # Eval ("
sname") %>'></asp:Label>
                <br/>
                性别:< asp: Label ID= "lblSex" runat= "server" Text= '<% # Eval ("
sex") %>'></asp:Label>
                <br/>
                年龄:< asp: Label ID= "lblAge" runat= "server" Text= '<% # Eval ("
age") %>'></asp:Label>
                <br/>
                部门: < asp: Label ID= "lblDept" runat= "server" Text= '<% # Eval ("
dept") %>'></asp:Label>
                <br/>
                <asp:Button ID="btnDelete" runat="server" Text="删除" Command-
Name="delete"/>
                <asp:Button ID="btnEdit" runat="server" Text="编辑" CommandName
="edit"/>
              <asp:Button ID="btnBack" runat="server" Text="返回" CommandName="
cancel"/>
            </SelectedItemTemplate>
        </asp:DataList>
        <br/>
             < asp: SqlDataSource ID = " SqlDataSource1 " runat = " server "
ConnectionString="<%$ ConnectionStrings:DemoConnectionString %>" SelectCom-
mand="SELECT * FROM [student]"></asp:SqlDataSource>
    </div>
</form>
```

(6)为控件添加事件,并编辑代码如下。

```
//单击 CommandName 为 edit 的按钮触发 EditCommand 事件,并通过委托调用该方法
protected void DataList1_EditCommand(object source, DataListCommandEventArgs e)
{
    DataList1.SelectedIndex=-1;
    DataList1.EditItemIndex=e.Item.ItemIndex;
    DataList1.DataBind();
}
//单击 CommandName 为 cancel 的按钮触发 CancelCommand 事件,并通过委托调用该方法
protected void DataList1_CancelCommand(object source, DataListCommandEventArgs e)
{
    DataList1.SelectedIndex=-1;
    DataList1.EditItemIndex=-1;
    DataList1.DataBind();
}
//单击 CommandName 为 delete 的按钮触发 DeleteCommand 事件,并通过委托调用该方法
protected void DataList1_DeleteCommand(object source, DataListCommandEventArgs e)
{
    string dataKey=DataList1.DataKeys[e.Item.ItemIndex].ToString();
    Response.Write("要删除的学生学号为:"+dataKey);
}
//单击模板的任意按钮触发 ItemCommand 事件,并通过委托调用该方法
protected void DataList1_ItemCommand(object source, DataListCommandEventArgs e)
{
    //单击 CommandName 为 select 的按钮执行下列命令
    if (e.CommandName=="select")
    {
        DataList1.EditItemIndex=-1;
        DataList1.SelectedIndex=e.Item.ItemIndex;
        DataList1.DataBind();
    }
}
//单击 CommandName 为 update 的按钮触发 UpdateCommand 事件,并通过委托调用该方法
protected void DataList1_UpdateCommand(object source, DataListCommandEventArgs e)
{
    Label lblSno=e.Item.FindControl("lblSno") as Label;
    TextBox txtName=e.Item.FindControl("txtSname") as TextBox;
    TextBox txtSex=e.Item.FindControl("txtSex") as TextBox;
    TextBox txtAge=e.Item.FindControl("txtAge") as TextBox;
    TextBox txtDept=e.Item.FindControl("txtDept") as TextBox;
    string strInfo=string.Format("要修改的学生信息<br>学号:{0}<br>姓名:{1}<br>性
别:{2}<br>年龄:{3}<br>院系:{4}<br>", lblSno.Text, txtName.Text, txtSex.Text,
txtAge.Text, txtDept.Text);
    Response.Write(strInfo);
}
```

(7) 运行网站,初始页面执行效果如图 10.19 所示;单击任意项的“详细信息”可查看该学生的详细信息,如图 10.20 所示;单击选中项的“编辑”按钮可对该项信息进行编辑,之后单击“修改”按钮获取各控件修改后的内容,如图 10.21 所示。

在上述示例中添加数据库修改的相关代码即可完成学生信息的修改与删除。

图 10.19　DataList 控件应用页面演示

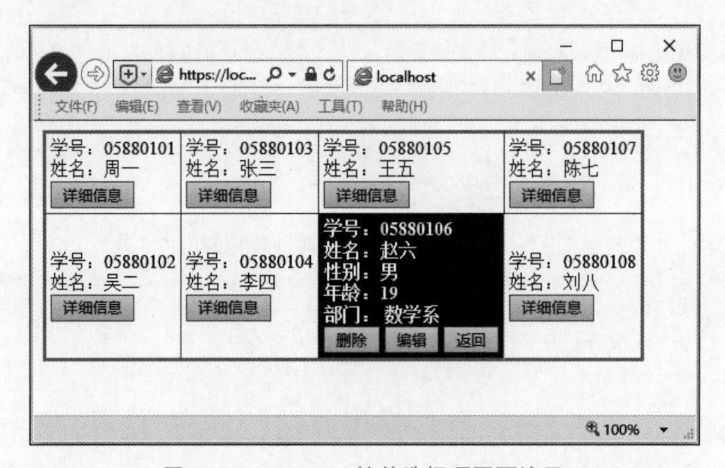

图 10.20　DataList 控件选择项页面演示

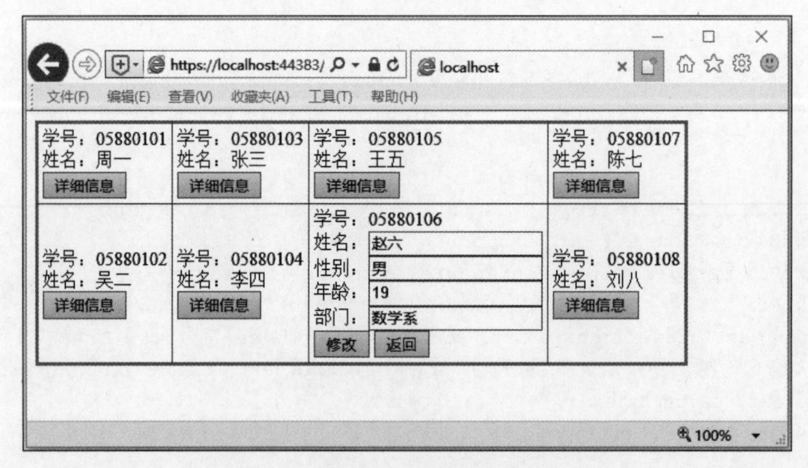

图 10.21　DataList 控件编辑项页面演示

10.4.4　网格视图控件 GridView

GridView 控件又称网格视图控件，在工具箱中图标为“ GridView”，封装在 System.

Web.UI.Control.WebControl 命名空间的 GridView 类中。GridView 是 ASP.NET 中功能比较丰富的控件之一，可以通过数据源控件自动绑定和显示数据，并以表格的形式显示数据的内容，能够通过配置数据源控件对 GridView 中的数据进行选择、排序、分页、编辑和删除等操作。GridView 控件的常用属性如表 10.4 所示。

表 10.4　GridView 控件的常用属性

属　性　名	说　　明
Adapter	获取控件的浏览器特定适配器
AllowPaging	获取或设置一个值，指示是否启用分页功能
AllowSorting	获取或设置一个值，指示是否启用排序功能
Caption	获取或设置要在一个 HTML 标题元素中呈现的文本 GridView 控件
CaptionAlign	获取或设置 HTML 标题元素的对齐方式
Columns	获取 GridView 控件的列字段
Controls	获取数据绑定复合控件内的子控件的集合
DataKeys	获取 GridView 控件每行的键值
DataMember	获取或设置控件绑定表的名称
DataSource	获取或设置绑定到控件的数据源
EditIndex	获取或设置要编辑行的索引
EmptyDataText	获取或设置空数据项的显示内容
ShowFooter	获取或设置一个值，指示是否显示页脚行
ShowHeader	获取或设置一个值，指示是否显示标题行

GridView 控件支持内置格式，单击“自动套用格式”连接可以选择 GridView 中的默认格式，如图 10.22 所示。

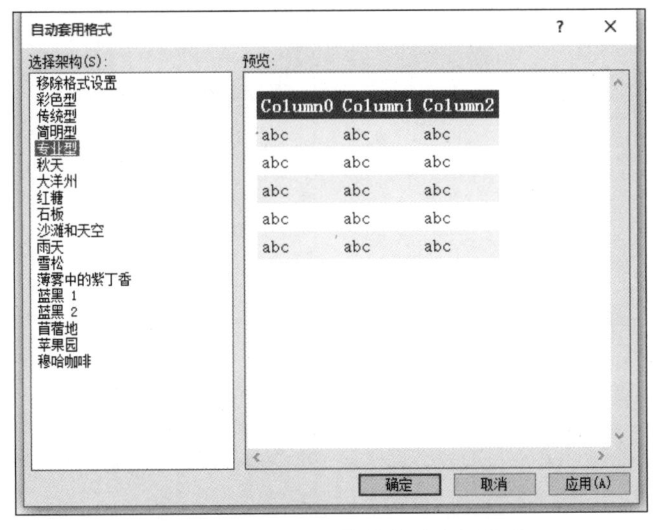

图 10.22　GridView 控件自动套用格式

GridView 以表格为表现形式，通过配置相应的属性能够编辑相应行的样式，可以选择

"编辑列"选项编写相应列的样式，如图 10.23 所示。

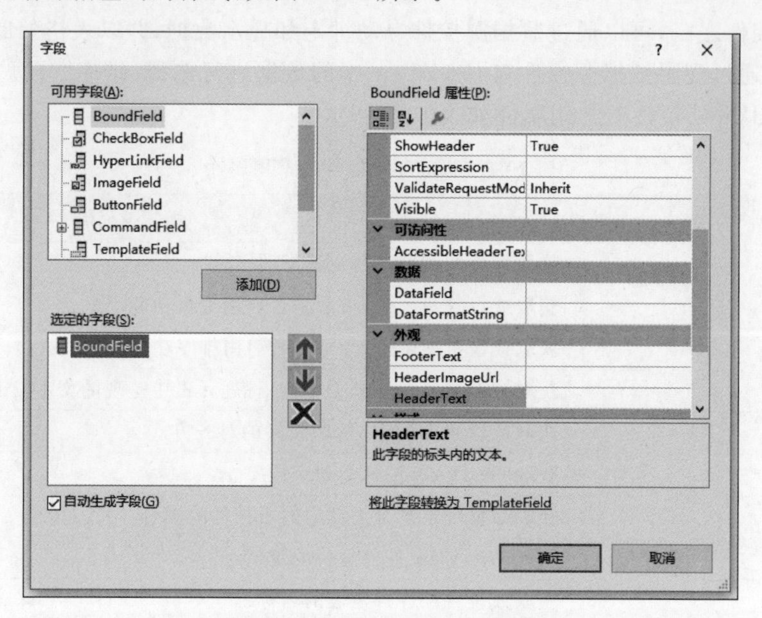

图 10.23　GridView 控件编辑列

使用 DataSourceID 进行数据绑定后，GridView 控件能够自动进行分页、选择等操作，如图 10.24 所示。GridView 控件能够自定义字段，单击"添加"按钮，可以选择相应类型的字段。在添加字段选项中支持多种类型的字段，如复选框、图片、单选框、超链接等，如图 10.25 所示。

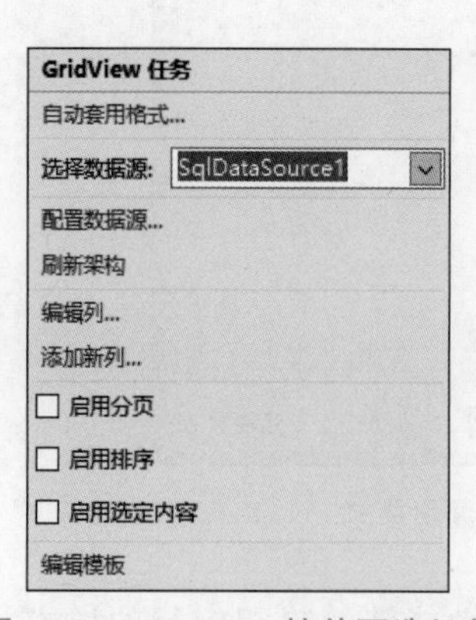

图 10.24　GridView 控件可选任务

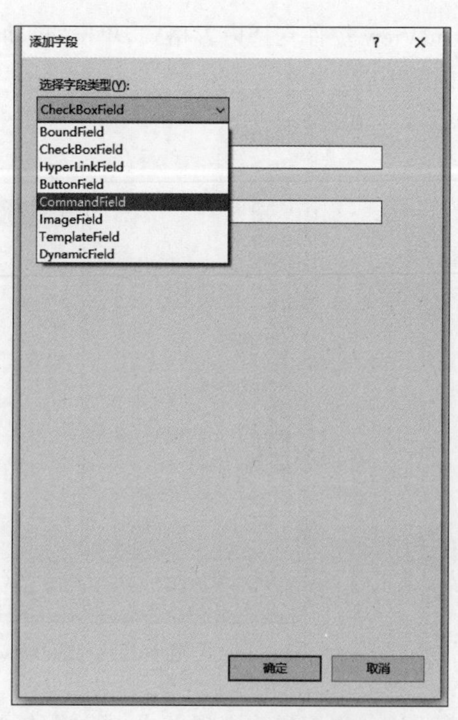

图 10.25　GridView 控件添加字段

GridView 支持多个事件，对 GridView 控件进行排序、选择等操作时，创建当前行或对当前行绑定数据，以及单击命令控件时都会引发事件。GridView 控件的常用事件如表 10.5 所示。

扫一扫

表 10.5　GridView 控件的常用事件

事件名	说明
RowCommand	在 GridView 控件中单击某个按钮时发生
PageIndexChanging	单击页导航按钮时发生，在 GridView 控件执行分页操作之前
PageIndexChanged	单击页导航按钮时发生，在 GridView 控件执行分页操作之后
SelectedIndexChanging	单击 GridView 控件内某一行的 Select 按钮时发生，但在 GridView 控件执行选择操作之前
SelectedIndexChanged	单击 GridView 控件内某一行的 Select 按钮时发生，但在 GridView 控件执行选择操作之后
Sorting	单击某个用于对列进行排序的超链接时发生，但在 GridView 控件执行排序操作之前
Sorted	单击某个用于对列进行排序的超链接时发生，但在 GridView 控件执行排序操作之后
RowDataBound	在 GridView 控件中的某个行被绑定到一个数据记录时发生
RowCreated	在 GridView 控件中创建新行时发生
RowDeleting	单击 GridView 控件内某一行的 Delete 按钮(其中 CommandName 属性设置为 Delete 的按钮)时发生，但在 GridView 控件从数据源删除记录之前
RowDeleted	单击 GridView 控件内某一行的 Delete 按钮时发生，但在 GridView 控件从数据源删除记录之后
RowEditing	单击 GridView 控件内某一行的 Edit 按钮(其中 CommandName 属性设置为 Edit 的按钮)时发生，但在 GridView 控件进入编辑模式之前
RowCancelingEdit	单击 GridView 控件内某一行的 Cancel 按钮(其中 CommandName 属性设置为 Cancel 的按钮)时发生，但在 GridView 控件退出编辑模式之前
RowUpdating	单击 GridView 控件内某一行的 Update 按钮(其中 CommandName 属性设置为 Update 的按钮)时发生，但在 GridView 控件更新记录之前
RowUpdated	单击 GridView 控件内某一行的 Update 按钮时发生，但在 GridView 控件更新记录之后
DataBound	继承自 BaseDataBoundControl 控件，在 GridView 控件完成到数据源的绑定后发生

【示例 10-5】　在 chapter10 网站根目录下创建名为 example10-5 的网页，编写 C# 代码建立数据源，对 GridView 控件进行数据绑定，编辑字段，以网格视图的形式显示数据。

(1) 在 example10-5 页面添加 GridView 控件，设置其 ID 属性为默认值 GridView1。

(2) 单击 GridView 控件右上角的▶图标，选择“编辑列”，并显示“字段”窗口，添加字段如图 10.26 所示。

(3) 按如下源文件设置 GridView 控件及字段的相关属性值。

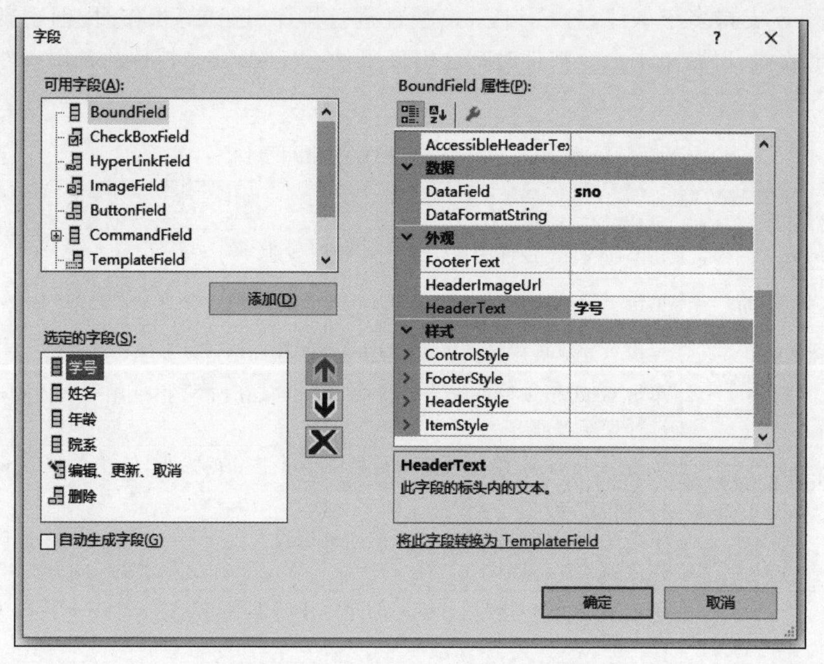

图 10.26　GridView 控件添加字段

```
<form id="form1" runat="server">
    <div>
        <asp:GridView ID="GridView1" runat="server" AllowSorting="True" Auto-
GenerateColumns="False" DataKeyNames="sno" Height="185px" OnRowCancelingEdit
="GridView1_RowCancelingEdit" OnRowDeleting="GridView1_RowDeleting" OnRowEd-
iting="GridView1_RowEditing" OnRowUpdating="GridView1_RowUpdating"  Width="
478px">
            <Columns>
                <asp:BoundField DataField="sno" HeaderText="学号" ReadOnly="True"/>
                <asp:BoundField DataField="sname" HeaderText="姓名"/>
                <asp:BoundField DataField="age" HeaderText="年龄"/>
                <asp:BoundField DataField="dept" HeaderText="院系"/>
                <asp:CommandField InsertVisible="False" ShowEditButton="True">
                <ItemStyle Wrap="False"/>
                </asp:CommandField>
                <asp:ButtonField ButtonType="Button" CommandName="delete" Text="删
除"/>
            </Columns>
        </asp:GridView>
    </div>
</form>
```

（4）为控件添加事件,并编辑代码如下。

```
protected void Page_Load(object sender, EventArgs e)
{
    if (!Page.IsPostBack)
        BindGridView();
}
```

```
void BindGridView()
{
    string conStr=@"server=.\sqlexpress;database=demo;trusted_connection=true;";
    DataSet ds=new DataSet();
    SqlConnection con=new SqlConnection(conStr);
    SqlDataAdapter sda=new SqlDataAdapter("select sno, sname, age, dept from student", con);
    sda.Fill(ds, "tb_student");
    GridView1.DataSource=ds.Tables["tb_student"];
    GridView1.DataBind();
    con.Dispose();
    sda.Dispose();
}
protected void GridView1_RowEditing(object sender, GridViewEditEventArgs e)
{
    GridView1.EditIndex=e.NewEditIndex;
    BindGridView();
}
protected void GridView1_RowUpdating(object sender, GridViewUpdateEventArgs e)
{
    string conStr=@"server=.\sqlexpress;database=demo;trusted_connection=true;";
    SqlConnection con=new SqlConnection(conStr);
    try
    {
        con.Open();
        SqlCommand cmd=new SqlCommand();
        cmd.Connection=con;
        cmd.CommandText="update student set sname=@Sname,age=@Age,dept=@Dept where sno=@Sno";
        cmd.Parameters.AddWithValue("@Sno", GridView1.DataKeys[e.RowIndex].Value.ToString());
        cmd.Parameters.AddWithValue("@Sname", ((TextBox)GridView1.Rows[e.RowIndex].Cells[1].Controls[0]).Text);
        cmd.Parameters.AddWithValue("@Age", ((TextBox)GridView1.Rows[e.RowIndex].Cells[2].Controls[0]).Text);
        cmd.Parameters.AddWithValue("@Dept", ((TextBox)GridView1.Rows[e.RowIndex].Cells[3].Controls[0]).Text);
        cmd.ExecuteNonQuery();
        con.Close();
        con.Dispose();
        cmd.Dispose();
    }
    catch (Exception ex)
    {
        Response.Write("数据库更新出错:"+ex.ToString());
    }
    GridView1.EditIndex=-1;
    BindGridView();
}
protected void GridView1_RowCancelingEdit(object sender, GridViewCancelEditEventArgs e)
{
```

```
        GridView1.EditIndex=-1;
        BindGridView();
    }
    protected void GridView1_RowDeleting(object sender, GridViewDeleteEventArgs e)
    {
        //设置数据库连接
        string conStr = @" server = . \ sqlexpress; database = demo; trusted _ connection =
true;"; ;
        SqlConnection con=new SqlConnection(conStr);
        //执行删除行处理
        try
        {
            con.Open();
            String sql="delete from student where sno= '"+GridView1.DataKeys[e.RowIndex].
Value.ToString()+"'";
            SqlCommand cmd=new SqlCommand(sql, con);
            cmd.ExecuteNonQuery();
            con.Close();
            con.Dispose();
            cmd.Dispose();
            GridView1.EditIndex=-1;
            BindGridView();
        }
        catch (Exception ex)
        {
            Response.Write("数据库删除出错:"+ex.ToString());
        }
    }
```

(5) 运行网站,初始页面执行效果如图 10.27 所示,单击任意项的“编辑”按钮,可对该项信息进行编辑,单击“更新”按钮,可在数据库中修改对应内容,如图 10.28 所示。单击任意项的“删除”按钮,可删除该学生信息。

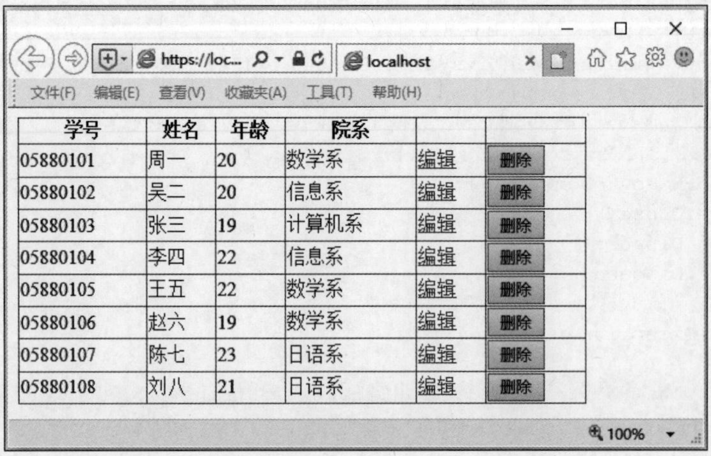
https://loc...
localhost
文件(F) 编辑(E) 查看(V) 收藏夹(A) 工具(T) 帮助(H)
学号 姓名 年龄 院系
05880101 周一 20 数学系 编辑 删除
05880102 吴二 20 信息系 编辑 删除
05880103 张三 19 计算机系 编辑 删除
05880104 李四 22 信息系 编辑 删除
05880105 王五 22 数学系 编辑 删除
05880106 赵六 19 数学系 编辑 删除
05880107 陈七 23 日语系 编辑 删除
05880108 刘八 21 日语系 编辑 删除
100%

图 10.27 GridView 控件应用页面演示

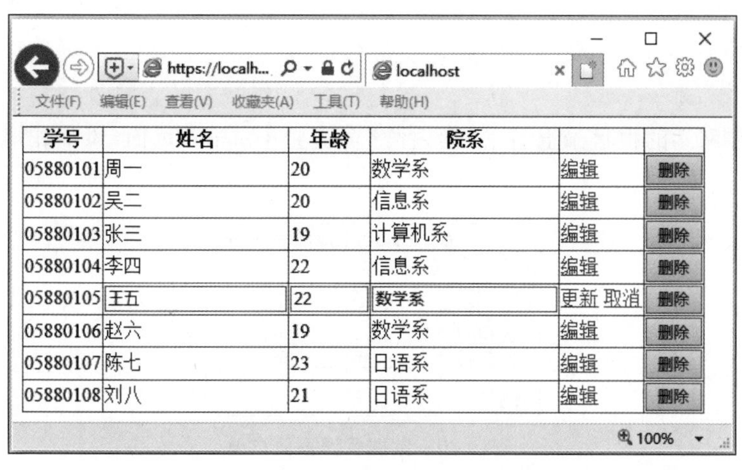

学号	姓名	年龄	院系		
05880101	周一	20	数学系	编辑	删除
05880102	吴二	20	信息系	编辑	删除
05880103	张三	19	计算机系	编辑	删除
05880104	李四	22	信息系	编辑	删除
05880105	王五	22	数学系	更新 取消	删除
05880106	赵六	19	数学系	编辑	删除
05880107	陈七	23	日语系	编辑	删除
05880108	刘八	21	日语系	编辑	删除

图 10.28 GridView 控件编辑应用页面演示

综合实验十 XML 文件数据的绑定

主要任务

创建 ASP.NET 应用程序，使用 GridView 数据控件绑定 XML 文件。

实验步骤

步骤 1：在 VS 2019 菜单上选择“文件”→“新建”→“项目”命令。

步骤 2：创建 ASP.NET 应用程序，命名为“综合实验十”。

步骤 3：在网站的根目录上右击，添加 XML 文件，命名为 bookstore.xml。

步骤 4：编辑 bookstore.xml 文件，代码如下。

```
<?xml version="1.0" encoding="utf-8" ?>
<bookstore>
    <book  genre="natural science" ISBN="9787301327685">
      <No>1001</No>
      <title>芯片战争</title>
      <author>脑极体</author>
      <price>59.00</price>
    </book>
  <book genre="social scince" ISBN="9787545527100">
      <No>1002</No>
      <title>故宫院长说故宫</title>
      <author>李文儒</author>
      <price>68.00</price>
    </book>
  <book genre="natural science" ISBN="9787302469131">
    <No>1003</No>
    <title>数字信号处理教程</title>
    <author>程佩青</author>
```

```
    <price>69.00</price>
  </book>
</bookstore>
```

步骤 5：在网站的根目录上右击，创建“综合实验十.aspx”页面，添加相应控件，具体如图 10.29 所示。

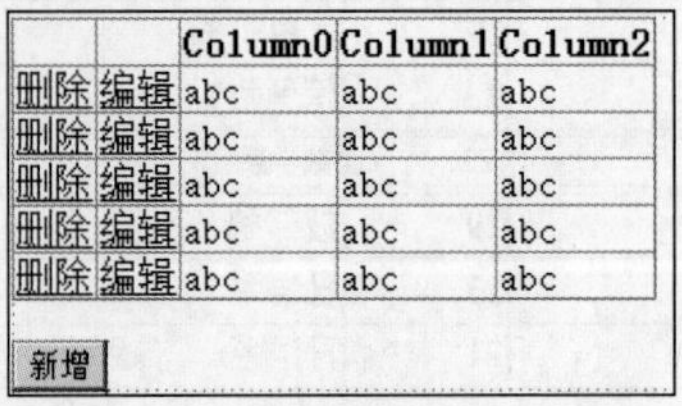

		Column0	Column1	Column2
删除	编辑	abc	abc	abc
删除	编辑	abc	abc	abc
删除	编辑	abc	abc	abc
删除	编辑	abc	abc	abc
删除	编辑	abc	abc	abc

新增

图 10.29　数据管理页面布局

步骤 6：按如下源文件设置控件的相关属性值。

```
<form id="form1" runat="server">
        <div>
            <asp:GridView ID="GridView1" runat="server" DataKeyNames="No" On-
RowCancelingEdit="GridView1_RowCancelingEdit" OnRowDeleting="GridView1_Row-
Deleting" OnRowEditing="GridView1_RowEditing" OnRowUpdating="GridView1_RowUp-
dating">
                <Columns>
                    <asp:ButtonField CommandName="delete" Text="删除"/>
                    <asp:TemplateField ShowHeader="False">
                        <EditItemTemplate>
                            <asp:LinkButton ID="LinkButton2" runat="server" Com-
mandName="update">确认</asp:LinkButton>
                             
                            <asp:LinkButton ID="LinkButton3" runat="server" Com-
mandName="Cancel">取消</asp:LinkButton>
                        </EditItemTemplate>
                        <FooterTemplate>
                            <br/>
                        </FooterTemplate>
                        <ItemTemplate>
                              <asp:LinkButton ID="LinkButton1" runat="server"
CausesValidation="false" CommandName="edit" Text="编辑"></asp:LinkButton>
                        </ItemTemplate>
                    </asp:TemplateField>
                </Columns>
            </asp:GridView>
            <br/>
             <asp:Button ID="btnInsert" runat="server" OnClick="btnInsert_
Click" Text="新增"/>
            <br/>
        </div>
</form>
```

步骤 7：编辑“综合实验十.aspx.cs”，代码如下。

```
using System;
using System.Data;
using System.Web.UI.WebControls;
using System.Xml;

namespace 综合实验十
{
    public partial class 综合实验十: System.Web.UI.Page
    {
        DataSet ds;
        protected void Page_Load(object sender, EventArgs e)
        {
            ds=new DataSet();
            if (! IsPostBack)
            {
                GridViewDataBind("bookstore.xml" ,-1);
            }
        }
        protected void GridView1_RowDeleting(object sender, GridViewDeleteEven-
tArgs e)
        {
            ds.ReadXml(Server.MapPath("bookstore.xml"));
            DataRow row=ds.Tables[0].Rows[e.RowIndex];
            ds.Tables[0].Rows.Remove(row);
            WriteXmlFile(ds, "bookstore.xml");
            GridViewDataBind("bookstore.xml" ,-1);
        }
        protected void GridView1_RowEditing(object sender, GridViewEditEventArgs e)
        {
            GridViewDataBind("bookstore.xml", e.NewEditIndex);
        }
        protected void GridView1_RowUpdating(object sender, GridViewUpdateEventArgs e)
        {
            DataSet ds=new DataSet();
            ds.ReadXml(Server.MapPath("bookstore.xml"));
            int count=ds.Tables[0].Rows.Count;
            DataRow row;
            if (! btnInsert.Visible)
            {
                row=ds.Tables[0].NewRow();
                ds.Tables[0].Rows.Add(row);
                int No=Convert.ToInt32(ds.Tables[0].Rows[count-2]["No"])+1;
                row["No"]=No;
                btnInsert.Visible=true;
            }
            else
            {
                row=ds.Tables[0].Rows[count-1];
            }
            row["title"]=e.NewValues[0];
            row["author"]=e.NewValues[1];
            row["price"]=e.NewValues[2];
            row["genre"]=e.NewValues[3];
```

```
            row["ISBN"]=e.NewValues[4];
            WriteXmlFile(ds, "bookstore.xml");
            GridViewDataBind("bookstore.xml", -1);
        }
        protected void btnInsert_Click(object sender, EventArgs e)
        {
            btnInsert.Visible=false;
            GrirdViewBind("bookstore.xml");
        }
        protected void GridView1_RowCancelingEdit(object sender,
GridViewCancelEditEventArgs e)
        {
            GridViewDataBind("bookstore.xml", -1);
        }
        /// <summary>
        /// 将 DataSet 数据存到 XML 文件中
        /// </summary>
        /// <param name="ds">DataSet 数据表</param>
        /// <param name="xmlFilePath">文件要保存到的虚拟路径</param>
        public void WriteXmlFile(DataSet ds, string xmlFilePath)
        {
            // string filePath=Server.MapPath(xmlFilePath);
            if (ds.Tables[0].Rows.Count >0)
            {
                 string xmlstring="<? xml version='1.0' encoding='utf-8' ?>
<bookstore></bookstore>";
                XmlDocument xml=new XmlDocument();
                xml.LoadXml(xmlstring);
                XmlDocumentFragment xdf=xml.CreateDocumentFragment();
                for (int i=0; i<ds.Tables[0].Rows.Count; i++)
                {
                    string xmlst =string.Format("<book genre='{0}' ISBN='{1}'>
<No>{2}</No><title>{3}</title><author>{4}</author><price>{5}</price></
book>",ds.Tables[0].Rows[i]["genre"],ds.Tables[0].Rows[i]["ISBN"], ds.Tables
[0].Rows[i]["No"], ds.Tables[0].Rows[i]["title"], ds.Tables[0].Rows[i]["
author"], ds.Tables[0].Rows[i]["price"]);
                    xdf.InnerXml=xmlst;
                    xml.DocumentElement.AppendChild(xdf);
                }
                xml.Save(Server.MapPath(xmlFilePath));
            }
        }
        /// <summary>
        /// 将 XML 文件中的数据绑定到 GridView 控件
        /// </summary>
        /// <param name="xmlFileName">XML 文件名</param>
        /// <param name="editIndex">GridView 控件中编辑项索引</param>
        private void GridViewDataBind(string xmlFileName, int editIndex)
        {
            DataSet ds2=new DataSet();
            ds2.ReadXml(Server.MapPath(xmlFileName));
            GridView1.DataSource=ds2.Tables[0];
            GridView1.EditIndex=editIndex;
```

```
            GridView1.DataBind();
        }
        /// <summary>
        /// 将 XML 文件中的数据绑定到 GridView 控件,并创建新增数据
        /// </summary>
        /// <param name="xmlFileName">XML 文件名</param>
        private void GrirdViewBind(string xmlFileName)
        {
            ds.ReadXml(Server.MapPath(xmlFileName));
            int count=ds.Tables[0].Rows.Count;
            int No=Convert.ToInt32(ds.Tables[0].Rows[count-1]["No"])+1;
            ds.Tables[0].Rows.Add(No);
            GridView1.DataSource=ds.Tables[0];
            GridView1.EditIndex=count;
            GridView1.DataBind();
        }
    }
}
```

步骤 8：运行"综合实验十"页面，初始显示图书信息如图 10.30 所示。单击"新增"按钮，增加新行，可添加数据并保存，如图 10.31 所示。单击任一行中的"删除"按钮，可删除该条记录；单击任一行中的"编辑"按钮，可编辑并保存除 No 外的其他字段值，如图 10.32 所示。

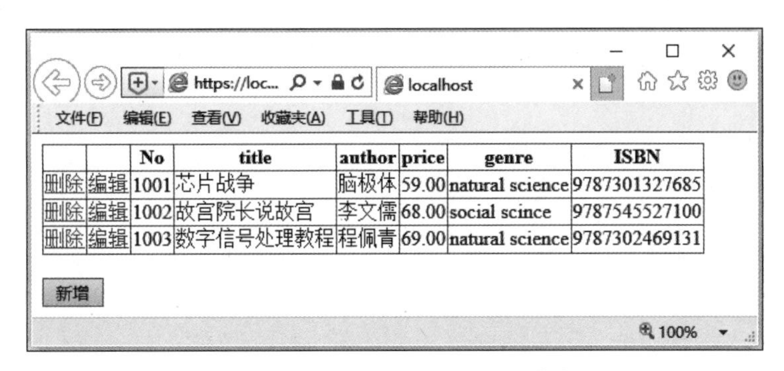

		No	title	author	price	genre	ISBN
删除	编辑	1001	芯片战争	脑极体	59.00	natural science	9787301327685
删除	编辑	1002	故宫院长说故宫	李文儒	68.00	social scince	9787545527100
删除	编辑	1003	数字信号处理教程	程佩青	69.00	natural science	9787302469131

图 10.30　显示图书信息运行测试

https://localhost:44383/综合实验十.aspx

文件(F) 编辑(E) 查看(V) 收藏夹(A) 工具(T) 帮助(H)

		No	title	author	price	genre	ISBN
删除	编辑	1001	芯片战争	脑极体	59.00	natural science	9787301327685
删除	编辑	1002	故宫院长说故宫	李文儒	68.00	social scince	9787545527100
删除	编辑	1003	数字信号处理教程	程佩青	69.00	natural science	9787302469131
删除	确认 取消	1004	数据结构	严蔚敏	32.00	natural science	9787302023685

图 10.31　添加图书功能运行测试

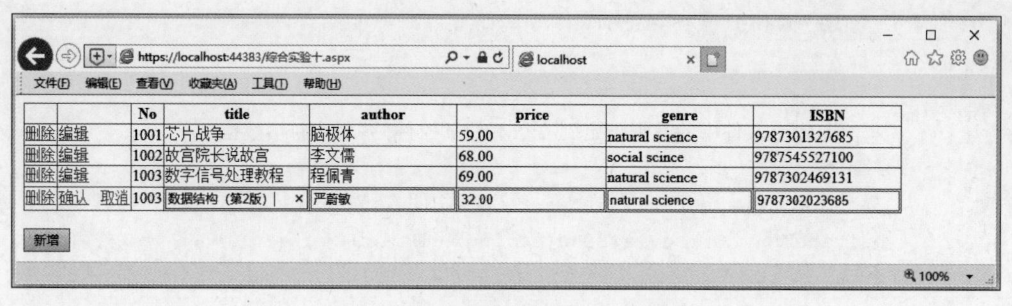

图 10.32　编辑图书功能运行测试

Web系统中的三层结构

本章学习目标

- 了解项目分层的意义
- 熟练掌握三层架构
- 熟练掌握实体层的使用方法
- 熟练掌握 ASP.NET 中三层架构的搭建

本章首先介绍项目中分层的意思，然后对三层架构中的数据访问层、业务逻辑层和表现层进行详细讲解，对实体层的作用进行说明，最后在 ASP.NET 中对三层架构的搭建进行详细讲解。

11.1 三层架构

随着软件工程的不断进步和规范化以及面向对象思想的广泛应用，为满足软件在封装性、复用性、扩展性等方面的要求，三层架构体系应运而生。三层架构在传统的双层结构模型中引入了新的中间层，提供业务规则处理、数据存取、合法性校验等操作，有效地实现了页面与数据库的分离，增强了应用程序的灵活性、可移植性和安全性。

11.1.1 项目结构分层的意义

在早期的 Web 开发中只有静态的 HTML 页面，有了数据库后，产生了所谓的动态页面。编码时会把所有代码都写在页面上，包括数据库连接、事务控制、接收参数、各种校验、各种逻辑以及各种 HTML、JS、CSS 代码等。一个页面一般有成千上万条代码，出现问题需要修改时很难定位错误，往往需要从头到尾地逐个排查，费时费工。

借鉴生活、生产中分工协作、流水线作业的理念将页面和逻辑拆开，页面只负责显示，逻辑都放在后台，这便是项目分层的起源，经过不断发展产生了 MVC 架构、三层架构等经典架构。项目的分层结构主要有如下意义。

1. 降低耦合性，提升程序可维护性

分层后上一层只依赖于下一层，如果测试下一层没有问题，那么问题就只有可能在本层了，便于发现和改正 BUG，体现了“高内聚，低耦合”的思想，各层间通过接口解耦，接口与实现分离，从而可以简单地替换等，使得软件开发有条理、有秩序，一目了然，可以清晰识别程序架构的框架。

2. 简化问题复杂性，提升程序可复用性

各层次分工明确，将一个复杂问题拆分为简单问题，劳动成本减少，分层后各层之间具有互不依赖的内部实现，可实现即插即用，如将 SQL Server 数据库换为 Oracle 数据库，只修改数据访问层即可，无须对其他层做任何改动。

3. 增强团队协作性，提升代码规范性

团队合作开发，可以提高工作效率，在开发前规定好各层的接口，每层可以独立开发和维护。在人员的分配方面，可实现技术强者负责重要的开发工作，简单重复性的工作安排新手完成，大大地提高了开发效率，使项目具有固定的语言开发的风格。

项目分层也不是越多越好，过多的分层会限制开发人员与客户对系统的理解，影响客户与开发人员的交流，在性能、复杂性等难度上带来不良影响，降低可靠性和稳定性。

11.1.2 什么是三层架构

对三层架构结构最简单的理解是在客户端与数据库之间加入了一个“中间层”，即业务逻辑层。三层架构的应用程序将业务规则、数据访问、合法性校验等工作放到中间层进行处理，通常情况下客户端不直接与数据库进行交互，而是通过业务逻辑层建立连接，再经由中间层与数据库进行交互。所谓的三层架构，不是指物理上的三层，而是指逻辑上的三层；不是简单地放置三台机器就是三层架构，三个层也可放置一台机器上；不仅 B/S 应用程序可以使用三层架构，C/S 应用程序同样可以使用三层架构。

11.1.3 三层架构中每层的作用

三层架构（3-Tier Architecture）是将整个网站的业务应用划分为用户界面层（User Interface Layer）、业务逻辑层（Business Logic Layer）和数据访问层（Data Access Layer）三层，每一层的作用如下。

• 用户界面层

用户界面层（表示层）位于最外层，最接近用户，即 ASPX 或 HTML 页面，用于显示数据和接收用户输入的数据，为用户提供一种交互式操作的界面。

• 业务逻辑层

业务逻辑层是系统架构中体现核心价值的部分，处于数据访问层与表示层中间，在数据交换中起到承上启下的作用。对于数据访问层而言，它是调用者；对于表示层而言，它是被调用者，项目的依赖与被依赖关系都体现在业务逻辑层上。

• 数据访问层

数据访问层也称为持久层，其功能主要是负责数据的访问，可以读写数据库文件、二进制文件、文本文档或 XML 文档等，用于与数据库进行交互，存取数据，简单的说法就是实现对数据表的 Select、Insert、Update、Delete 等操作。

三层架构中各层之间调用关系如图 11.1 所示。

三层架构将网站分为用户界面层、业务逻辑层和数据访问层后，具有如下优缺点。

用户界面层

用户请求　反馈数据

业务逻辑层

用户请求　反馈数据

数据访问层

图 11.1　三层架构中各层的调用关系

1. 三层架构的优点

- 开发人员只关注整个架构中的某一层，可以用新的实现替换原有层的实现，降低层与层之间的依赖。
- 有利于标准化，实现各层逻辑的复用。
- 架构更加明确，极大地降低了维护成本和维护时间。

2. 三层架构的缺点

- 可能降低系统的性能。在单层架构中可以直接访问数据库，获取相应的数据，三层架构必须通过业务逻辑层完成，性能有所降低。
- 可能导致级联的修改，如果在表示层中需要增加一个功能，为保证其设计符合分层式结构，可能需要在相应的业务逻辑层和数据访问层中都增加相应的代码。
- 可能增加开发成本。

11.1.4　三层架构与实体层

扫一扫

实体层（Entity Layer）也可以称作模型层，通常是一个类库文件，类库中的每一类对应数据库中的一张数据表，起到封装数据的作用，但并不属于三层架构中的一层。

数据处理时通常使用变量作参数，但参数声明往往比较烦琐，容易因参数类型匹配不一致而导致出错，可以根据数据类型定义各个实体类进行数据的封装，类中的属性与数据库表中的字段对应，通过访问器属性获取或设置实体类里的成员值，便于程序的维护和扩展。

在三层架构中实体层的作用是在层与层之间传递数据。实体层与三层架构之间的依赖关系如图 11.2 所示。

11.2　三层架构的应用

扫一扫

在 ASP.NET 解决方案下创建的三层架构项目包含三个类库文件，分别对应业务逻辑层、数据访问层和实体层，具体为 Business 类库对应业务逻辑层、DataAccess 类库对应数据访问层、Entity 类库对应实体层，Web UI 网站对应表示层，解决方案具体如图 11.3 所示。

扫一扫

【示例 11-1】　在 E 盘 ASP.NET 项目代码目录中创建 chapter11 子目录，将其作为解决方案根目录，添加类库文件和网站，搭建三层架构项目，对 Demo 数据库内 student 表中的学生信息进行添加。

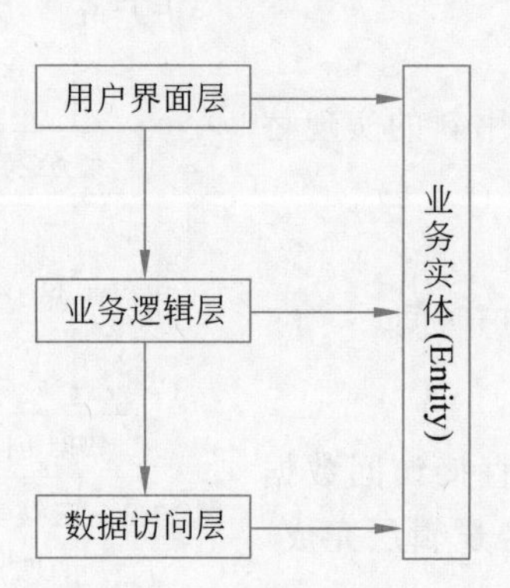

图 11.2 实体层与三层架构之间的依赖关系

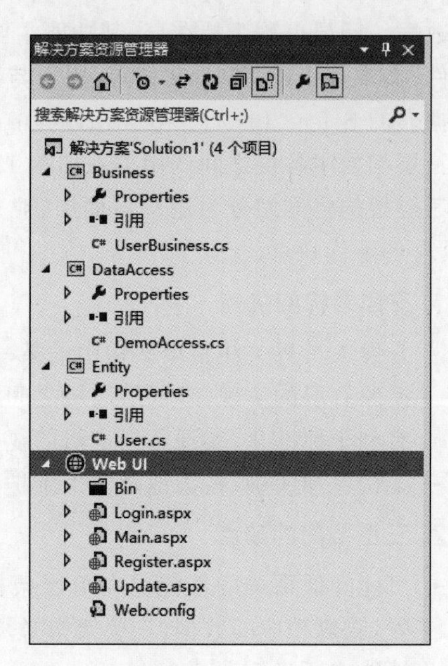

图 11.3 解决方案下的三层架构项目

(1) 在 VS 2019 菜单中选择"空白解决方案",如图 11.4 所示,设置解决方案名称并选择保存位置,如图 11.5 所示。

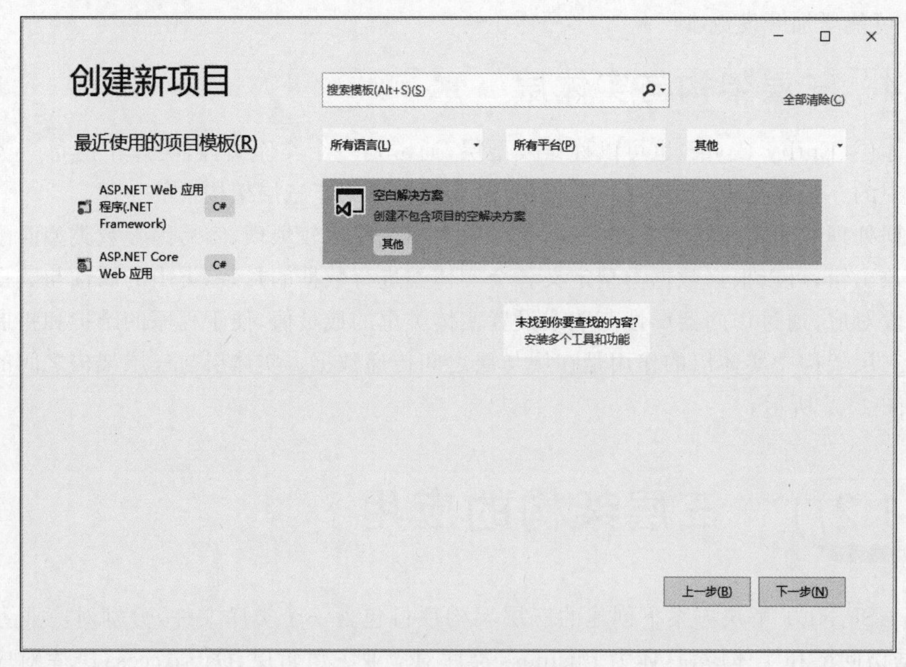

图 11.4 解决方案的创建

(2) 右击"解决方案资源管理器",从弹出的快捷菜单中选择"添加"→"新建项目",在"添加新建项目"窗口中选择"类库",如图 11.6 所示,设置类库名称为 DataAccess,如图 11.7

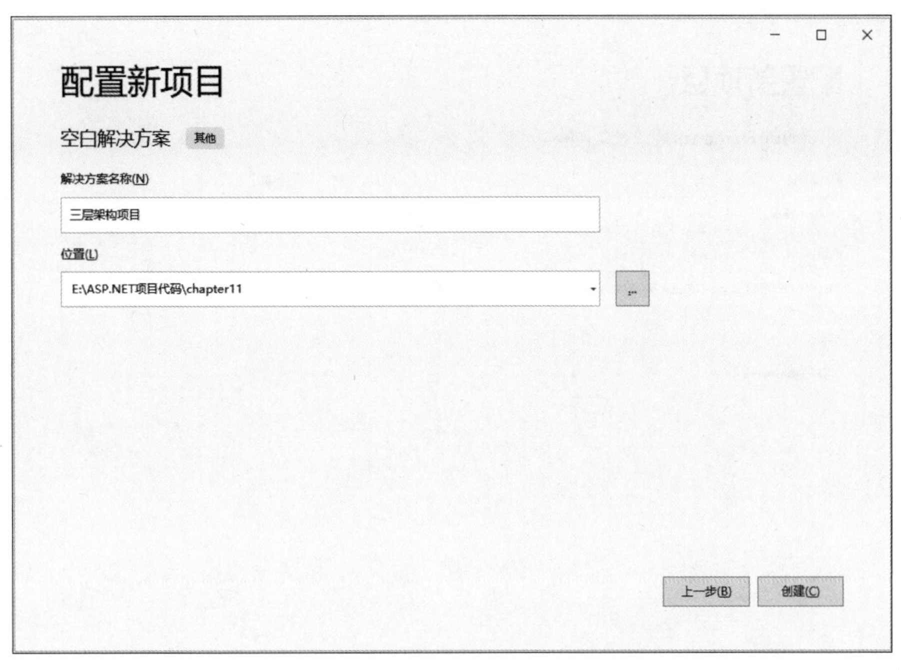

图 11.5　解决方案名称和保存位置的设置

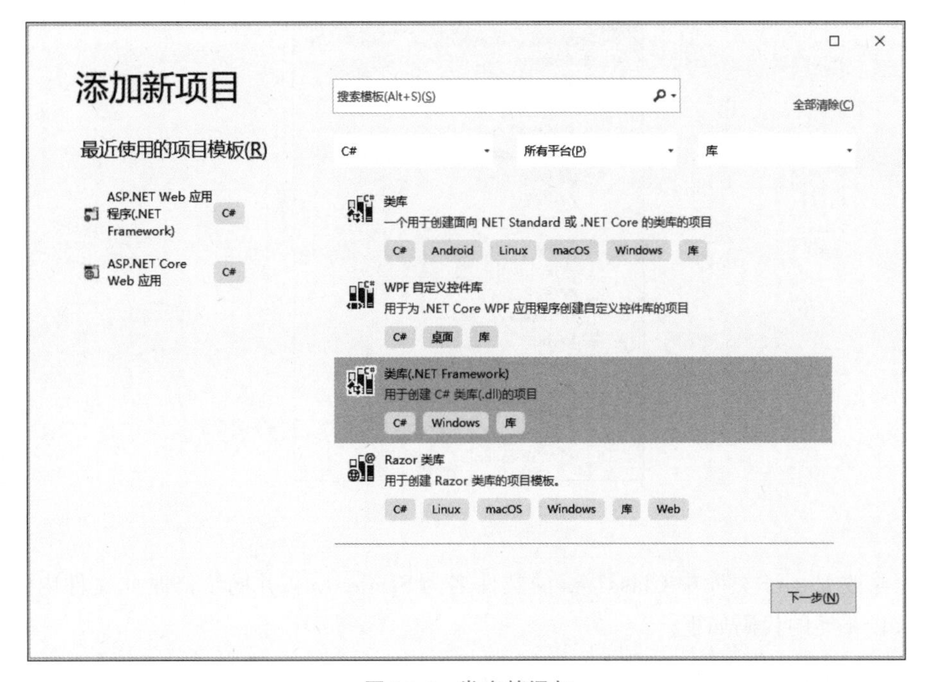

图 11.6　类库的添加

所示。

(3) 按步骤(2)的方法添加类库 Business 和 Entity，添加空白网站 WebUI，如图 11.8 所示。

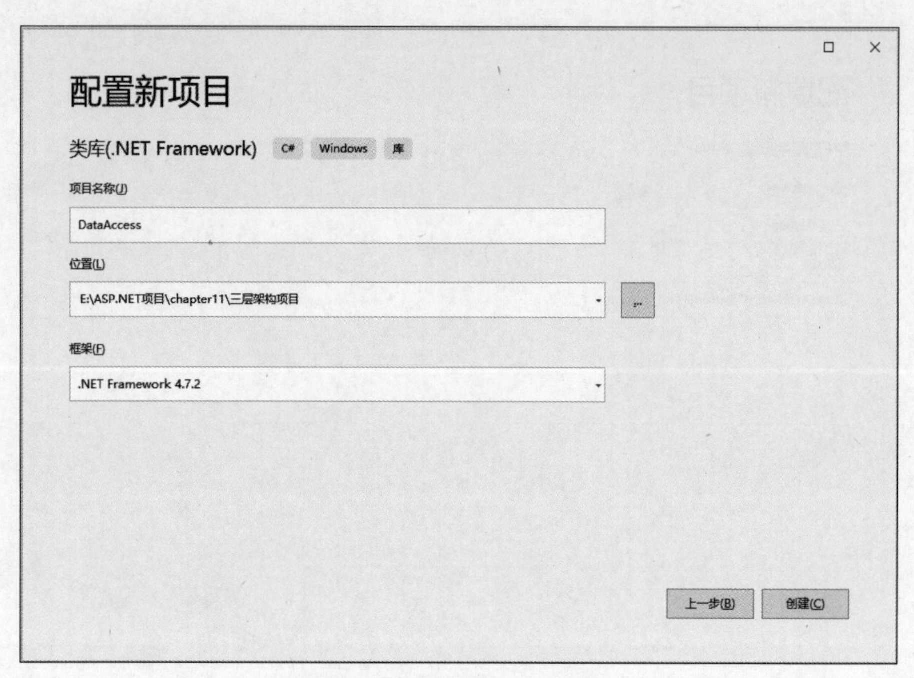

图 11.7　类库名称的设置

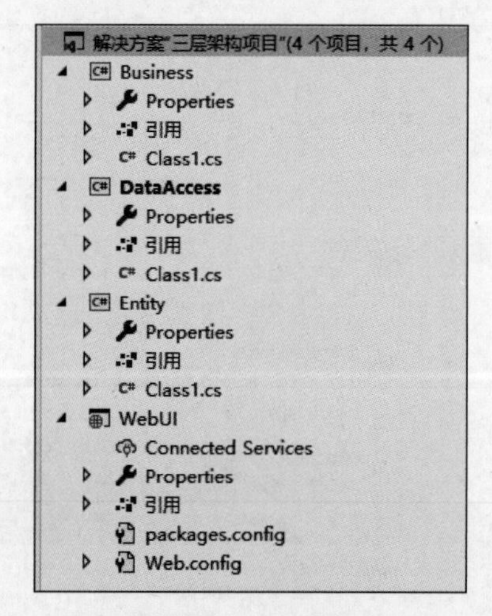

图 11.8　项目资源图

（4）修改 Entity 类库中 Class1.cs 的文件名为 Student.cs，并同步修改此文件内的类名，编写 Student 类的代码如下。

```
public class Student
{
    string sno;
    public string Sno
    {
        get { return sno; }
        set { sno=value; }
```

```
        }
        string sname;
        public string Sname
        {
            get { return sname; }
            set { sname=value; }
        }
        string sex;
        public string Sex
        {
            get { return sex; }
            set { sex=value; }
        }
        int age;
        public int Age
        {
            get { return age; }
            set { age=value; }
        }
        string dept;
        public string Dept
        {
            get { return dept; }
            set { dept=value; }
        }
    }
```

(5) 修改 DataAccess 类库中 Class1.cs 的文件名为 DataAccess.cs,并同步修改此文件内的类名,编写代码如下。

```
    using System.Data.SqlClient;
    //添加对 Entity 类库所属命名空间的引用
    using Entity;
    namespace DataAccess
    {
        static public class DemoDA
        {
            static SqlConnection con;
            static SqlCommand cmd;
            static  DemoDA()
            {
                con=new SqlConnection(@"server=.\sqlexpress;database=demo;trusted
    _connection=true;");
                cmd=new SqlCommand();
                cmd.Connection=con;
            }
            public static int InsertStudentInfo(Student stu)
            {
                cmd.CommandText="insert into student values (@Sno, @Sname, @Sex, @Age, @
    Dept)";
                cmd.Parameters.Clear();
                cmd.Parameters.AddWithValue("@Sno", stu.Sno);
```

```
            cmd.Parameters.AddWithValue("@Sname", stu.Sname);
            cmd.Parameters.AddWithValue("@Sex", stu.Sex);
            cmd.Parameters.AddWithValue("@Age", stu.Age);
            cmd.Parameters.AddWithValue("@Dept", stu.Dept);
            con.Open();
            int n=cmd.ExecuteNonQuery();
            con.Close();
            return n;
        }
    }
}
```

(6) 修改 Business 类库中 Class1.cs 的文件名为 Business.cs,并同步修改此文件内的类名,编写代码如下。

```
using Entity;
using DataAccess;
namespace Business
{
    static public class StudentBusiness
    {
        public static bool AddStudentInfo(Student stu)
        {
            int n=DemoDA.InsertStudentInfo(stu);
            if (n==1)
                return true;
            else
                return false;
        }
    }
}
```

(7) 在 WebUI 网站根目录下添加 Web 窗体"添加学生信息.aspx",并添加相关控件,如图 11.9 所示。

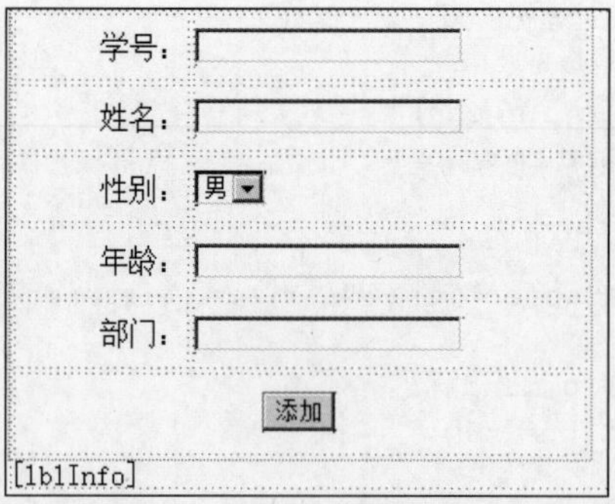

学号:
姓名:
性别: 男
年龄:
部门:
添加
[lblInfo]

图 11.9 WebUI 页面设计

(8) 为页面内的控件设置相关属性,源文件代码如下。

```
<form id="form1" runat="server">
    <div>
        <div>
            <table>
                <tr>
                    <td>学号:</td>
                    <td>
                        <asp:TextBox ID="txtNum" runat="server"></asp:TextBox>
                    </td>
                </tr>
                <tr>
                    <td>姓名:</td>
                    <td>
                        <asp:TextBox ID="txtName" runat="server"></asp:TextBox>
                    </td>
                </tr>
                <tr>
                    <td>性别:</td>
                    <td>
                        <asp:DropDownList ID="ddlSex" runat="server">
                            <asp:ListItem Selected="True">男</asp:ListItem>
                            <asp:ListItem>女</asp:ListItem>
                        </asp:DropDownList>
                    </td>
                </tr>
                <tr>
                    <td>年龄:</td>
                    <td>
                        <asp:TextBox ID="txtAge" runat="server"></asp:TextBox>
                    </td>
                </tr>
                <tr>
                    <td>部门:</td>
                    <td>
                        <asp:TextBox ID="txtDept" runat="server"></asp:TextBox>
                    </td>
                </tr>
                <tr>
                    <td colspan="2">
                        <asp:Button ID="btnInsert" runat="server" OnClick="
btnInsert_Click" Text="添加" />
 </td>
                </tr>
            </table>
        </div>
        <asp:Label ID="lblInfo" runat="server" Text=""></asp:Label>
    </div>
</form>
```

（9）为页面控件添加事件代码。“添加学生信息.cs”文件代码如下。

```
using Business;
using Entity;
```

```
namespace WebUI
{
    public partial class 添加学生信息 : System.Web.UI.Page
    {
        protected void Page_Load(object sender, EventArgs e)
        {
        }
        Student stu=new Student();
        protected void btnInsert_Click(object sender, EventArgs e)
        {
            stu.Sno=txtNum.Text;
            stu.Sname=txtName.Text;
            stu.Sex=ddlSex.SelectedValue;
            stu.Age=int.Parse(txtAge.Text);
            stu.Dept=txtDept.Text;
            if (StudentBusiness.AddStudentInfo(stu))
            {
                lblInfo.Text="添加学生信息成功!";
            }
            else
            {
                lblInfo.Text="添加学生信息失败!";
            }
        }
    }
}
```

（10）右击 DataAccess 类库，从弹出的快捷菜单中选择“添加引用”，在“引用管理器”窗口选择引用 Entity 类库，如图 11.10 所示。采取同样的方法为 Business 类库添加对 Entity 和 DataAccess 的引用，为 WebUI 网站添加对 Entity 和 DataAccess 的引用。

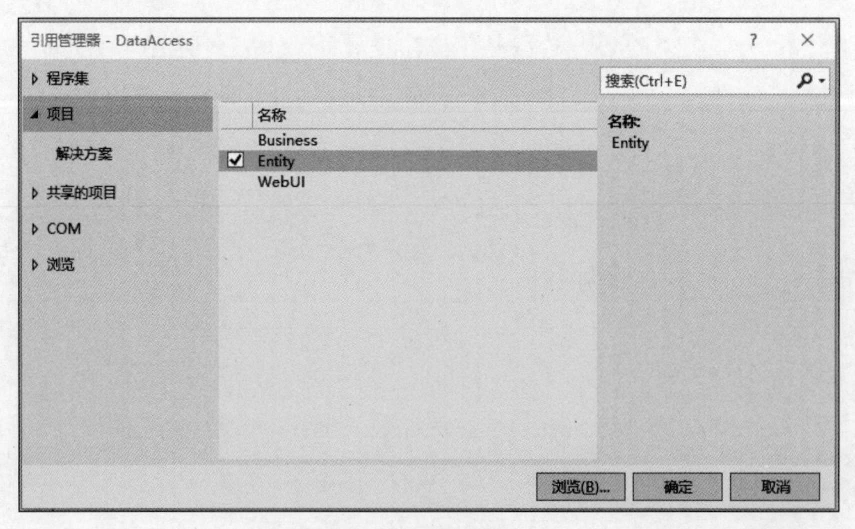

图 11.10　项目添加引用

（11）右击 Entity 类库，从弹出的快捷菜单中选择“生成”，为类库生成动态链接库文件(.dll)，采取同样的方法为 DataAccess 类库和 Business 类库分别进行“生成”操作。

(12) 运行网站,执行效果如图 11.11 所示。

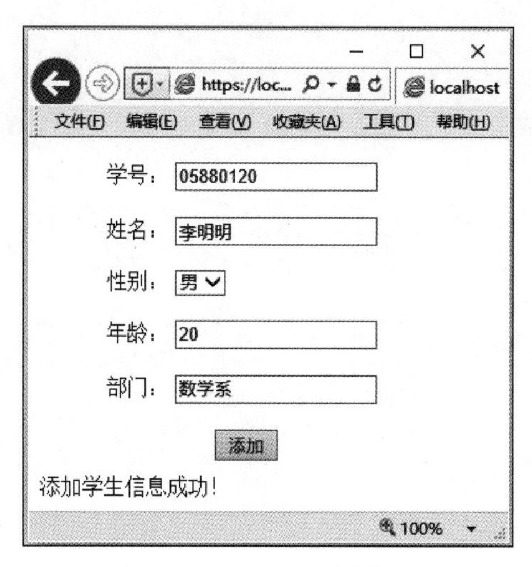

图 11.11 WebUI 页面演示

第12章 美妆网的设计与实现

本章学习目标

- 了解网站创建的业务流程
- 熟练掌握网站的创建方法
- 了解网站中数据库的创建方法
- 熟练掌握三层架构中类库的引用关系

本章首先对网站的业务逻辑进行分析，其次详细讲解网站中数据库的设计、网站层次的划分，并对三层架构类中代码进行设计，最后对网站进行页面设计和后台代码实现。

12.1 网站功能

美妆网是一个在线选购化妆品的网站，提供以图像和文字为主的界面，向用户展示化妆品并实现在线选购，由管理员、一般用户/会员、浏览者三种用户组成。

12.1.1 管理员

(1) 管理员拥有网站最大管理权限，可以配置系统信息。

(2) 管理员可以管理会员，对会员进行删除。

(3) 管理员可以更新产品信息，上传新产品，删除产品。

(4) 管理员可以查看和删除留言。

12.1.2 一般用户/会员

(1) 会员可以查看商品和账单结算信息。

(2) 会员可查看商品信息，包括编码、名称、类型、描述、商品状态和图片等。

(3) 会员可查询商品，也可实现关键字模糊查询。

（4）会员可查看购物车，显示所购商品的编码、名称、价格和数量，并显示总价格。

（5）会员可在购物车中删除商品，进行二次购买，更新产品及总产品的数量。

（6）会员可查看购物车，并可随时下订单。

（7）会员可查看订单。

（8）会员可进行留言或删除留言。

12.1.3 浏览者

（1）浏览者可查看化妆品。

（2）浏览者可查询化妆品信息。

12.2 网站业务流程

用户首先需要注册，通过用户名与密码验证登录主页面。用户业务流程如图 12.1 所示。管理员业务流程如图 12.2 所示。

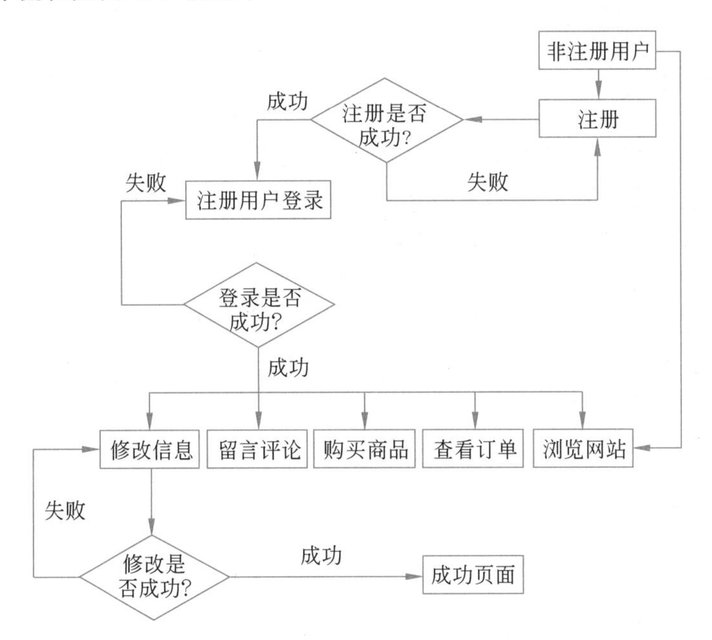

图 12.1 用户业务流程

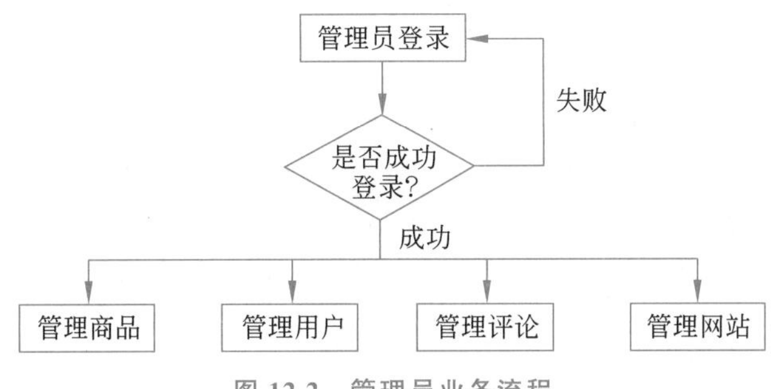

图 12.2 管理员业务流程

12.3 系统概要设计

将流程图中的各处理模块进一步分解，确定系统的层次结构关系，转换为单元功能模块。系统功能模块结构图如图 12.3 所示。

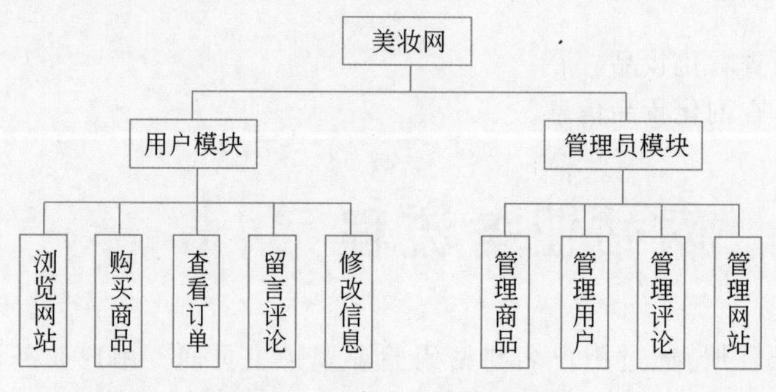

图 12.3　系统功能模块结构图

12.3.1　用户模块功能描述

(1) 浏览网站模块：包含热门商品浏览、新到商品浏览、商品分类浏览、按商品名称搜索、商品详细信息浏览等功能。

(2) 购买商品模块：包含添加购物车、修改购物车、结账等功能。

(3) 修改信息模块：包含注册新用户、登录、修改密码、个人信息管理等功能。

(4) 查看订单模块：包含订单查询等功能。

(5) 留言评论模块：包含留言、发表评论等功能。

用户模块用例图如图 12.4 所示。

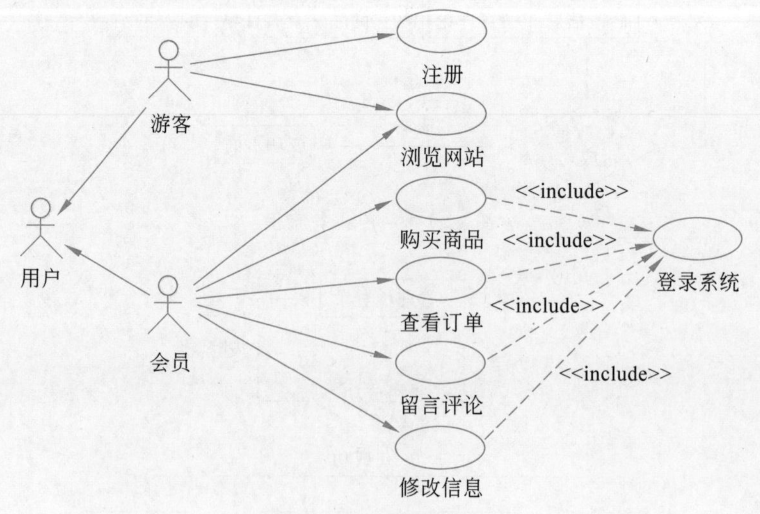

图 12.4　用户模块用例图

12.3.2 管理员模块功能描述

(1) 管理用户模块：登录、查询用户、删除用户等功能。

(2) 管理商品模块：添加、修改、删除商品信息等功能。

(3) 管理网站模块：网站信息的更新和维护等功能。

(4) 管理评论模块：回复、删除、评论等功能。

管理员模块用例图如图 12.5 所示。

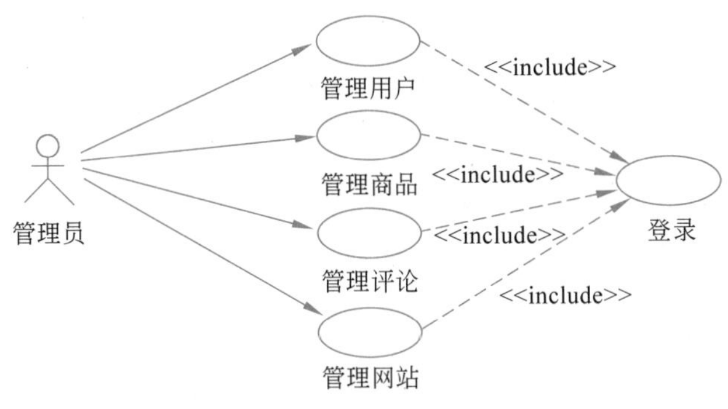

图 12.5 管理员模块用例图

12.4 数据库设计

12.4.1 概念设计

概念设计是将需求分析得到的用户需求抽象为信息结构的过程，可以将用户的数据要求清晰明确地表达出来，建立面向问题的数据模型。

管理员信息实体属性图如图 12.6 所示。

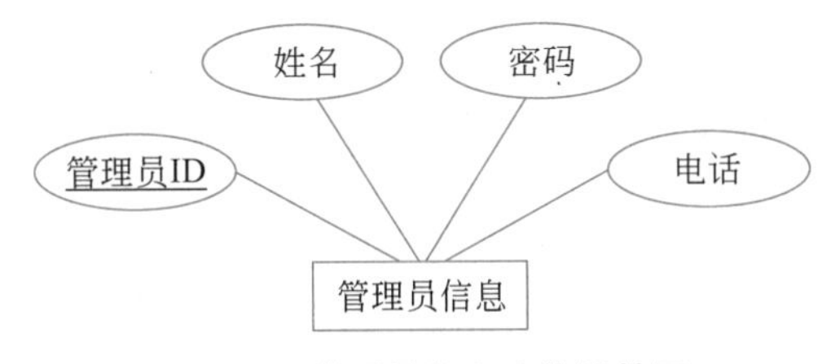

图 12.6 管理员信息实体属性图

用户信息实体属性图如图 12.7 所示。

商品信息实体属性图如图 12.8 所示。

订单信息实体属性图如图 12.9 所示。

购物车信息实体属性图如图 12.10 所示。

留言信息实体属性图如图 12.11 所示。

订单详情信息实体属性图如图 12.12 所示。

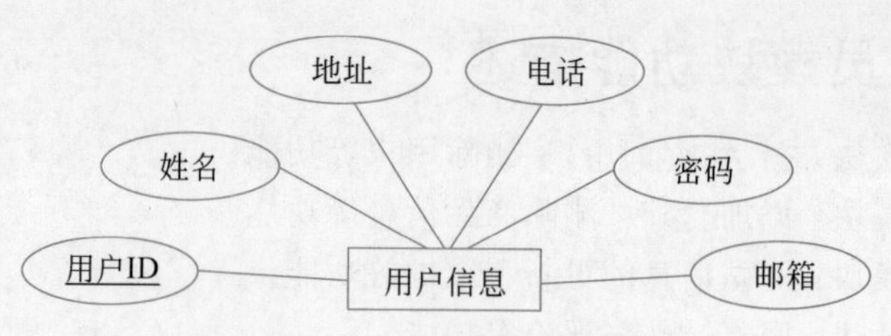

图 12.7　用户信息实体属性图

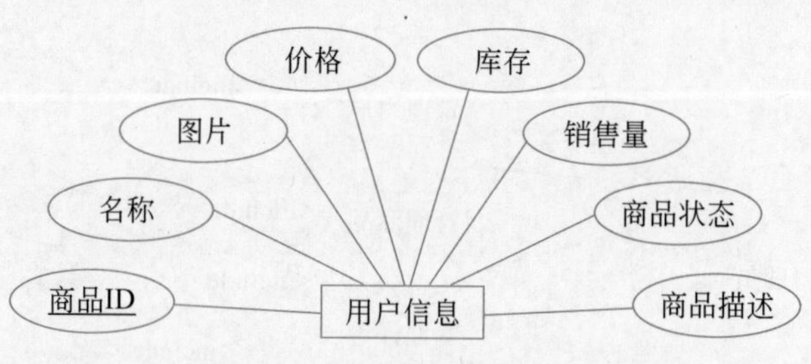

图 12.8　商品信息实体属性图

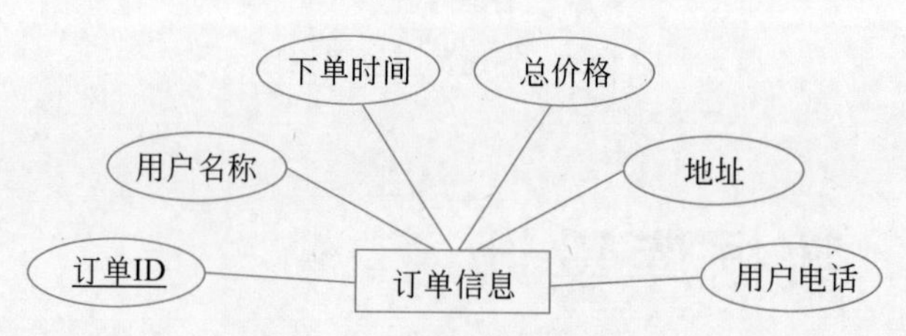

图 12.9　订单信息实体属性图

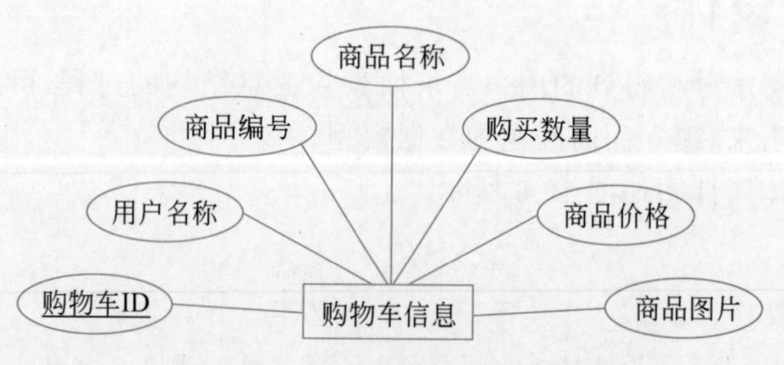

图 12.10　购物车信息实体属性图

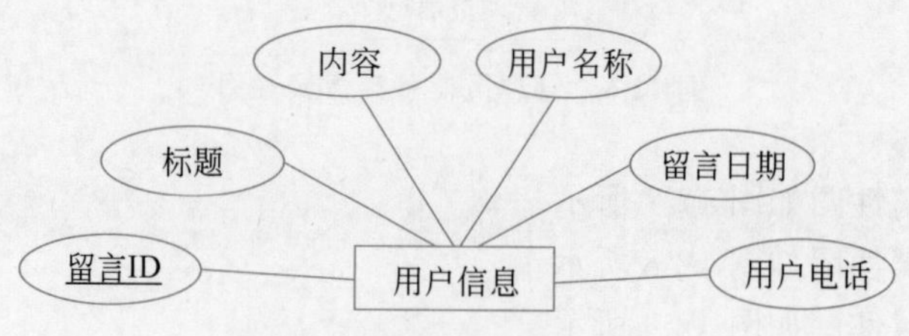

图 12.11　留言信息实体属性图

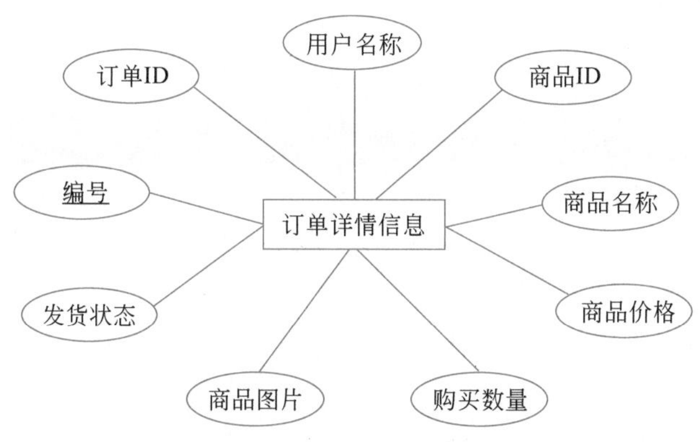

图 12.12 订单详情信息实体属性图

12.4.2 逻辑设计

对 12.4.1 节的属性图关系模型进行转化，将实体及实体间关系转换成为关系模型，实体转换出的关系模式如下。

用户信息表（用户 ID，密码，姓名，邮箱，电话，地址）

管理员信息表（管理员 ID，密码，姓名，电话）

商品信息表（商品 ID，名称，图片，价格，库存，销售量，商品状态，商品描述）

订单信息表（订单 ID，用户名称，下单时间，总价格，地址，用户电话）

购物车信息表（购物车 ID，用户名称，商品编号，商品名称，商品价格，商品图片，购买数量）

订单详情信息表（编号，订单 ID，用户名称，商品 ID，商品名称，商品价格，购买数量，商品图片，发货状态）

留言信息表（留言 ID，标题，内容，用户名称，留言日期，用户电话）

12.4.3 物理设计

从逻辑设计上转换实体以及实体之间的关系模式，形成数据库中表以及各表之间的关系。数据库物理设计用于确定存储结构、存取方法、存取路径、数据的存放位置等信息。

用户信息表如表 12-1 所示。

表 12-1 用户信息表

字段名	说明	类型	可否为空	主键	外键
uid	用户 ID	int	否	是	否
uname	姓名	varchar	否	否	否
password	密码	varchar	否	否	否
address	地址	varchar	否	否	否
tell	电话	varchar	否	否	否
email	邮箱	varchar	否	否	否

商品信息表如表 12-2 所示。

表 12-2　商品信息表

字段名	说明	类型	可否为空	主键	外键
pid	商品 ID	int	否	是	否
pname	名称	varchar	否	否	否
photp	图片	varchar	否	否	否
price	价格	decimal	否	否	否
pnums	库存	int	否	否	否
salenums	销售量	int	否	否	否
state	商品状态	text	否	否	否
mess	商品描述	text	否	否	否

购物车信息表如表 12-3 所示。

表 12-3　购物车信息表

字段名	说明	类型	可否为空	主键	外键
cid	购物车 ID	int	否	是	否
uname	用户名称	varchar	否	否	否
pid	商品编号	int	否	否	是
pname	商品名称	varchar	否	否	否
price	商品价格	decimal	否	否	否
nums	购买数量	int	否	否	否
photo	商品图片	varchar	否	否	否

管理员信息表如表 12-4 所示。

表 12-4　管理员信息表

字段名	说明	类型	可否为空	主键	外键
aid	管理员 ID	int	否	是	否
uname	姓名	varchar	否	否	否
password	密码	int	否	否	否
tel	电话	varchar	否	否	否

订单详情信息表如表 12-5 所示。

表 12-5　订单详情信息表

字段名	说明	类型	可否为空	主键	外键
id	编号	int	否	是	否
uname	用户名称	varchar	否	否	否

续表

字段名	说明	类型	可否为空	主键	外键
oid	订单 ID	int	否	否	是
pid	商品 ID	int	否	否	是
pname	商品名称	varchar	否	否	否
price	商品价格	decimal	否	否	否
nums	购买数量	int	否	否	否
photo	商品图片	varchar	否	否	否
state	发货状态	varchar	否	否	否

留言信息表如表 12-6 所示。

表 12-6　留言信息表

字段名	说明	类型	可否为空	主键	外键
mid	留言 ID	int	否	是	否
title	标题	varchar	否	否	否
mess	内容	text	否	否	否
uname	用户名称	varchar	否	否	否
messdate	留言日期	datetime	否	否	否
tel	用户电话	int	否	否	否

订单信息表如表 12-7 所示。

表 12-7　订单信息表

字段名	说明	类型	可否为空	主键	外键
oid	订单 ID	int	否	是	否
uname	用户名称	varchar	否	否	否
orderTime	下单时间	datetime	否	否	否
allPrice	总价格	decimal	否	否	否
address	地址	varchar	否	否	否
tel	用户电话	int	否	否	否

12.5 系统详细设计

系统的模块设计是在系统架构的基础上，通过精化架构、分析用例、设计模块标识设计元素，发现设计元素的行为细节，精化设计元素的定义，以确保用例实现的更新。系统的模

块设计根据业务内容的不同分为管理员系统、用户登录系统。

12.5.1 用户模块设计

用户模块主要包含已注册的用户登录、网站信息浏览、商品信息浏览、商品详情、购物车查看、留言评论、游客注册等模块类。用户模块类图如图 12.13 所示。

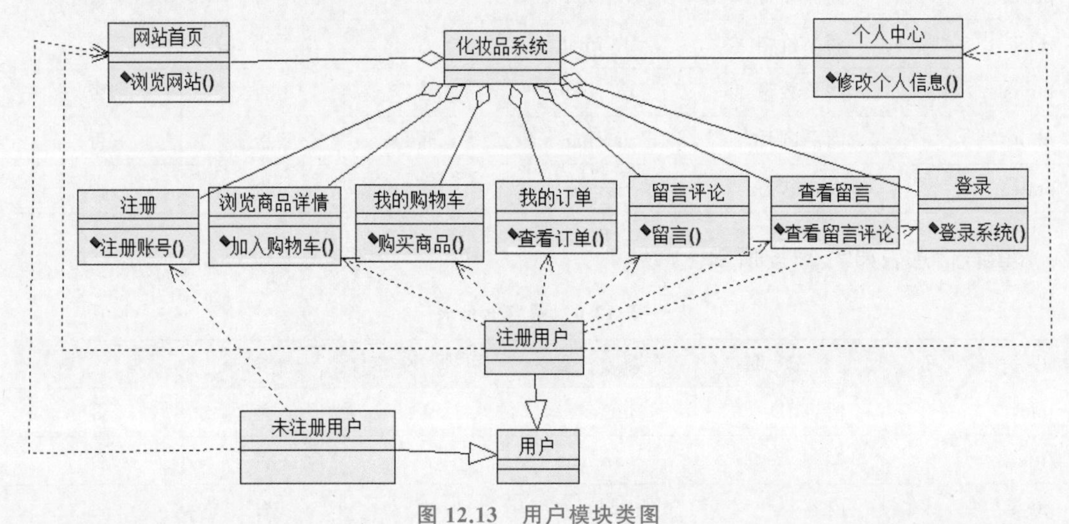

图 12.13　用户模块类图

1. 会员登录

会员登录涉及 UserLogin、UserBusiness、DA、UserEntity 等类库，具体如图 12.14 所示。

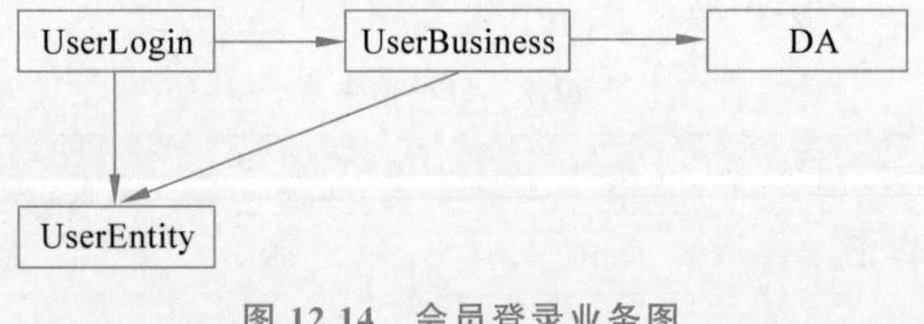

图 12.14　会员登录业务图

(1) UserLogin 类：UserLogin 调用 UserBusiness 类中的 UserAndPWD()方法，实现用户登录相应功能。

(2) UserBusiness 类：UserBusiness 类中的 UserAndPWD()方法负责验证用户名、密码是否正确的业务逻辑，调用 DA 类中的 GetOneData()方法。

(3) DA 类：DA 类中的 GetOneData()方法负责在 SQL Server 数据库中取出 UserBusiness 类要验证的数据。

(4) UserEntity 类：UserEntity 类实现对数据库中的 tb_user 表的面向对象化处理，实现数据的封装。

2. 浏览商品信息

浏览商品信息涉及 UserProduct、ProductBusiness、DA、ProductEntity 等类库，具体如

图 12.15 所示。

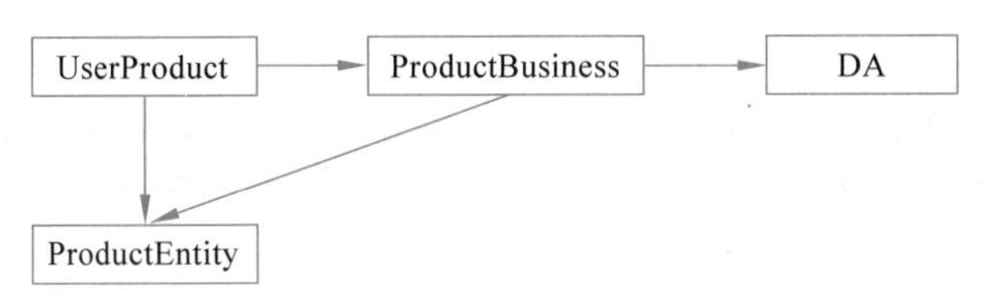

图 12.15 浏览商品信息业务图

(1) UserProduct 类：用 Datalist 绑定商品信息，调用 ProductBusiness 类中的 SelectProductByPname()方法查询商品信息，单击商品图片可进入商品详情页面。

(2) ProductBusiness 类：ProductBusiness 类中的 SelectProductByPname()方法调用 DA 类中的 GetDataSet()方法，实现按商品名称查找商品。

(3) DA 类：DA 类中的 GetDataSet()方法负责在 SQL Server 数据库中取出 UserProduct()方法要显示的商品信息数据。

(4) ProductEntity 类：ProductEntity 类负责将数据库中的 tb_product 表进行面向对象化处理，实现数据封装。

3. 商品详情

商品详情信息涉及 UserProductDetail、CartBusiness、OrdersBusiness、DA、ProductEntity、OrdersEntity 等类库，具体如图 12.16 所示。

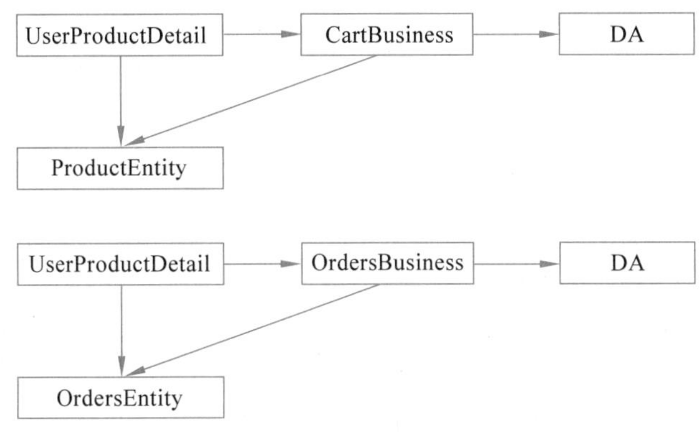

图 12.16 商品详情业务图

(1) UserProductDetail 类：UserProductDetail 类调用 CartBusiness 类中的 InsertCart()方法，将商品添加到购物车。

(2) CartBusiness 类：CartBusiness 类中的 InsertCart()方法调用 DA 类中的 ExcuteSql()方法，将商品添加到购物车。

(3) OrdersBusiness 类：OrdersBusinessl 类调用 DA 类中的 ExcuteSql()方法实现订单生成。

(4) DA 类：DA 类中的 ExcuteSql()方法负责对 SQL Server 数据库中的数据进行增加、删除、修改操作，使方法 InsertCart()、InsertOrders()的功能得以实现。

(5) ProductEntity 类和 OrdersEntity 类：ProductEntity 及 OrdersEntity 类负责将数

据库中的 tb_product、tb_orders 表进行面向对象化的处理，实现数据封装。

4. 购物车

购物车类涉及 UserCart、CartBusiness、OrdersBusiness、OrderDetailBusiness、DA、CartEntity、OrdersEntity、OrderDetailsEntity 等类库，具体如图 12.17 所示。

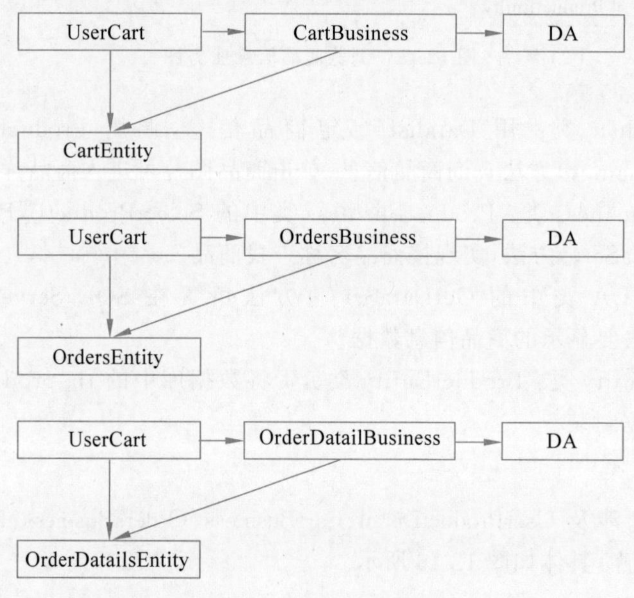

图 12.17　购物车业务图

5. 留言评论

留言评论模块涉及 UserAddMessage、MessageBusiness、DA、MessageEntity 等类库，具体如图 12.18 所示。

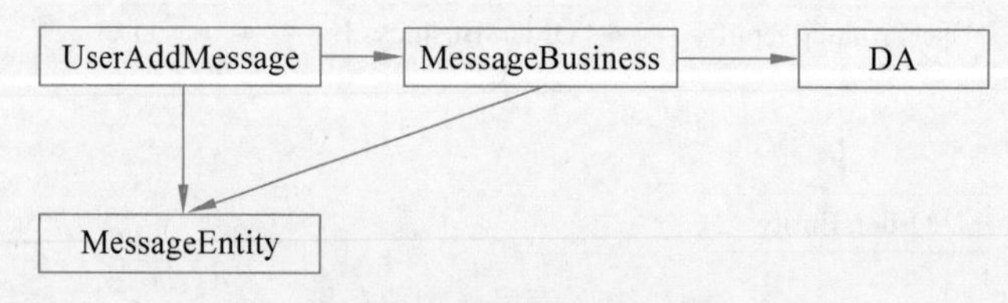

图 2.18　留言评论业务图

(1) UserAddMessage 类：UserMessage 类调用 MessageBusiness 类中的 InsertMessage()方法新增留言。

(2) MessageBusiness 类：MessageBusiness 类调用 DA 类中的 ExcuteSql()方法新增留言。

(3) DA 类：DA 类中的 ExcuteSql()方法负责在 SQL Server 数据库中对数据进行增加、删除、修改操作，使方法 InsertMessage()得以实现。

6. 游客

游客模块涉及 UserRegister、UserBusiness、DA、UserEntity 等类库，具体如图 12.19 所示。

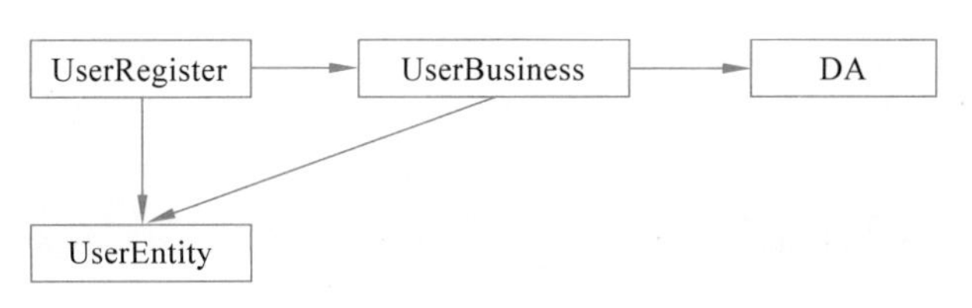

图 12.19 用户注册业务图

(1) UserRegister 类：UserRegister 类调用 UserBusiness 类中的 InsertUser()方法，实现新用户注册。

(2) UserBusiness 类：UserBusiness 类调用 DA 类中的 ExcuteSql()方法实现新用户的注册。

(3) DA 类：DA 类中的 ExcuteSql()方法负责在 SQL Server 数据库中对数据进行增加、删除、修改操作，使 InsertUser()得以实现。

(4) UserEntity 类：UserEntity 类实现对数据库中的 tb_user 表的面向对象化处理，实现数据的封装。

12.5.2 管理员模块设计

管理员模块主要进行管理员登录、用户信息的管理、商品管理、留言管理等功能。管理员功能模块类图如图 12.20 所示。

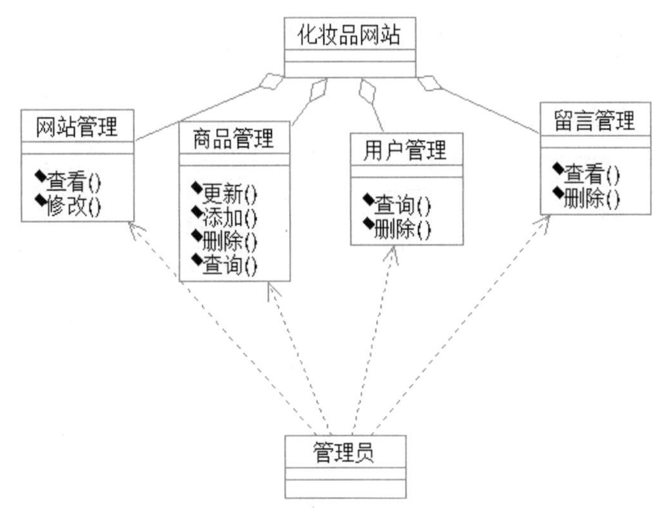

图 12.20 管理员功能模块类图

管理员模块挑选代表性功能进行详细叙述如下。

1. 登录功能

登录模块涉及 AdminLogin、AdminBusiness、DA、AdminEntity 等类库，具体如图 12.21 所示。

(1) AdminLogin 类：AdminLogin 类调用 AdminBusiness 类中的 AdminAndPWD()方法验证管理员登录信息。

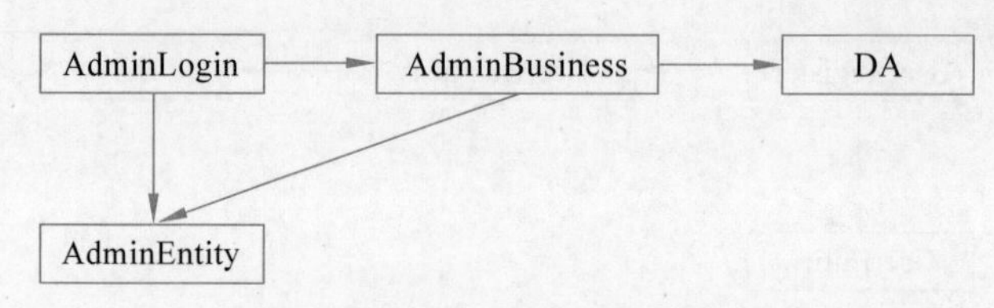

图 12.21 管理员登录业务图

(2) AdminBusiness 类：AdminBusiness 类中的 AdminAndPWD()方法调用 DA 类中的 GetOneData()方法验证管理员登录信息。

(3) DA 类：DA 类中的 GetOneData()方法对 SQL Server 数据库中的数据进行查询操作，使 AdminAndPWD()方法得以实现。

(4) AdminEntity 类：AdminEntity 类实现对数据库中的 tb_admin 表进行面向对象化处理，实现数据的封装。

2. 更新商品信息功能

更新商品信息模块涉及 AdminUpdateProduct、ProductBusiness、DA、ProductEntity 等类库，具体如图 12.22 所示。

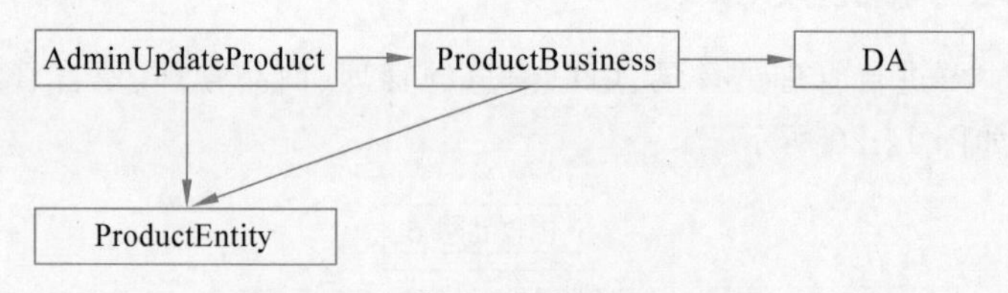

图 12.22 管理员更新商品业务图

(1) AdminUpdateProduct 类：AdminUpdateProduct 类调用 ProductBusiness 类中的 InsertProduct()方法进行商品更新。

(2) ProductBusiness 类：ProductBusiness 类调用 DA 类中的 ExcuteSql()方法进行商品添加。

(3) DA 类：DA 类中的 ExcuteSql()方法对 SQL Server 数据库中的数据进行更新操作，使 InsertProduct()方法得以实现。

(4) ProductEntity 类：ProductEntity 类实现对数据库中的 tb_product 表进行面向对象化处理，实现数据的封装。

3. 删除商品信息功能

删除商品信息模块涉及 AdminDeleteProduct、ProductBusiness、DA、ProductEntity 等类库，具体如图 12.23 所示。

(1) AdminDeleteProduct 类：AdminDeleteProduct 类调用 ProductBusiness 类中的 DeleteProductByPid()方法进行商品删除。

(2) ProductBusiness 类：ProductBusiness 类调用 DA 类中的 ExcuteSql()方法进行商品删除。

(3) DA 类：DA 类中的 ExcuteSql()方法对 SQL Server 数据库中的数据进行删除操作，使 DeleteProductByPid()方法得以实现。

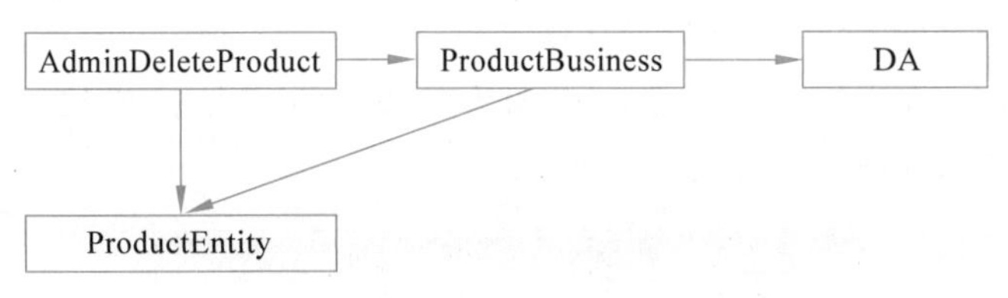

图 12.23　管理员删除商品业务图

(4) ProductEntity 类:ProductEntity 类实现对数据库中的 tb_product 表进行面向对象化处理,实现数据的封装。

4. 新增商品信息功能

新增商品信息模块涉及 AdminAddProduct、ProductBusiness、DA、ProductEntity 等类库,具体如图 12.24 所示。

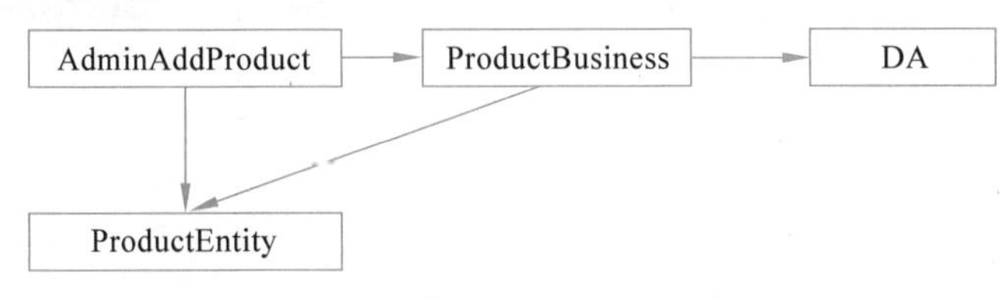

图 12.24　管理员添加商品业务图

(1) AdminAddProduct 类: AdminAddProduct 类调用 ProductBusiness 类中的 UpdateProductByPid()方法进行商品添加。

(2) ProductBusiness 类:ProductBusiness 类调用 DA 类中的 ExcuteSql()方法进行商品添加。

(3) DA 类: DA 类中的 ExcuteSql()方法对 SQL Server 数据库中的数据进行更新操作,使 UpdateProductByPid()方法得以实现。

(4) ProductEntity 类: ProductEntity 类实现对数据库中的 tb_product 表进行面向对象化处理,实现数据的封装。

12.6　网站建立

在 E 盘 ASP.NET 项目代码目录中创建 chapter12 子目录,将其作为解决方案根目录,在 VS 2019 菜单中创建"Visual Stidio 解决方案"并命名为"美妆网",按照三层架构框架添加类库 Entity(实体层)、DataAccess(数据访问层)、Business(业务逻辑层)和 Web UI 网站(表现层)。

添加类库及网站后,解决方案布局如图 12.25 所示。

为各类库添加引用关系如下。

(1) DataAccess 引用 Entity。

(2) Business 引用 Entity 和 DataAccess。

(3) Web UI 网站引用 Entity 和 Business。

按 12.5 节中的详细设计在类库 Entity、DataAccess、Business 中添加类文件，在 Web UI 网站中添加 Web 页面及文件，完成网站基本架构及文件的搭建。类库文件结构如图 12.26 所示。网站页面结构如图 12.27 所示。

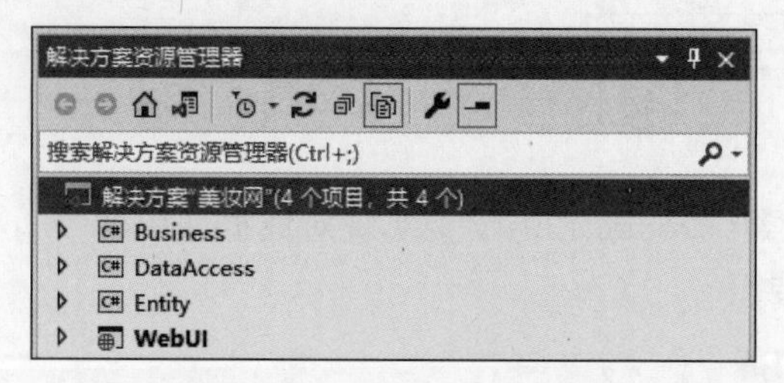

图 12.25 "美妆网"解决方案

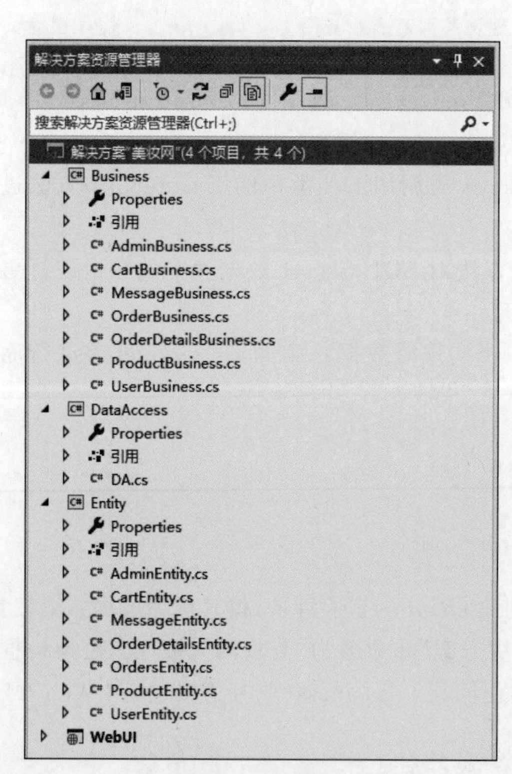

图 12.26 类库文件结构

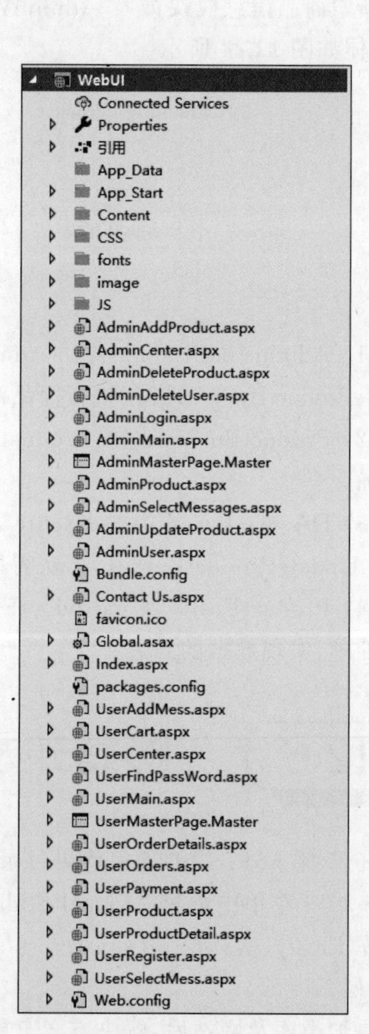

图 12.27 网站页面结构

12.7 类库代码实现

12.7.1 实体层 Entity 设计

实体层(Entity Layer)包含7个类,分别与数据库中的数据表对应,代码设计如下。

(1) 用户表对应 UserEntity 类,编辑 UserEntity.cs 代码如下。

```
public class UserEntity
{
    private int uid;
    public int Uid
    {
        get { return uid; }
        set { uid=value; }
    }
    private string name;
    public string Name
    {
        get { return name; }
        set { name=value; }
    }
    private string password;
    public string Password
    {
        get { return password; }
        set { password=value; }
    }
    private string address;
    public string Address
    {
        get { return address; }
        set { address=value; }
    }
    private string tel;
    public string Tel
    {
        get { return tel; }
        set { tel=value; }
    }
    private string email;
    public string Email
    {
        get { return email; }
        set { email=value; }
    }
}
```

(2) 商品表对应 ProductEntity 类,编辑 ProductEntity.cs 代码如下。

```
public class ProductEntity
{
    private int pid;
    public int Pid
    {
        get { return pid; }
        set { pid=value; }
    }
    private string pname;
    public string Pname
    {
        get { return pname; }
        set { pname=value; }
    }
    private string photo;
    public string Photo
    {
        get { return photo; }
        set { photo=value; }
    }
    private decimal price;
    public decimal Price
    {
        get { return price; }
        set { price=value; }
    }
    private int pnums;
    public int Pnums
    {
        get { return pnums; }
        set { pnums=value; }
    }
    private int salenums;
    public int Salenums
    {
        get { return salenums; }
        set { salenums=value; }
    }
    private string mess;
    public string Mess
    {
        get { return mess; }
        set { mess=value; }
    }
    private string state;
    public string State
    {
        get { return state; }
        set { state=value; }
    }
}
```

(3) 购物车表对应 CartEntity 类，编辑 CartEntity.cs 代码如下。

```
public class CartEntity
{
    private int cid;
    public int Cid
    {
        get { return cid; }
        set { cid=value; }
    }
    private string uname;
    public string Uname
    {
        get { return uname; }
        set { uname=value; }
    }
    private int pid;
    public int Pid
    {
        get { return pid; }
        set { pid=value; }
    }
    private string pname;
    public string Pname
    {
        get { return pname; }
        set { pname=value; }
    }
    private decimal price;
    public decimal Price
    {
        get { return price; }
        set { price=value; }
    }
    private int nums;
    public int Nums
    {
        get { return nums; }
        set { nums=value; }
    }
    private string photo;
    public string Photo
    {
        get { return photo; }
        set { photo=value; }
    }
}
```

（4）管理员表对应 AdminEntity 类，编辑 AdminEntity.cs 代码如下。

```
public class AdminEntity
{
    private int aid;
    public int Aid
    {
```

```
        get { return aid; }
        set { aid=value; }
    }
    private string aname;
    public string Aname
    {
        get { return aname; }
        set { aname=value; }
    }
    private string password;
    public string Password
    {
        get { return password; }
        set { password=value; }
    }
    private string tel;
    public string Tel
    {
        get { return tel; }
        set { tel=value; }
    }
}
```

（5）订单表对应 OrdersEntity 类，编辑 OrdersEntity.cs 代码如下。

```
public class OrdersEntity
{
    private int oid;
    public int Oid
    {
        get { return oid; }
        set { oid=value; }
    }
    private string uname;
    public string Uname
    {
        get { return uname; }
        set { uname=value; }
    }
    private DateTime orderTime;
    public DateTime OrderTime
    {
        get { return orderTime; }
        set { orderTime=value; }
    }
    private decimal allPrice;
    public decimal AllPrice
    {
        get { return allPrice; }
        set { allPrice=value; }
    }
    private string address;
    public string Address
```

```
    {
        get { return address; }
        set { address=value; }
    }
    private string tel;
    public string Tel
    {
        get { return tel; }
        set { tel=value; }
    }
}
```

（6）留言板表对应 MessageEntity 类，编辑 MessageEntity.cs 代码如下。

```
public class MessageEntity
{
    private int mid;
    public int Mid
    {
        get { return mid; }
        set { mid=value; }
    }
    private string title;
    public string Title
    {
        get { return title; }
        set { title=value; }
    }
    private string mess;
    public string Mess
    {
        get { return mess; }
        set { mess=value; }
    }
    private string uname;
    public string Uname
    {
        get { return uname; }
        set { uname=value; }
    }
    private DateTime messDate;
    public DateTime MessDate
    {
        get { return messDate; }
        set { messDate=value; }
    }
}
```

（7）订单信息表对应 OrderDetailsEntity 类，编辑 OrderDetailsEntity.cs 代码如下。

```
public class OrderDetailsEntity
{
    private int id;
    public int Id
```

```
        {
            get { return id; }
            set { id=value; }
        }
        private int oid;
        public int Oid
        {
            get { return oid; }
            set { oid=value; }
        }
        private string uname;
        public string Uname
        {
            get { return uname; }
            set { uname=value; }
        }
        private int pid;
        public int Pid
        {
            get { return pid; }
            set { pid=value; }
        }
        private string pname;
        public string Pname
        {
            get { return pname; }
            set { pname=value; }
        }
        private decimal price;
        public decimal Price
        {
            get { return price; }
            set { price=value; }
        }
        private int nums;
        public int Nums
        {
            get { return nums; }
            set { nums=value; }
        }
        private string photo;
        public string Photo
        {
            get { return photo; }
            set { photo=value; }
        }
        private string states;
        public string States
        {
            get { return states; }
            set { states=value; }
        }
    }
```

12.7.2 数据访问层 DataAccess 设计

数据访问层包含 1 个 DA 类，编写 DA.cs 代码如下。

```
using System.Data;
using System.Data.SqlClient;
namespace DataAccess
{
    public static class DA
    {
        static SqlConnection conn=null;
        static SqlCommand cmd=null;
        static SqlDataAdapter sda=null;
        static DataSet ds=null;
        static DA()
        {
            conn=new SqlConnection(@"server=.\sqlexpress;database=Cosmetics;
trusted_connection=true;");
            cmd=new SqlCommand();
            cmd.Connection=conn;
        }
        /// <summary>
        /// 对数据库进行增加、删除、修改操作
        /// </summary>
        /// <param name="sqlText">sql 命令文本</param>
        /// <param name="commandType">命令类型</param>
        /// <param name="paraNames">参数名数组</param>
        /// <param name="paraValues">参数值数组</param>
        /// <returns></returns>
        public static int ExecuteSql(string sqlText, CommandType commandType,
string[] paraNames, object[] paraValues)
        {
            int count;
            if (conn.State !=ConnectionState.Open)
                conn.Open();
            cmd.CommandType=commandType;
            cmd.CommandText=sqlText;
            if (paraNames !=null)
            {
                for (int i=0; i<paraNames.Length; i++)
                    cmd.Parameters.AddWithValue(paraNames[i], paraValues[i]);
            }
            count=cmd.ExecuteNonQuery();
            cmd.Parameters.Clear();
            conn.Close();
            return count;
        }
        //返回一个值的查询,聚组函数
        public static object GetOneData(string sqlText, CommandType commandType,
string[] paraNames, object[] paraValues)
        {
            object result;
```

```
            if (conn.State ! =ConnectionState.Open)
                conn.Open();
            cmd.CommandType=commandType;
            cmd.CommandText=sqlText;
            if (paraNames ! =null)
            {
                for (int i=0; i<paraNames.Length; i++)
                    cmd.Parameters.AddWithValue(paraNames[i], paraValues[i]);
            }
            result=cmd.ExecuteScalar();
            cmd.Parameters.Clear();
            conn.Close();
            return result;
        }
        //查询一个数据集并返回该数据集(离线模式)
         public static DataSet GetDataSet (string sqlText, string tableName,
CommandType commandType, string[] paraNames, object[] paraValues)
        {
            cmd.CommandType=commandType;
            cmd.CommandText=sqlText;
            if (paraNames ! =null)
            {
                for (int i=0; i<paraNames.Length; i++)
                    cmd.Parameters.AddWithValue(paraNames[i], paraValues[i]);
            }
            sda=new SqlDataAdapter(cmd);
            ds=new DataSet();
            sda.Fill(ds, tableName);
            cmd.Parameters.Clear();
            return ds;
        }
    }
}
```

12.7.3 业务逻辑层 Business 设计

业务逻辑层关注网站业务规则的制定、业务流程的实现等与业务需求有关的系统设计，包含 7 个类，分别对应数据库中每一张数据表的操作，代码设计如下。

(1) 操作用户表对应业务逻辑 UserBusiness 类，编辑 UserBusiness.cs 代码如下。

```
using DataAccess;
using Entity;
using System.Data;
namespace Business
{
    public class UserBusiness
    {
        //添加新用户
        public int InsertUser(UserEntity ue)
        {
            string sqlText="insert into tb_user values(@name,@password,@address,@
tel,@email)";
```

```
            string[] paras={"@name", "@password", "@address", "@tel", "@email" };
            object[] values={ue.Name, ue.Password, ue.Address, ue.Tel, ue.Email };
            int i=DA.ExecuteSql(sqlText, CommandType.Text, paras, values);
            return i;
        }
        //判断用户名和密码是否匹配
        public int UserAndPWD(UserEntity ue)
        {
            string sqlText="select count(*) from tb_user where name=@name and 
password=@password";
            string[] paras={ "@name", "@password" };
            object[] values={ ue.Name, ue.Password };
             int i=Convert.ToInt32(DA.GetOneData(sqlText, CommandType.Text, 
paras, values));
            return i;
        }
        //判断用户名和手机号码是否匹配
        public int UserAndTel(UserEntity ue)
        {
            string sqlText="select count(*) from tb_user where name=@name and 
tel=@tel";
            string[] paras={ "@name", "@tel" };
            object[] values={ ue.Name, ue.Tel };
             int i=Convert.ToInt32(DA.GetOneData(sqlText, CommandType.Text, 
paras, values));
            return i;
        }
        //根据用户昵称,查询用户详细信息
        public DataSet SelectUserByUname(UserEntity ue, string tableName)
        {
            string sqlText="select * from tb_user where Name=@Name";
            string[] paras={ "@Name" };
            object[] values={ ue.Name};
            DataSet ds=new DataSet();
              ds=DA.GetDataSet(sqlText, tableName, CommandType.Text, paras, 
values);
            return ds;
        }
        public DataSet SelectUser(string tableName)
        {
            string sqlText="select * from tb_user";
            DataSet ds=new DataSet();
            ds=DA.GetDataSet(sqlText, tableName, CommandType.Text, null, null);
            return ds;
        }
        //按照用户编号删除信息
        public int DeleteUserByUid(UserEntity ue)
        {
            string sqlText="delete from tb_user where uid=@uid";
            string[] paras={ "@uid" };
            object[] values={ ue.Uid };
            int i=DA.ExecuteSql(sqlText, CommandType.Text, paras, values);
            return i;
```

```
        }
        public int DeleteUserByPart(string sql)
        {
            string sqlText="delete from tb_user where 1>1"+sql;
            int i=DA.ExecuteSql(sqlText, CommandType.Text, null, null);
            return i;
        }
        //根据用户名称查询信息
        public DataSet SelectUserByName(UserEntity ue,string tableName)
        {
            string sqlText="select * from tb_user where name like @name";
            string[] paras={ "@name" };
            object[] values={ "%"+ue.Name+"%" };
            DataSet ds=new DataSet();
            ds=DA.GetDataSet(sqlText, tableName, CommandType.Text, paras, values);
            return ds;
        }
        //根据用户编号更新用户信息
        public int UpdateUserByUid(UserEntity ue)
        {
             string sqlText ="update tb_user set name= @ name, password= @ password,
 address=@address,tel=@tel,email=@email where uid=@uid";
            string[] paras={ "@name", "@password", "@address", "@tel", "@email","@
uid"};
            object[] values={ue.Name,ue.Password,ue.Address,ue.Tel,ue.Email,
 ue.Uid};
            int i=DA.ExecuteSql(sqlText, CommandType.Text, paras, values);
            return i;
        }
        //根据编号查询用户
        public DataSet SelectUserByUid(UserEntity ue, string tableName)
        {
            string sqlText="select * from tb_user where uid=@uid";
            string[] paras={ "@uid" };
            object[] values={ ue.Uid };
            DataSet ds=new DataSet();
            ds=DA.GetDataSet(sqlText, tableName, CommandType.Text, paras, values);
            return ds;
        }
    }
}
```

(2) 操作商品表对应业务逻辑 ProductBusiness 类，编辑 ProductBusiness.cs 代码如下。

```
using DataAccess;
using Entity;
using System.Data;
namespace Business
{
    public class ProductBusiness
    {
        public int InsertProduct(ProductEntity pe)
        {
```

```
            string sqlText="insert into tb_product values(@pname,@photo,@price,@
pnums,@salenums,@mess,@state)";
            string[] paras={ "@pname", "@photo", "@price", "@pnums","@salenums","
@mess","@state" };
            object[] values={ pe.Pname, pe.Photo, pe.Price, pe.Pnums,pe.Salenums,
pe.Mess,pe.State };
            int i=DA.ExecuteSql(sqlText, CommandType.Text, paras, values);
            return i;
        }
        public DataSet SelectProduct(string tableName)
        {
            string sqlText="select * from tb_product";
            DataSet ds=new DataSet();
            ds=DA.GetDataSet(sqlText, tableName, CommandType.Text, null, null);
            return ds;
        }
        public int DeleteProductByPid(ProductEntity pe)
        {
            string sqlText="delete from tb_product where pid=@pid";
            string[] paras={ "@pid" };
            object[] values={ pe.Pid };
            int i=DA.ExecuteSql(sqlText, CommandType.Text, paras, values);
            return i;
        }
        public int DeleteProductByPname(ProductEntity pe)
        {
            string sqlText="delete from tb_product where pname=@pname";
            string[] paras={ "@pname" };
            object[] values={ pe.Pname };
            int i=DA.ExecuteSql(sqlText, CommandType.Text, paras, values);
            return i;
        }
        public int DeleteProductByPart(string sql)
        {
            string sqlText="delete from tb_product where 1>1"+sql;
            int i=DA.ExecuteSql(sqlText, CommandType.Text, null, null);
            return i;
        }
        public DataSet SelectProductByPid(ProductEntity pe, string tableName)
        {
            string sqlText="select * from tb_product where pid=@pid";
            string[] paras={ "@pid" };
            object[] values={ pe.Pid };
            DataSet ds=new DataSet();
            ds=DA.GetDataSet(sqlText, tableName, CommandType.Text, paras, values);
            return ds;
        }
        public int UpdateProductByPid(ProductEntity pe)
        {
            string sqlText="update tb_product set pname=@pname,photo=@photo,
price=@price,pnums=@pnums,salenums=@salenums,mess=@mess,state=@state where
pid=@pid";
```

```
            string[] paras={ "@pname", "@photo", "@price", "@pnums", "@salenums", "
@mess","@state","@pid" };
            object[] values={ pe.Pname, pe.Photo, pe.Price, pe.Pnums, pe.Salenums,
pe.Mess,pe.State,pe.Pid };
            int i=DA.ExecuteSql(sqlText, CommandType.Text, paras, values);
            return i;
        }
        public DataSet SelectProductByPname(ProductEntity pe, string tableName)
        {
            string sqlText="select * from tb_product where pname like @pname";
            string[] paras={ "@pname" };
            object[] values={ "%"+pe.Pname+"%" };
            DataSet ds=new DataSet();
            ds=DA.GetDataSet(sqlText, tableName, CommandType.Text, paras, values);
            return ds;
        }
    }
}
```

(3) 操作购物车表对应业务逻辑 CartBusiness 类，编辑 CartBusiness.cs 代码如下。

```
using DataAccess;
using Entity;
using System.Data;
namespace Business
{
    public class CartBusiness
    {
        //将商品信息添加到购物车
        public int InsertCart(CartEntity ce)
        {
            string sqlText="insert into tb_cart values(@uname,@pid,@pname,@price,
@nums,@photo)";
            string[] paras={ "@uname", "@pid", "@pname", "@price", "@nums", "@photo" };
            object[] values={ ce.Uname, ce.Pid, ce.Pname, ce.Price, ce.Nums, ce.Photo };
            int i=DA.ExecuteSql(sqlText, CommandType.Text, paras, values);
            return i;
        }
        //删除购物车中已生成订单的商品信息
        public int DeleteCartByCid(CartEntity ce)
        {
            string sqlText="delete from tb_cart where cid=@cid";
            string[] paras={ "@cid" };
            object[] values={ ce.Cid };
            int i=DA.ExecuteSql(sqlText, CommandType.Text, paras, values);
            return i;
        }
    }
}
```

(4) 操作管理员表对应业务逻辑 AdminBusiness 类，编辑 AdminBusiness.cs 代码如下。

```
using DataAccess;
using Entity;
using System.Data;
namespace Business
{
    public class AdminBusiness
    {
        //判断管理员名称和密码是否匹配
        public int AdminAndPWD(AdminEntity ae)
        {
            string sqlText="select count(*) from tb_admin where aname=@aname and password=@password";
            string[] paras={ "@aname", "@password" };
            object[] values={ ae.Aname, ae.Password };
            int i=Convert.ToInt32(DA.GetOneData(sqlText, CommandType.Text, paras, values));
            return i;
        }
        //根据用户编号更新用户信息
        public int UpdateAdminByAid(AdminEntity ae)
        {
            string sqlText="update tb_admin set aname=@aname,password=@password,tel=@tel where aid=@aid";
            string[] paras={ "@aname", "@password", "@tel","@aid" };
            object[] values={ae.Aname,ae.Password,ae.Tel,ae.Aid};
            int i=DA.ExecuteSql(sqlText, CommandType.Text, paras, values);
            return i;
        }
        public DataSet SelectAdminByAid(AdminEntity ae, string tableName)
        {
            string sqlText="select * from tb_Admin where aid=@aid";
            string[] paras={ "@aid" };
            object[] values={ ae.Aid };
            DataSet ds=new DataSet();
            ds=DA.GetDataSet(sqlText, tableName, CommandType.Text, paras, values);
            return ds;
        }
    }
}
```

(5) 操作订单表对应业务逻辑 OrdersBusiness 类，编辑 OrdersBusiness.cs 代码如下。

```
using DataAccess;
using Entity;
using System.Data;
namespace Business
{
    public class OrdersBusiness
    {
        //向订单总表插入信息
        public int InsertOrders(OrdersEntity oe)
```

```
        {
            string sqlText ="insert into tb_orders values (@uname, @orderTime, @allPrice,@address,@tel)";
            string[] paras={ "@uname", "@orderTime", "@allPrice", "@address", "@tel" };
            object[] values={ oe.Uname, oe.OrderTime, oe.AllPrice, oe.Address, oe.Tel };
            int i=DA.ExecuteSql(sqlText, CommandType.Text, paras, values);
            return i;
        }
        //订单总表信息降序排列
        public DataSet SelectOrderByDesc(string tableName)
        {
            string sqlText="select * from tb_orders order by oid desc";
            DataSet ds=new DataSet();
            ds=DA.GetDataSet(sqlText, tableName, CommandType.Text, null, null);
            return ds;
        }
    }
}
```

(6) 操作留言板表对应业务逻辑 MessageBusiness 类，编辑 MessageBusiness.cs 代码如下。

```
using DataAccess;
using Entity;
using System.Data;
namespace Business
{
    public class MessageBusiness
    {
        //发表留言
        public int InsertMessage(MessageEntity me)
        {
            string sqlText ="insert into tb_message values (@title, @mess, @uname, @messDate)";
            string[] paras={ "@title", "@mess", "@uname", "@messDate" };
            object[] values={ me.Title, me.Mess, me.Uname, me.MessDate };
            int i=DA.ExecuteSql(sqlText, CommandType.Text, paras, values);
            return i;
        }
        //删除留言
        public int DeleteMessageByMid(MessageEntity me)
        {
            string sqlText="delete from tb_message where mid=@mid";
            string[] paras={ "@mid" };
            object[] values={ me.Mid };
            int i=DA.ExecuteSql(sqlText, CommandType.Text, paras, values);
            return i;
        }
    }
}
```

(7) 操作订单信息表对应业务逻辑 OrderDetailsBusiness 类，编辑 OrderDetailsBusiness.cs

代码如下。

```
using DataAccess;
using Entity;
using System.Data;
namespace Business
{
    public class OrderDetailsBusiness
    {
        //添加订单详情信息
        public int InsertOrderDetail(OrderDetailsEntity ode)
        {
            string sqlText="insert into tb_orderDetails values(@oid,@uname,
@pid,@pname,@price,@nums,@photo,@states)";
            string[] paras={ "@oid", "@uname", "@pid", "@pname", "@price",
"@nums", "@photo", "@states" };
            object[] values={ ode.Oid, ode.Uname, ode.Pid, ode.Pname, ode.Price,
ode.Nums, ode.Photo, ode.States };
            int i=DA.ExecuteSql(sqlText, CommandType.Text, paras, values);
            return i;
        }
        //按照编号详情删除订单信息
        public int DeleteOrderDetailsByid(OrderDetailsEntity ode)
        {
            string sqlText="delete from tb_orderDetails where id=@id";
            string[] paras={ "@id" };
            object[] values={ ode.Id };
            int i=DA.ExecuteSql(sqlText, CommandType.Text, paras, values);
            return i;
        }
    }
}
```

12.8 系统页面设计

12.8.1 游客模块的实现

1. 主页面设计

网站以浅蓝为主色调,导航菜单上方是网站新上市商品宣传图片,以菜单控件构建导航条,包含网站首页、浏览商品、我的购物车、我的订单、留言评论、查看留言和个人中心等内容项,Index 页面设计如图 12.28 所示。

Index.aspx 源视图主要代码如下。

```
<%@Page Language="C#" AutoEventWireup="true" CodeFile=" WebDefault.aspx.cs"
Inherits=" WebDefault" %>
<html xmlns="http://www.w3.org/1999/xhtml">
<head runat="server">
```

图 12.28 Index 主页面设计

```
    <title>index</title>
</head>
<body>
    <form id="form1" runat="server">
    <div>
            <table >
            <asp:Menu ID="Menu1" runat="server" Font-Bold="True" Font-Italic=
"False"
                ForeColor="Black" Height="49px" Orientation="Horizontal" Width=
"1374px">
                        <DynamicItemTemplate> <%# Eval("Text") %>
</DynamicItemTemplate>
                <Items>
                        <asp:MenuItem Text="网站首页" Value="网站首页"
NavigateUrl="~/UserMain.aspx"></asp:MenuItem>
                        <asp:MenuItem Text="浏览商品" Value="浏览商品"
NavigateUrl="~/UserProduct.aspx"></asp:MenuItem>
                       <asp:MenuItem Text="我的购物车" Value="我的购物车"
NavigateUrl="~/UserCart.aspx"></asp:MenuItem>
                    <asp:MenuItem Text="我的订单" Value="我的订单" NavigateUrl=
"~/UserOrders.aspx"></asp:MenuItem>
                    <asp:MenuItem Text="留言评论" Value="留言评论" NavigateUrl=
"~/UserAddMess.aspx"></asp:MenuItem>
                    <asp:MenuItem Text="查看留言" Value="查看留言" NavigateUrl=
"~/UserSelectMess.aspx"></asp:MenuItem>
```

```
                              <asp:MenuItem Text="个人中心" Value="个人中心"
NavigateUrl="~/UserCenter.aspx"></asp:MenuItem>
                        </Items>
          </asp:Menu>
               <td>  用户名<br/>
            <asp:TextBox ID="txt" runat="server" Height="32px" Width="192px">
</asp:TextBox>  <br/>
   密码<br/>
           <asp:TextBox ID="txtPassword" runat="server" Height="31px" TextMode=
"Password" Width="192px"></asp:TextBox><br/>
         <asp:Button ID="btnLogin" runat="server" onclick="btnLogin_Click" Text=
"登录"/>
         <asp:Button ID="btnCancel" runat="server" onclick="btnCancel_Click"
Text="取消"/>
              <br/>忘记密码了?<asp:LinkButton ID="LinkButton12" runat="server"
                    PostBackUrl="~/UserFindPassWord.aspx">点击这里</asp:
LinkButton>
               找回<br/>还不是会员?现在就去<asp:LinkButton ID="LinkButton9"
runat="server"  PostBackUrl="~/UserRegister.aspx">注册</asp:LinkButton>吧!
<br/>
                </td>
               <td>
                        <asp:DataList ID="dlNewProducts" runat="server"
DataSourceID="SqlDataSource1"  RepeatDirection="Horizontal" Height="262px"
Width="819px"  onitemcommand="dlNewProducts_ItemCommand">
                    <ItemTemplate>
                         <asp:ImageButton ID="ImageButton1" runat="server"
CommandArgument='<%#Eval("pid")%>' Height="185px"  ImageUrl='<%#Eval
("photo")%>' Width="203px" CommandName="详情"/>
                          <asp:Label ID="Label1" runat="server" Text='<%#
Eval("pname")%>'></asp:Label>
                    </ItemTemplate>
                  </asp:DataList>
              </td>
               </tr>
               <tr>
                 <td></td>
                 <td>
                    <asp:SqlDataSource ID="SqlDataSource1" runat="server"
   ConnectionString="<%$ ConnectionStrings:CosmeticsConnectionString %>"
SelectCommand="select top 5 * from tb_product order by pid desc">
                     </asp:SqlDataSource>
                       </td>
                         </tr>
                    <tr>
                       <td class="style13">
                              化 妆 品 分 类</td>
                           <td>
                              <strong>销 量 排 行</strong></td>
                         </tr><tr>
                           <td>
                             <asp:LinkButton ID="LinkButton3" runat="server"
```

```
    PostBackUrl="~/UserProduct.aspx">洁面</asp:LinkButton>
                                <br/><br/>
                                    <asp:LinkButton ID="LinkButton4" runat=
"server"
  PostBackUrl="~/UserProduct.aspx">爽肤水</asp:LinkButton>
                                <br/><br/>
                                <asp:LinkButton ID="LinkButton5" runat="server"
  PostBackUrl="~/UserProduct.aspx">精华液</asp:LinkButton>
                                <br/><br/>
                                <asp:LinkButton ID="LinkButton6" runat="server"
    PostBackUrl="~/UserProduct.aspx">乳液面霜</asp:LinkButton>
                                <br/><br/>
                                <asp:LinkButton ID="LinkButton7" runat="server"
    PostBackUrl="~/UserProduct.aspx">隔离防晒</asp:LinkButton>
                                <br/><br/>
                                <asp:LinkButton ID="LinkButton8" runat="server"
  PostBackUrl="~/UserProduct.aspx">面膜</asp:LinkButton>
                                <br/><br/>
                                </td>
                            <td>
                                <asp:DataList ID="dlTopSaleProcucts" runat=
"server"  DataSourceID =" SqlDataSource2 " RepeatDirection =" Horizontal "
 onitemcommand="dlTopSaleProcucts_ItemCommand">
                               <ItemTemplate>
                               <asp:ImageButton ID="ImageButton2" runat="server"
 CommandArgument='<%#Eval("pid")%>' Height="163px"   ImageUrl='<%#Eval
 ("photo")%>' Width="192px" CommandName="详情"/>
                                     <br/><br/>
                               <asp:Label ID="Label2" runat="server" Text=
'<%#Eval("pname")%>'></asp:Label>
                                </ItemTemplate>
                                </asp:DataList>
                            </td>
                        </tr><tr>
                            <td> </td>
                            <td>
                                <asp:SqlDataSource ID="SqlDataSource2" runat=
"server"  ConnectionString="<%$ ConnectionStrings:CosmeticsConnectionString %>"
 SelectCommand="select top 5 * from tb_product  where [state] is not null order by
 salenums desc">
                                </asp:SqlDataSource>
                            </td>
                        </tr>
                    </table>
                Copyright 2022.Website name All rights reserved.
            </form>
    </body>
</html>
```

Index.cs 部分 C#代码如下。

```
using Business;
using Entity;
```

```
using System;
using System.Web.UI.WebControls;
namespace WebUI
{
    public partial class Index : System.Web.UI.Page
    {
        protected void Page_Load(object sender, EventArgs e)
        {
        }
        protected void btnLogin_Click(object sender, EventArgs e)
        {
            if (txt.Text.Trim()=="" || txtPassword.Text.Trim()=="")
            {
                Response.Write("<script>alert('用户名和密码不能为空!')</script>");
            }
            else
            {
                UserEntity ue=new UserEntity();
                ue.Name=txt.Text.Trim();
                ue.Password=txtPassword.Text.Trim();
                UserBusiness ub=new UserBusiness();
                int count=ub.UserAndPWD(ue);
                if (count >0)
                {
                    Session["user"]=txt.Text.Trim();
                    Response.Write("<script>alert('登录成功'),location.href='
UserMain.aspx'</script>");
                }
                else
                {
                    Response.Write("<script>alert('用户名和密码不匹配!请重新输
入!')</script>");
                }
            }
        }
        protected void btnCancel_Click(object sender, EventArgs e)
        {
            txt.Text="";
            txtPassword.Text="";
        }

            protected void dlNewProducts _ ItemCommand ( object source,
DataListCommandEventArgs e)
        {
            if (e.CommandName=="详情")
            {
                    Response. Redirect ( " UserProductDetail. aspx pid =" + e.
CommandArgument.ToString());
            }
        }
            protected void dlTopSaleProcucts _ ItemCommand ( object source,
DataListCommandEventArgs e)
```

```
            {
                if (e.CommandName=="详情")
                {
                    Response.Redirect("UserProductDetail.aspx? pid="+e.CommandArgu-
ment.ToString());
                }
            }
        }
    }
```

2. 注册模块功能设计

单击"注册"按钮,跳转到 UserRegister 注册页面,填写相应信息,即可完成注册。用户注册页面设计如图 12.29 所示。

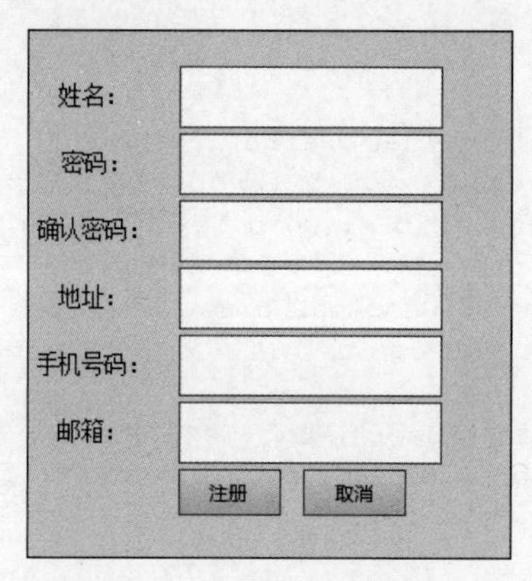

图 12.29　用户注册页面设计

UserRegister.cs 对应的部分 C#代码如下。

```
using Business;
using Entity;
using System;
namespace WebUI
{
    public partial class UserRegister : System.Web.UI.Page
    {
        protected void Page_Load(object sender, EventArgs e)
        {
        }
        protected void btnRegister_Click1(object sender, EventArgs e)
        {
            //获取最新的用户信息
            UserEntity ue=new UserEntity();
            ue.Name=txtName.Text.Trim();
            ue.Password=txtPassowrd.Text.Trim();
            ue.Address=txtAddress.Text.Trim();
            ue.Tel=txtTel.Text.Trim();
            ue.Email=txtEmail.Text.Trim();
            UserBusiness ub=new UserBusiness();
```

```
            int i=ub.InsertUser(ue);
            if (i >0)
            {
                Response.Write("<script>alert('注册成功!'),location.href='Index.
aspx'</script>");
            }
        }
        protected void btnCancel_Click(object sender, EventArgs e)
        {
            txtName.Text="";
            txtPassowrd.Text="";
            txtRePassword.Text="";
            txtAddress.Text="";
            txtTel.Text="";
        }
    }
}
```

12.8.2 会员模块的实现

1. 用户登录

在 Index 页面的会员中心输入正确的用户名和密码即可登录网站。用户登录页面设计如图 12.30 所示。

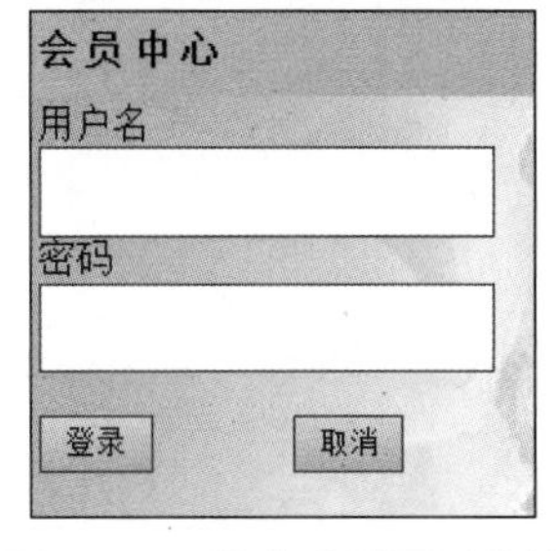

图 12.30 用户登录页面设计

2. 浏览商品模块

用户浏览商品，了解商品基本信息。单击商品图片或商品名称进入 UserProduct 商品详情页面，可根据需要查找商品。浏览商品页面设计如图 12.31 所示。

UserProduct.cs 对应的部分 C# 代码如下。

```
using Business;
using Entity;
using System;
using System.Data;
using System.Web.UI.WebControls;
public partial class UserProduct : System.Web.UI.Page
{
    protected void Page_Load(object sender, EventArgs e)
    {
        if (IsPostBack==false)
        {
            Bind();
        }
    }
    public void Bind()
    {
        ProductBusiness pb=new ProductBusiness();
        DataSet ds=new DataSet();
        ds=pb.SelectProduct("p");
        DataList1.DataSource=ds.Tables["p"];
```

图 12.31 浏览商品页面设计

```
            DataList1.DataBind();
        }
        protected void DataList1_ItemCommand(object source, DataListCommandEventArgs e)
        {
            if (e.CommandName=="详情" || e.CommandName=="详情 2")
            {
                Response.Redirect("UserProductDetail.aspx pid="+e.CommandArgument.
ToString());
            }
        }
        protected void btnSearch_Click(object sender, EventArgs e)
        {
            if (txtName.Text.Trim()=="")
            {
                Bind();
            }
            else
            {
                ProductEntity pe=new ProductEntity();
                pe.Pname=txtName.Text.Trim();
                ProductBusiness pb=new ProductBusiness();
                DataSet ds=new DataSet();
                ds=pb.SelectProductByPname(pe, "P");
                DataList1.DataSource=ds.Tables["p"];
                DataList1.DataBind();
```

```
            }
        }
    }
```

3. 商品详情模块

在浏览商品页面,单击商品图片或商品名称进入 UserProductDetail 商品详情页面,用户可以选择直接购买商品或将商品加入购物车。商品详情页面设计如图 12.32 所示。

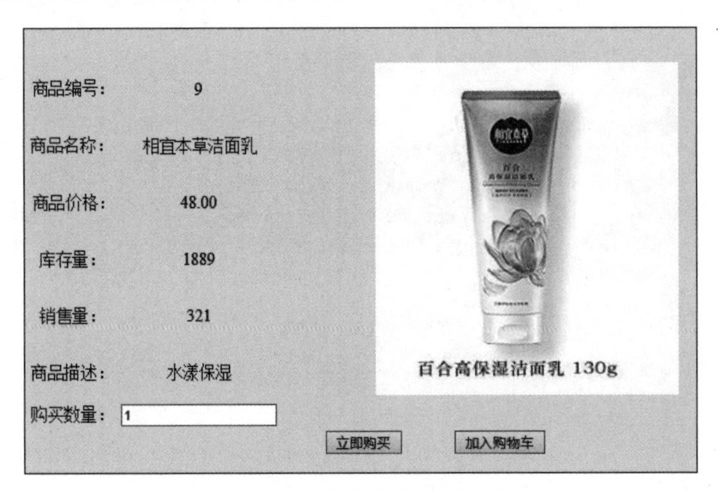

图 12.32 商品详情页面设计

UserProductDetail.cs 对应的部分 C#代码如下。

```
using Business;
using Entity;
using System;
using System.Data;
public partial class UserProductDetail : System.Web.UI.Page
{
    protected void Page_Load(object sender, EventArgs e)
    {
        if (IsPostBack==false)
        {
            lblPid.Text=Request.QueryString["pid"].ToString();
            ProductEntity pe=new ProductEntity();
            pe.Pid=Convert.ToInt32(lblPid.Text);
            ProductBusiness pb=new ProductBusiness();
            DataSet ds=new DataSet();
            ds=pb.SelectProductByPid(pe, "p");
            lblPname.Text=ds.Tables["p"].Rows[0][1].ToString();
            Image1.ImageUrl=ds.Tables["p"].Rows[0][2].ToString();
            lblPrice.Text=ds.Tables["p"].Rows[0][3].ToString();
            lblPnums.Text=ds.Tables["p"].Rows[0][4].ToString();
            lblSaleNums.Text=ds.Tables["p"].Rows[0][5].ToString();
            lblMess.Text=ds.Tables["p"].Rows[0][6].ToString();
        }
    }
```

```
        //将商品添加到购物车
        protected void btnInsertShopCart_Click(object sender, EventArgs e)
        {
            try
            {
                if (Session["user"]==null)
                {
                    Response.Write("<script>alert('登录后才能购买商品!'),location.
    href='Index.aspx'</script>");
                }
                else
                {
                    CartEntity ce=new CartEntity();
                    ce.Uname=Session["User"].ToString();
                    ce.Pid=Convert.ToInt32(lblPid.Text);
                    ce.Pname=lblPname.Text;
                    ce.Price=Convert.ToDecimal(lblPrice.Text);
                    ce.Nums=Convert.ToInt32(txtNums.Text.Trim());
                    ce.Photo=Image1.ImageUrl;
                    CartBusiness cb=new CartBusiness();
                    int count=cb.InsertCart(ce);
                    if (count >0)
                    {
                        Response.Write("<script>alert('成功添加购物车!'),location.
    href='UserCart.aspx'</script>");
                    }
                }
            }
            catch
            {
                Response.Write("<script>alert('信号差啊!请稍后再试!')</script>");
            }
        }
        //确定购买商品
        protected void btnInsertOrders_Click(object sender, EventArgs e)
        {
            if (Session["user"]==null)
            {
                Response.Write("<script>alert('登录后才能购买商品!'),location.href=
    'Index.aspx'</script>");
            }
            else
            {
                if (Convert.ToInt32(txtNums.Text) >Convert.ToInt32(lblPnums.Text))
                {
                    Response.Write("<script>alert('库存不足,不能购买!')</script>");
                }
                else
                {
                    decimal allprice=0;
                    OrdersEntity oe=new OrdersEntity();
                    oe.Uname=Session["user"].ToString();
                    oe.OrderTime=DateTime.Now;
```

```
                oe.AllPrice=Convert.ToInt32(txtNums.Text.Trim()) * Convert.
ToDecimal(lblPrice.Text);
                oe.Address=txtAddress.Text;
                oe.Tel=txtTel.Text;
                OrdersBusiness ob=new OrdersBusiness();
                int i=ob.InsertOrders(oe);
                if (i >0)
                {
                    OrderDetailsEntity ode=new OrderDetailsEntity();
                    DataSet ds=new DataSet();
                    ds=ob.SelectOrderByDesc("orders");
                    ode.Oid=Convert.ToInt32(ds.Tables["orders"].Rows[0][0]);
                    ode.Uname=Session["user"].ToString();
                    ode.Pid=Convert.ToInt32(lblPid.Text);
                    ode.Pname=lblPname.Text;
                    ode.Price=Convert.ToDecimal(lblPrice.Text);
                    ode.Nums=Convert.ToInt32(txtNums.Text.Trim());
                    ode.Photo=Image1.ImageUrl;
                    ode.States="已付款";
                    OrderDetailsBusiness odb=new OrderDetailsBusiness();
                    int count=odb.InsertOrderDetail(ode);
                    if (count >0)
                    {
                  Response.Write("<script>location.href='UserPayment.aspx'</
script>");
                    }
                    Session["user"]=oe.Uname;
                    Session["price"]=oe.AllPrice;
                }
            }
        }
    }
}
```

4. 购物车模块

添加商品到购物车页面 UserCart，购物车显示具体购买信息。购物车页面设计如图 12.33 所示。

图 12.33 购物车页面设计

UserCart.cs 对应的部分 C#代码如下。

```
using Business;
using Entity;
using System;
using System.Data;
using System.Web.UI.WebControls;
public partial class UserCart : System.Web.UI.Page
{
    protected void Page_Load(object sender, EventArgs e)
    {
        if (Session["user"]==null)
        {
       Response.Write("<script>alert('请先登录'),location.href='Index.aspx'</
script>");
        }
        else
        {
            if (IsPostBack==false)
            {
                UserEntity ue=new UserEntity();
                ue.Name=Session["user"].ToString();
                UserBusiness ub=new UserBusiness();
                DataSet ds=new DataSet();
                ds=ub.SelectUserByName(ue, "user");
                txtAddress.Text=ds.Tables["user"].Rows[0][3].ToString();
                txtTel.Text=ds.Tables["user"].Rows[0][4].ToString();
            }
        }
    }
    protected void btnBuy_Click(object sender, EventArgs e)
    {
        decimal allPrice=0;
        for (int i=0; i<GridView1.Rows.Count; i++)
        {
            CheckBox cbx1=(CheckBox)GridView1.Rows[i].FindControl("cbxID");
            if (cbx1.Checked==true)
            {
                 allPrice=allPrice+Convert.ToDecimal(GridView1.Rows[i].Cells[5].
Text)*Convert.ToInt32(GridView1.Rows[i].Cells[6].Text);
            }
        }
        if (allPrice>0)
        {
            OrdersEntity oe=new OrdersEntity();
            oe.Uname=Session["user"].ToString();
            oe.OrderTime=DateTime.Now;
            oe.AllPrice=allPrice;
            oe.Address=txtAddress.Text.Trim();
            oe.Tel=txtTel.Text.Trim();
            OrdersBusiness ob=new OrdersBusiness();
            ob.InsertOrders(oe);
            OrderDetailsEntity ode=new OrderDetailsEntity();
```

```
            ode.Uname=Session["user"].ToString();
            Session["user"]=ode.Uname;
            DataSet ds=new DataSet();
            ds=ob.SelectOrderByDesc("order");
            ode.Oid=Convert.ToInt32(ds.Tables["order"].Rows[0][0]);
            for (int i=0; i<GridView1.Rows.Count; i++)
            {
                CheckBox cbx2=(CheckBox)GridView1.Rows[i].FindControl("cbxID");
                if (cbx2.Checked==true)
                {
                  Response.Write("<script>location.href('UserPayment.aspx')</
script>");
                    ode.Pid=Convert.ToInt32(GridView1.Rows[i].Cells[3].Text);
                    ode.Pname=GridView1.Rows[i].Cells[4].Text;
                    ode.Price=Convert.ToDecimal(GridView1.Rows[i].Cells[5].Text);
                    ode.Nums=Convert.ToInt32(GridView1.Rows[i].Cells[6].Text);
                    ode.Photo=GridView1.Rows[i].Cells[7].Text;
                    ode.States="已付款";
                    OrderDetailsBusiness odb=new OrderDetailsBusiness();
                    odb.InsertOrderDetail(ode);
                    CartEntity ce=new CartEntity();
                    ce.Cid=Convert.ToInt32(GridView1.Rows[i].Cells[1].Text);
                    CartBusiness cb=new CartBusiness();
                    cb.DeleteCartByCid(ce);
                }
            }
            Response.Write("<script>alert('购买成功,订单已生成!'),location.href('
userOrders.aspx')</script>");
        }
        else
        {
            Response.Write("<script>alert('请选择商品!')</script>");
        }
        Session["price"]=allPrice;
    }
}
```

5. 支付模块

在支付页面 UserPayment 选择支付方式后，会出现具体的账单信息，核对订单信息后自动对其账户进行扣款。支付页面设计如图 12.34 所示。

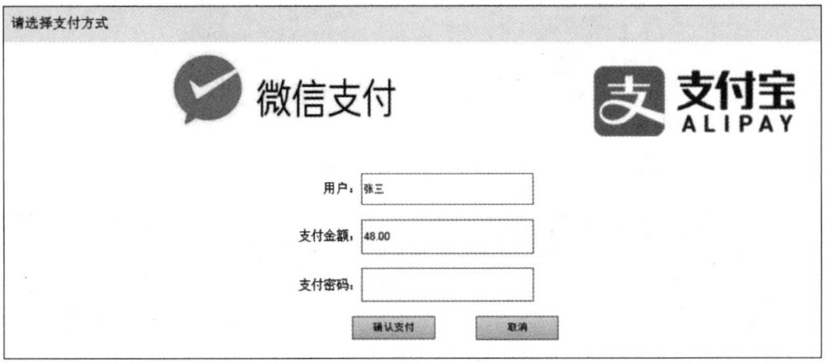

图 12.34 支付页面设计

UserPayment.cs 对应的部分 C# 代码如下。

```
using System;
using System.Web.UI;
public partial class UserPayment: System.Web.UI.Page
{
    protected void Page_Load(object sender, EventArgs e)
    {
    }
    protected void imgbtnPay1_Click(object sender, ImageClickEventArgs e)
    {
        Panel1.Visible=true;
        txtPrice.Text=Session["price"].ToString();
        txtID.Text=Session["user"].ToString();
    }
    protected void btnConfirmPay_Click(object sender, EventArgs e)
    {
        Response. Write ( " < script > alert (' 支 付 成 功! '), location. href ('
UserOrderDetails.aspx')</script>");
    }
    protected void imgbtnPay2_Click(object sender, ImageClickEventArgs e)
    {
        Panel1.Visible=true;
        txtPrice.Text=Session["price"].ToString();
        txtID.Text=Session["user"].ToString();
    }
    protected void btnCancelPay_Click(object sender, EventArgs e)
    {
        txtPassword.Text="";
    }
}
```

6. 订单模块

完成商品购买后，在订单页面 UserOrders 可查看订单详情。订单页面设计如图 12.35 所示。

图 12.35　订单页面设计

UserOrders.cs 对应的部分 C＃代码如下。

```
public partial class UserOrders : System.Web.UI.Page
{
    protected void Page_Load(object sender, EventArgs e)
    {
        if (Session["user"]==null)
        {
         Response.Write("<script>alert('请先登录'),location.href='Index.aspx'
</script>");
        }
    }
}
```

7. 留言模块

在留言页面 UserAddMess 可以对商品进行留言评论。留言页面设计如图 12.36 所示。

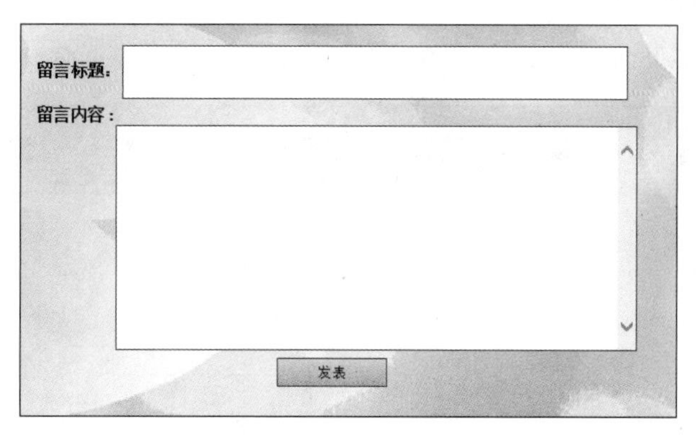

图 12.36　留言页面设计

UserAddMess.cs 对应的部分 C＃代码如下。

```
public partial class UserAddMess : System.Web.UI.Page
{
    protected void Page_Load(object sender, EventArgs e)
    {
        if (Session["user"]==null)
        {
      Response.Write("<script>alert('请先登录'),location.href='Index.aspx'</
script>");
        }
    }
    protected void btnSendMess_Click(object sender, EventArgs e)
    {
        MessageEntity me=new MessageEntity();
        me.Title=txtTitle.Text.Trim();
        me.Mess =txtMess .Text;
        me.Uname=Session["user"].ToString();
        me.MessDate=DateTime.Now;
        MessageBusiness mb=new MessageBusiness();
        int i=mb.InsertMessage(me);
```

```
            if (i >0)
            {
                Response.Write("<script>alert('留言成功!')</script>");
            }
            Panel1.Visible=true;
            lblTitle.Text=txtTitle.Text.Trim();
            lblMess.Text=txtMess.Text.Trim();
            lblMessDate.Text=DateTime.Now.ToString();
            lblUname.Text=Session["user"].ToString();
        }
        protected void btnDelMess_Click(object sender, EventArgs e)
        {
            Panel1.Visible=false;
        }
    }
```

8. 个人中心模块

可以在个人中心页面 UserCenter 修改个人信息。个人中心页面设计如图 12.37 所示。

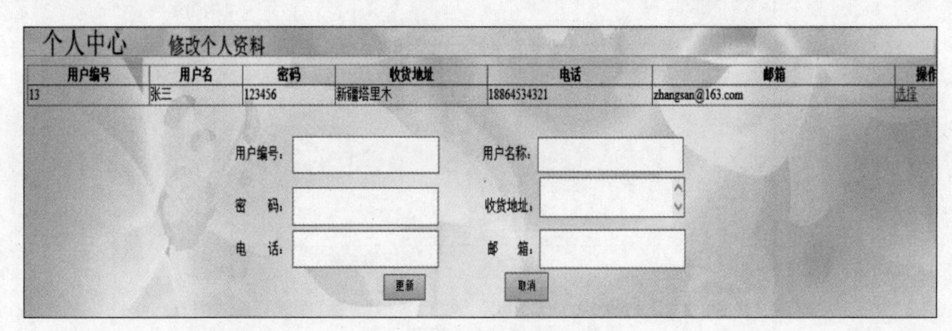

图 12.37　个人中心页面设计

UserCenter.cs 对应的部分 C#代码如下。

```
public partial class UserCenter : System.Web.UI.Page
{
    protected void Page_Load(object sender, EventArgs e)
    {
        if (Session["user"]==null)
        {
            Response.Write("<script>alert('请先登录!'),location.href='Index.
aspx'</script>");
        }
    }
    protected void btnUpdateInfo_Click(object sender, EventArgs e)
    {
        try
        {
            UserEntity ue=new UserEntity();
            ue.Uid=Convert.ToInt32(txtNo.Text.Trim());
            ue.Name=txtName.Text.Trim();
            ue.Password=txtPassword.Text.Trim();
```

```
                ue.Address=txtAdd.Text.Trim();
                ue.Tel=txtTel.Text.Trim();
                ue.Email=txtEmail.Text.Trim();
                UserBusiness ub=new UserBusiness();
                int count=ub.UpdateUserByUid(ue);
                if (count >0)
                {
                    Response.Write("<script>alert('信息修改成功!')</script>");
                    GridView1.DataBind();
                    txtNo.Text="";
                    txtName.Text="";
                    txtPassword.Text="";
                    txtAdd.Text="";
                    txtTel.Text="";
                    txtEmail.Text="";
                }
            }
            catch (Exception)
            {
                Response.Write("<script>alert('服务器异常,请稍后再试')</script>");
            }
        }
        protected void GridView1_RowCommand(object sender, GridViewCommandEventArgs e)
        {
            if (e.CommandName=="选择")
            {
                UserEntity ue=new UserEntity();
                ue.Uid=Convert.ToInt32(e.CommandArgument);
                UserBusiness ub=new UserBusiness();
                DataSet ds=new DataSet();
                ds=ub.SelectUserByUid(ue, "tb_user");
                if (ds.Tables["tb_user"].Rows.Count >0)
                {
                    txtNo.Text=ds.Tables["tb_user"].Rows[0][0].ToString();
                    txtName.Text=ds.Tables["tb_user"].Rows[0][1].ToString();
                    txtPassword.Text=ds.Tables["tb_user"].Rows[0][2].ToString();
                    txtAdd.Text=ds.Tables["tb_user"].Rows[0][3].ToString();
                    txtTel.Text=ds.Tables["tb_user"].Rows[0][4].ToString();
                    txtEmail.Text=ds.Tables["tb_user"].Rows[0][5].ToString();
                }
            }
        }
        protected void btnCancel_Click(object sender, EventArgs e)
        {
            txtNo.Text="";
            txtName.Text="";
            txtPassword.Text="";
            txtAdd.Text="";
            txtTel.Text="";
            txtEmail.Text="";
        }
    }
```

12.8.3 管理员模块的实现

1. 管理员登录

在管理员登录页面 AdminLogin 输入用户名、密码，验证通过后登录到后台管理系统对网站进行管理。管理员登录页面设计如图 12.38 所示。

图 12.38 管理员登录页面设计

AdminLogin.cs 对应的部分 C＃代码如下。

```
public partial class AdminLogin : System.Web.UI.Page
{
    protected void Page_Load(object sender, EventArgs e)
    {
        if (! IsPostBack)
        {
            txtName.Focus();
        }
    }
    protected void btnLogin_Click(object sender, EventArgs e)
    {
        AdminEntity ae=new AdminEntity();
        ae.Aname=txtName.Text.Trim();
        ae.Password=txtPassword.Text.Trim();
        AdminBusiness ab=new AdminBusiness();
        int count=ab.AdminAndPWD(ae);
        if (count >0)
        {
            Session["admin"]=txtName.Text.Trim();
            Response.Write("<script>alert('登录成功'),location.href='AdminMain.
aspx'</script>");
        }
        else
        {
            Response.Write("<script>alert('用户名和密码不匹配,请重新输入!')</
script>");
```

```
        }
    }
    protected void btnCancel_Click(object sender, EventArgs e)
    {
        txtName.Text="";
        txtPassword.Text="";
    }
}
```

2. 添加商品信息模块

在管理员添加商品页面 AdminAddProduct 可添加新商品，包括商品图片、商品名称、商品价格、商品描述等信息。管理员添加商品页面设计如图 12.39 所示。

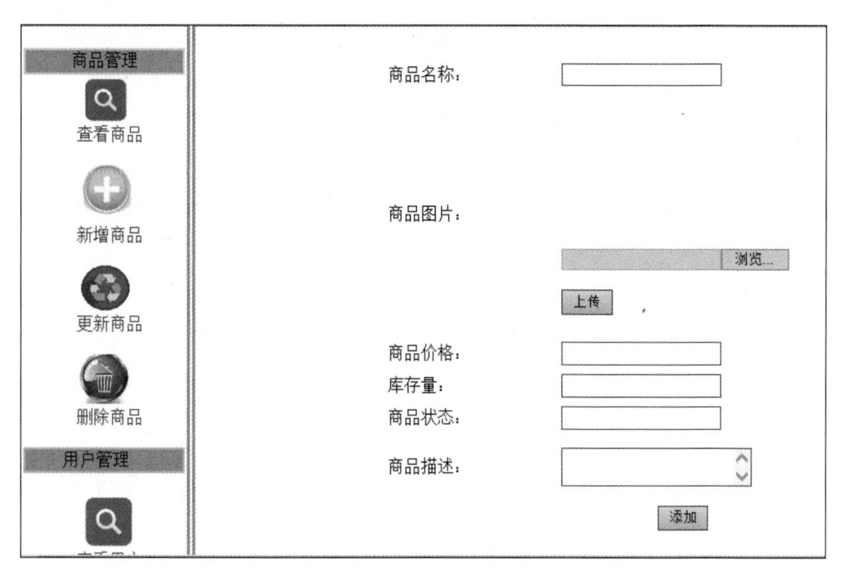

图 12.39　管理员添加商品页面设计

AdminAddProduct.cs 对应的部分 C＃代码如下。

```
public partial class AdminAddProduct : System.Web.UI.Page
{
    protected void Page_Load(object sender, EventArgs e)
    {
    }
    protected void btnImgUpload_Click(object sender, EventArgs e)
    {
        string savePath=Server.MapPath("~/image/");
        if (FileUpload1.HasFile)
        {
            savePath=savePath+FileUpload1.FileName;
            FileUpload1.SaveAs(savePath);
            Image1.ImageUrl="~/image/"+FileUpload1.FileName;
        }
    }
    protected void btnInsertProduct_Click(object sender, EventArgs e)
```

```
    {
        try
        {
            ProductEntity pe=new ProductEntity();
            pe.Pname=txtPname.Text.Trim();
            pe.Photo=Image1.ImageUrl;
            pe.Price=Convert.ToDecimal(txtPrice.Text.Trim());
            pe.Pnums=Convert.ToInt32(txtNums.Text.Trim());
            pe.Mess=txtMess.Text.Trim();
            pe.State=txtState.Text.Trim();
            ProductBusiness pb=new ProductBusiness();
            int count=pb.InsertProduct(pe);
            if (count >0)
            {
                Response.Write("<script>alert('添加成功!')</script>");
                txtPname.Text="";
                Image1.ImageUrl="";
                txtPrice.Text="";
                txtNums.Text="";
                txtState.Text="";
                txtMess.Text="";
            }
            else
            {
                Response.Write("<script>alert('添加失败!')</script>");
            }
        }
        catch (Exception)
        {
            Response.Write("<script>alert('数据输入格式有误,请重新输入!')</
script>");
        }
    }
    protected void btnClear_Click(object sender, EventArgs e)
    {
        txtPname.Text="";
        Image1.ImageUrl="";
        txtPrice.Text="";
        txtNums.Text="";
        txtState.Text="";
        txtMess.Text="";
    }
}
```

3. 更新商品信息模块

管理员可在 AdminUpdateProduct 页面查询特定商品并更新商品信息。管理员更新商品页面设计如图 12.40 所示。

AdminUpdateProduct.cs 对应的部分 C#代码如下。

```
public partial class AdminUpdateProduct : System.Web.UI.Page
{
    protected void Page_Load(object sender, EventArgs e)
```

编号	商品名称	价格	库存量	销售量	商品描述	商品状态	操作
7	相宜本草隔离乳	75.00	899	98	水嫩亮彩防晒乳	有货	选择
9	相宜本草洁面乳	48.00	1889	321	水样保湿	有货	选择
14	雅诗兰黛乳液	284.00	800	376	活肤原生乳液	有货	选择
15	雅诗兰黛红石榴活肤水	286.00	600	289	红石榴精华	有货	选择
123							

商品编号：

商品名称:

商品图片：

浏览...

上传

商品单价:

库存量:

销售量：

商品状态：

商品描述：

更新

图 12.40　管理员更新商品页面设计

```
    {
        if (TextBox2.Text.Trim()=="")
        {
            Bind();
        }
    }
    public void Bind()
    {
        ProductBusiness pb=new ProductBusiness();
        DataSet ds=new DataSet();
        ds=pb.SelectProduct("p");
        GridView1.DataSource=ds.Tables["p"];
        GridView1.DataBind();
    }
    protected void btnImgUpload_Click(object sender, EventArgs e)
    {
        string savePath=Server.MapPath("~/image/");
        if (FileUpload1.HasFile)
        {
            savePath=savePath+FileUpload1.FileName;
            FileUpload1.SaveAs(savePath);
            Image1.ImageUrl="~/image/"+FileUpload1.FileName;
        }
    }
    protected void GridView1_RowCommand(object sender, GridViewCommandEventArgs e)
    {
```

```
        if (e.CommandName=="选择")
        {
            ProductEntity pe=new ProductEntity();
            pe.Pid=Convert.ToInt32(e.CommandArgument);
            ProductBusiness pb=new ProductBusiness();
            DataSet ds=new DataSet();
            ds=pb.SelectProductByPid(pe, "product");
            if (ds.Tables["product"].Rows.Count >0)
            {
                txtPid.Text=ds.Tables["product"].Rows[0][0].ToString();
                txtPname.Text=ds.Tables["product"].Rows[0][1].ToString();
                Image1.ImageUrl=ds.Tables["product"].Rows[0][2].ToString();
                txtPrice.Text=ds.Tables["product"].Rows[0][3].ToString();
                txtNums.Text=ds.Tables["product"].Rows[0][4].ToString();
                txtSalenums.Text=ds.Tables["product"].Rows[0][5].ToString();
                txtMess.Text=ds.Tables["product"].Rows[0][6].ToString();
                txtState.Text=ds.Tables["product"].Rows[0][7].ToString();
            }
        }
    }
    protected void btnUpdateProduct_Click(object sender, EventArgs e)
    {
        try
        {
            ProductEntity pe=new ProductEntity();
            pe.Pid=Convert.ToInt32(txtPid.Text.Trim());
            pe.Pname=txtPname.Text.Trim();
            pe.Photo=Image1.ImageUrl;
            pe.Price=Convert.ToDecimal(txtPrice.Text);
            pe.Pnums=Convert.ToInt32(txtNums.Text.Trim());
            pe.Salenums=Convert.ToInt32(txtSalenums.Text.Trim());
            pe.State=txtState.Text.Trim();
            pe.Mess=txtMess.Text.Trim();
            ProductBusiness pb=new ProductBusiness();
            int count=pb.UpdateProductByPid(pe);
            if (count >0)
            {
                Response.Write("<script>alert('更新成功!')</script>");
                Bind();
                txtPid.Text="";
                txtPname.Text="";
                Image1.ImageUrl="";
                txtPrice.Text="";
                txtNums.Text="";
                txtSalenums.Text="";
                txtState.Text="";
                txtMess.Text="";
            }
            else
            {
                Response.Write("<script>alert('更新失败!')</script>");
            }
        }
```

```
        catch (Exception)
        {
            Response.Write("<script>alert('输入数据格式有误!')</script>");
        }
    }
    protected void btnSearch_Click(object sender, EventArgs e)
    {
        ProductEntity pe=new ProductEntity();
        pe.Pname=TextBox2.Text.Trim();
        ProductBusiness pb=new ProductBusiness();
        DataSet ds=new DataSet();
        ds=pb.SelectProductByPname(pe, "P");
        GridView1.DataSource=ds.Tables["p"];
        GridView1.DataBind();
    }
}
```

4. 删除商品模块

在 AdminDeleteProduct 页面可以删除已经停产、停售的产品。管理员删除商品页面设计如图 12.41 所示。

选择	商品编号	商品名称	商品单价	库存量	销售量	商品描述	商品状态	删除
□	7	相宜本草隔离乳	75.00	899	98	水嫩亮彩防晒乳	有货	删除
□	9	相宜本草洁面乳	48.00	1889	321	水漾保湿	有货	删除
□	14	雅诗兰黛乳液	284.00	800	376	活肤原生乳液	有货	删除
□	15	雅诗兰黛红石榴活肤水	286.00	600	289	红石榴精华	有货	删除
□	17	雅诗兰黛眼部修护	644.00	666	211	雅诗兰黛眼部修护	有货	删除
□	19	相宜本草精华液	67.00	578	433	相宜本草精华液	有货	删除
□	20	相宜本草保湿水	54.00	489	211	相宜本草保湿水	有货	删除
□	22	相宜本草眼霜	88.00	400	222	相宜本草眼霜	有货	删除
□	23	相宜本草面膜	66.00	300	156	相宜本草面膜	有货	删除
□	29	雅诗兰黛洗面奶	266.00	400	0	雅诗兰黛洗面奶	有货	删除

□全选　　批量删除

图 12.41　管理员删除商品页面设计

AdminDeleteProduct.cs 对应的部分 C# 代码如下。

```
public partial class AdminDeleteProduct : System.Web.UI.Page
{
    protected void Page_Load(object sender, EventArgs e)
    {
        if (IsPostBack==false)
        {
            Bind();
        }
```

```
        }
        public void Bind()
        {
            ProductBusiness pb=new ProductBusiness();
            DataSet ds=new DataSet();
            ds=pb.SelectProduct("product");
            GridView1.DataSource=ds.Tables["product"];
            GridView1.DataBind();
        }
        protected void btnDeleteProduct_Click(object sender, EventArgs e)
        {
            string sql="";
            for (int i=0; i<GridView1.Rows.Count; i++)
            {
                CheckBox cbx2=(CheckBox)GridView1.Rows[i].FindControl("cbxID");
                if (cbx2.Checked==true)
                {
                    sql=sql+" or pid="+GridView1.Rows[i].Cells[1].Text;
                }
            }
            ProductBusiness pb=new ProductBusiness();
            int count=pb.DeleteProductByPart(sql);
            if (count >0)
            {
                Response.Write("<script>alert('删除了"+count+"行')</script>");
                Bind();
            }
        }
        protected void cbxAll_CheckedChanged(object sender, EventArgs e)
        {
            for (int i=0; i<GridView1.Rows.Count; i++)
            {
                CheckBox cbx1=(CheckBox)GridView1.Rows[i].FindControl("cbxID");
                if (cbxAll.Checked==true)
                {
                    cbx1.Checked=true;
                }
                else
                {
                    cbx1.Checked=false;
                }
            }
        }
        protected void GridView1_RowCommand(object sender, GridViewCommandEventArgs e)
        {
            if (e.CommandName=="删除")
            {
                ProductEntity pe=new ProductEntity();
                pe.Pname=Convert.ToString(e.CommandArgument);
                ProductBusiness pb=new ProductBusiness();
                int count=pb.DeleteProductByPname(pe);
                if (count >0)
```

```
            {
                Response.Write("<script>alert('删除成功！')</script>");
                Bind();
            }
        }
    }
}
```

5. 查找用户模块

管理员可以在 AdminSelectUser 页面查看以及查找用户。管理员查看用户页面设计如图 12.42 所示。

查找用户：[ ] 搜索

用户编号	用户姓名	密码	地址	电话	电子邮箱
24	张三	*	大连	13236789090	zhangsan@qq.com
25	李可	*	北京	18842667890	like@163.com
26	王丹	*	上海	13074143678	wangdan@qq.com

图 12.42　管理员查看用户页面设计

AdminSelectUser.cs 对应的部分 C# 代码如下。

```
public partial class AdminSelectUser : System.Web.UI.Page
{
    protected void Page_Load(object sender, EventArgs e)
    {
        if (txtID.Text.Trim()=="")
        {
            Bind();
        }
    }
    public void Bind()
    {
        UserBusiness ub=new UserBusiness();
        DataSet ds=new DataSet();
        ds=ub.SelectUser("p");
        GridView1.DataSource=ds.Tables["p"];
        GridView1.DataBind();
    }
    protected void btnSearch_Click(object sender, EventArgs e)
    {
        UserEntity ue=new UserEntity();
        ue.Name=txtID.Text.Trim();
```

```
        UserBusiness ub=new UserBusiness();
        DataSet ds=new DataSet();
        ds=ub.SelectUserByName(ue, "p");
        GridView1.DataSource=ds.Tables["p"];
        GridView1.DataBind();
    }
}
```

6. 删除用户模块

对于严重影响网站正常使用的用户，管理员可在 AdminDeleteUser 页面对其进行删除。管理员删除用户页面设计如图 12.43 所示。

选择	用户编号	用户名称	密码	地址	电话	邮箱	删除
□	24	张三	*	辽宁大连	13236921159	zhangsan@qq.com	删除
□	26	王丹	*	上海	13074143678	wangdan@qq.com	删除
□	29	姜楠	*	海南海口	18842687926	jiangnan@163.com	删除
□	30	刘以	*	沈阳	12344	1@1b	删除

图 12.43　管理员删除用户页面设计

AdminDeleteUser.cs 对应的部分 C＃代码如下。

```
public partial class AdminDeleteUser : System.Web.UI.Page
{
    protected void Page_Load(object sender, EventArgs e)
    {
        if (IsPostBack==false)
        {
            Bind();
        }
    }
    public void Bind()
    {
        UserBusiness ub=new UserBusiness();
        DataSet ds=new DataSet();
        ds=ub.SelectUser("tb_user");
        GridView1.DataSource=ds.Tables["tb_user"];
        GridView1.DataBind();
    }
    protected void GridView1_RowCommand(object sender, GridViewCommandEventArgs e)
    {
        if (e.CommandName=="删除")
        {
            UserEntity ue=new UserEntity();
            ue.Uid=Convert.ToInt32(e.CommandArgument);
            UserBusiness ub=new UserBusiness();
```

```
            int count=ub.DeleteUserByUid(ue);
            if (count >0)
            {
                Response.Write("<script>alert('删除成功！')</script>");
                Bind();
            }
            else
            {
                Response.Write("<script>alert('删除失败！')</script>");
            }
        }
    }
}
```

第13章 学生档案管理系统的设计与实现

本章学习目标

- 了解系统创建的业务流程
- 了解系统中数据库的创建方法
- 掌握数据库中存储过程的创建方法
- 熟练掌握三层架构中类库的引用关系

本章首先对档案管理系统业务逻辑进行分析，然后详细讲解系统数据库的设计，对系统中项目层次的划分进行讲解，最后对系统进行页面设计和后台代码实现。

13.1 系统功能简介

学生档案管理系统包含学生基本档案管理、奖学金档案管理、借阅档案预约管理、奖学金申请管理、借阅管理等功能，提供了档案管理人员、教师及学生三种使用权限，可满足检索、借阅档案的需求。

系统采用三层架构进行设计，包含表示层、业务逻辑层、数据访问层和实体层。

(1) 在表示层中进行页面设计，调用业务逻辑层中的方法显示数据。

(2) 在业务逻辑层中调用存储过程，从数据访问层接收到返回值后传给表示层。

(3) 在数据访问层中访问数据库，将相关数据库命令执行的结果返回给业务逻辑层。

(4) 在实体层进行数据封装，在表示层和业务逻辑层之间进行数据传递。

13.2 系统业务流程

13.2.1 管理员权限业务流程

管理员通过账号、密码和验证码登录系统，登录过程中如有错误，会给出相应提示。通过验证后进入管理员管理主页面，可以进行学生信息管理、教师信息管理、基本档案信息管理、奖学金档案管理、借阅记录管理、借档预约管理和管理密码管理操作。管理员业务流程图如图13.1所示。管理员用例图如图13.2所示。

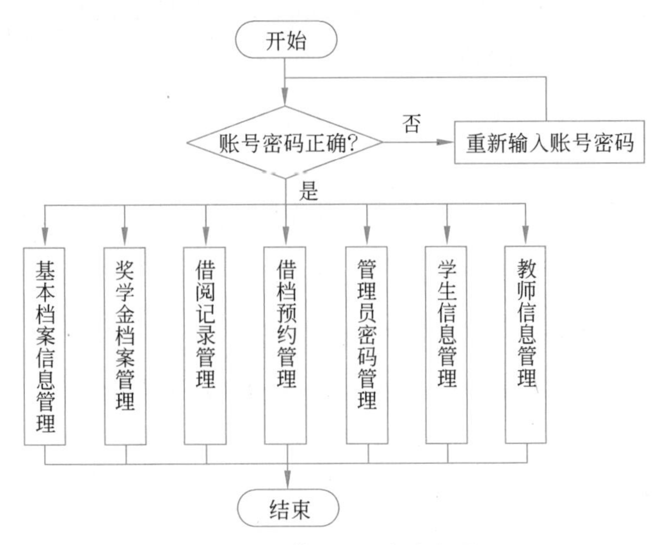

图13.1 管理员业务流程图

13.2.2 教师权限业务流程

教师通过教工号、密码和验证码登录系统，登录过程中如有错误会给出相应提示。通过验证后进入教师管理主页面，可进行个人信息修改、学生信息查看、奖学金申请审核以及借档预约申请提交等操作。教师权限业务流程图如图13.3所示。教师权限用例图如图13.4所示。

13.2.3 学生权限业务流程

学生通过学号、密码和验证码登录系统，登录过程中如有错误会给出相应提示。通过验证后进入学生管理主页面，可进行密码修改、个人档案查看、奖学金申请提交以及借档预约申请提交等操作。学生权限业务流程图如图13.5所示。学生权限用例图如图13.6所示。

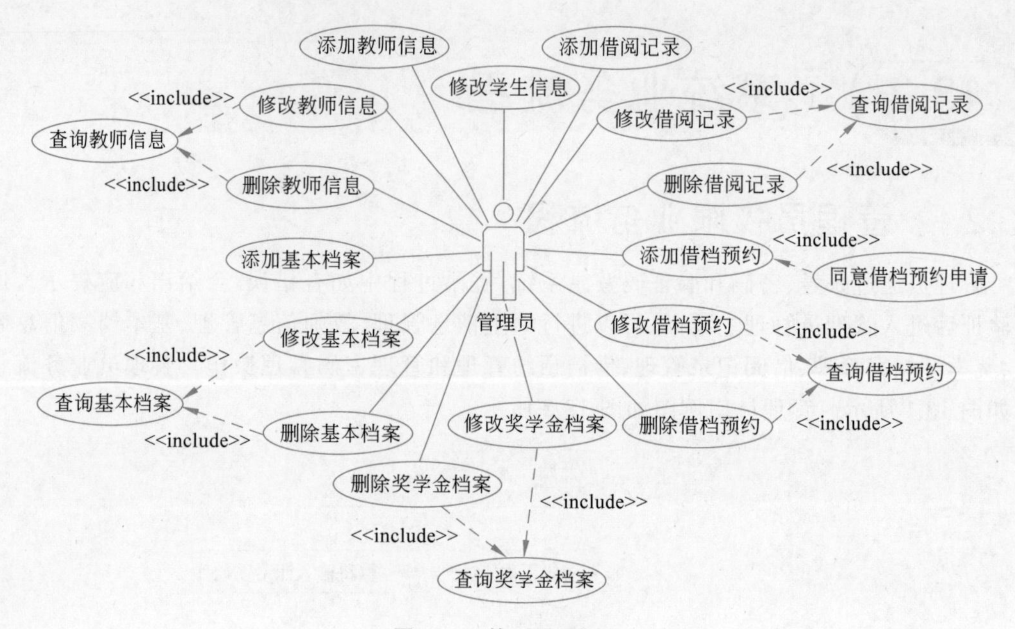

图 13.2 管理员用例图

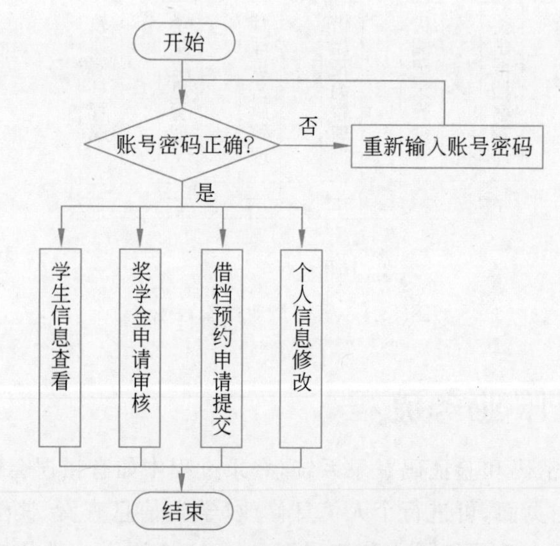

图 13.3 教师权限业务流程图

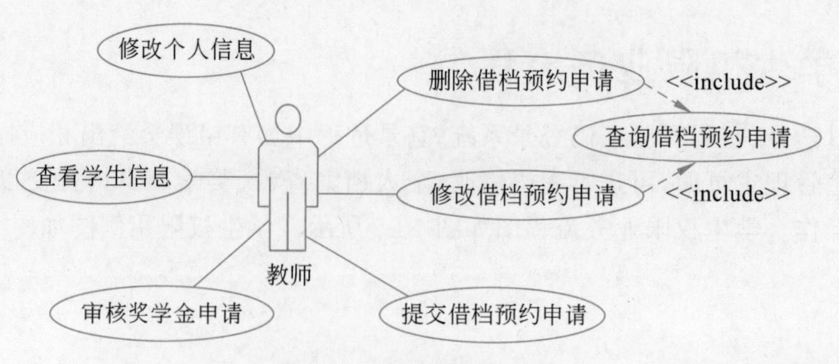

图 13.4 教师权限用例图

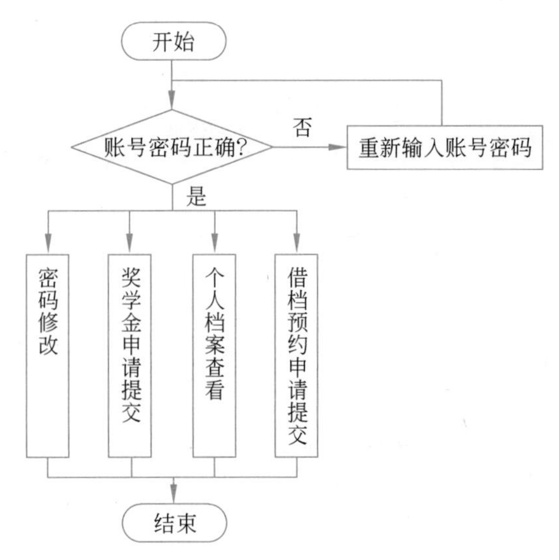

图 13.5　学生权限业务流程图

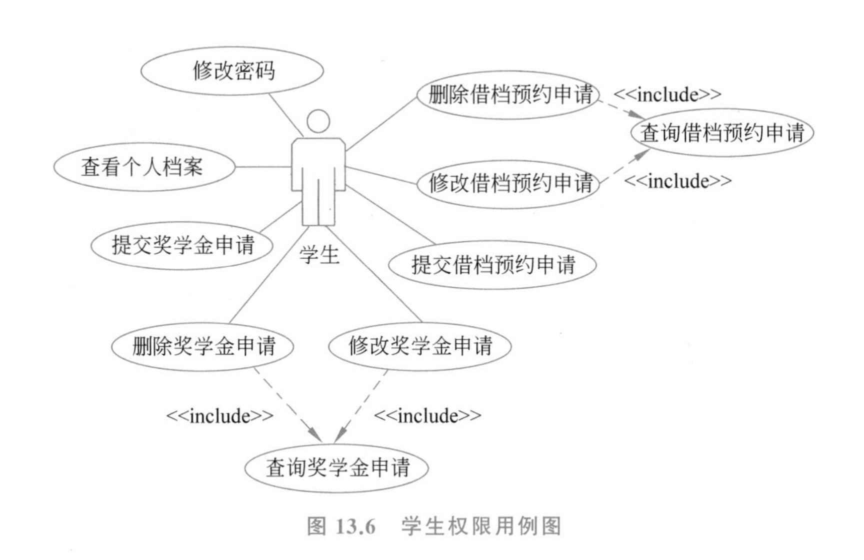

图 13.6　学生权限用例图

13.3　系统概要设计

将逻辑模型的各个处理模块进行分解，确定系统的层次结构关系，得到含义明确、职能单一的功能模块。系统模块的划分如图 13.7 所示。

13.3.1　概念设计

将需求分析的需求抽象化，形成概念模型。系统包含管理员信息属性、教师信息属性、学生基本档案属性、奖学金申请属性、借档预约属性、借阅记录属性等主要属性，其中管理员信息属性图如图 13.8 所示。

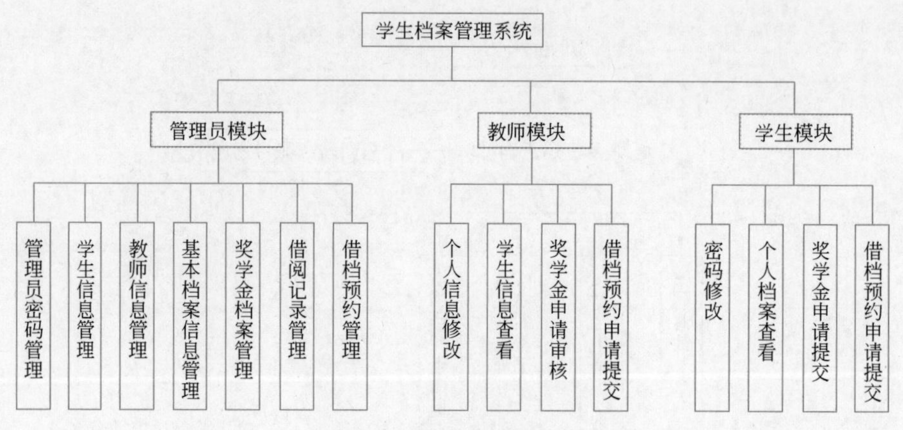

图 13.7　系统模块的划分

教师信息属性图如图 13.9 所示。

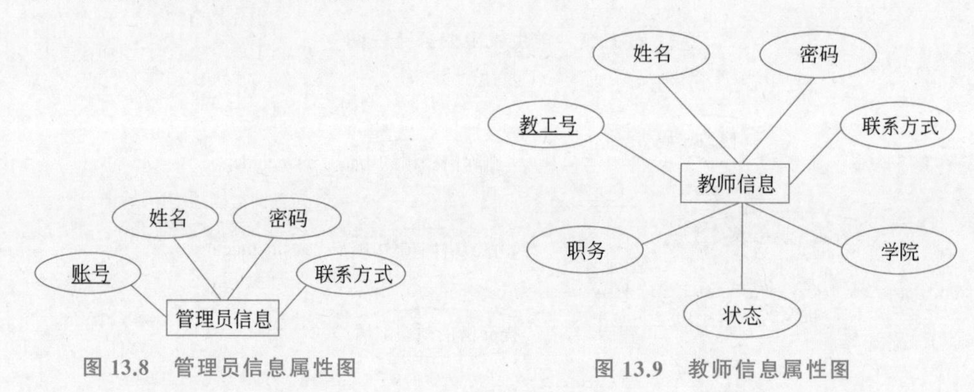

图 13.8　管理员信息属性图

图 13.9　教师信息属性图

学生基本档案属性图如图 13.10 所示。

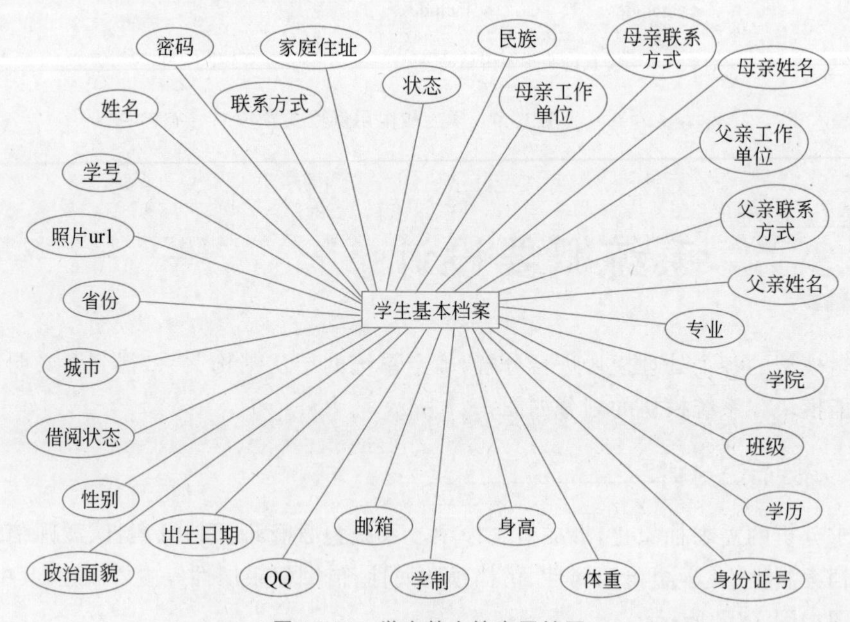

图 13.10　学生基本档案属性图

奖学金申请属性图如图 13.11 所示。

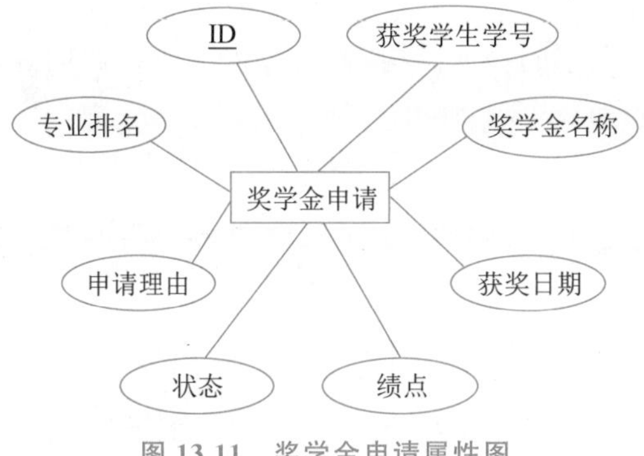

图 13.11　奖学金申请属性图

借档预约属性图如图 13.12 所示。

借阅记录属性图如图 13.13 所示。

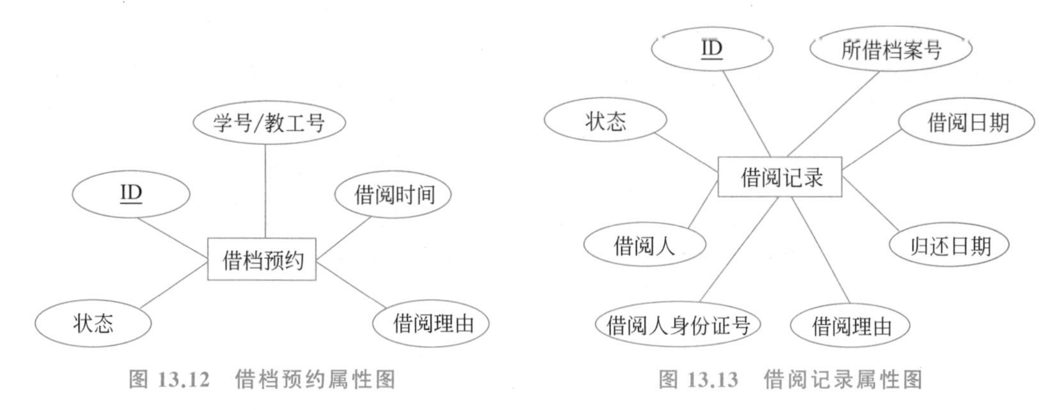

图 13.12　借档预约属性图

图 13.13　借阅记录属性图

13.3.2　逻辑设计

将概念设计得出的概念模型转化成数据模型，抽象体现对象之间的逻辑关系。主要属性图关系模型转化的关系模式如下。

管理员信息表(账号，姓名，密码，联系方式)

教师信息表(教工号，姓名，密码，联系方式，学院，职务，状态)

学生基本档案表(学号，姓名，密码，联系方式，家庭住址，状态，民族，母亲姓名，母亲联系方式，母亲工作单位，父亲姓名，父亲联系方式，父亲工作单位，学院，专业，班级，学历，身份证号，身高，体重，学制，邮箱，QQ，政治面貌，出生日期，性别，借阅状态，省份，城市，照片url)

奖学金申请表(ID，获奖学生学号，奖学金名称，获奖日期，绩点，专业排名，申请理由，状态)

借档预约表(ID，学号，教工号，借阅时间，借阅理由，状态)

借阅记录表(ID，所借档案号，借阅日期，归还日期，借阅理由，借阅人身份证号，借阅人，状态)

13.3.3 物理设计

为逻辑模型选定符合要求的物理结构，解决数据库的存储问题，包括存储结构、存取方法等信息。管理员信息表如表 13.1 所示。

表 13.1 管理员信息表

字段名	说明	类型	可否为空	主键	外键
adminid	管理员账号	int	否	是	否
adminname	管理员姓名	varchar	否	否	否
password	密码	varchar	否	否	否
tel	联系方式	varchar	是	否	否

学生基本档案表如表 13.2 所示。

表 13.2 学生基本档案表

字段名	说明	类型	可否为空	主键	外键
studentid	学号	int	否	是	否
studentname	学生姓名	varchar	否	否	否
tel	联系方式	varchar	否	否	否
address	家庭住址	varchar	否	否	否
password	密码	varchar	否	否	否
nation	民族	varchar	否	否	否
sex	性别	char	否	否	否
politics	政治面貌	varchar	否	否	否
QQ	QQ	varchar	否	否	否
email	邮箱	varchar	否	否	否
height	身高	char	否	否	否
weight	体重	char	否	否	否
ID	身份证号	varchar	否	否	否
father	父亲姓名	varchar	否	否	否
ftel	父亲联系方式	varchar	否	否	否
fjob	父亲工作单位	varchar	否	否	否
mother	母亲姓名	varchar	否	否	否
mtel	母亲联系方式	varchar	否	否	否
mjob	母亲工作单位	varchar	否	否	否
degree	学历	varchar	否	否	否

续表

字段名	说明	类型	可否为空	主键	外键
schoolid	学院 ID	int	否	否	是
majorid	专业 ID	int	否	否	是
classid	班级 ID	int	否	否	是
system	学制	char	否	否	否
authorizedid	状态 ID	int	否	否	是
photourl	照片 url	text	是	否	否
provinceid	省份 ID	int	否	否	是
cityid	城市 ID	int	否	否	是
lend	借阅状态	varchar	否	否	否

教师信息表如表 13.3 所示。

表 13.3 教师信息表

字段名	说明	类型	可否为空	主键	外键
teacherid	教工号	int	否	是	否
teachername	教师姓名	varchar	否	否	否
password	密码	varchar	否	否	否
tel	联系方式	varchar	否	否	否
authorizedid	状态 ID	int	否	否	是
job	职务	varchar	否	否	否
schoolid	学院 ID	int	否	否	是

奖学金表如表 13.4 所示。

表 13.4 奖学金表

字段名	说明	类型	可否为空	主键	外键
scholarshipid	奖学金 ID	int	否	是	否
scholarshipname	奖学金名称	varchar	否	否	否
money	金额	int	否	否	否

奖学金申请表如表 13.5 所示。

表 13.5 奖学金申请表

字段名	说明	类型	可否为空	主键	外键
sapplicationid	奖学金申请 ID	int	否	是	否
studentid	学号	int	否	否	是

续表

字段名	说明	类型	可否为空	主键	外键
scholarshipid	奖学金 ID	int	否	否	是
date	获奖日期	varchar	否	否	否
reason	申请理由	text	否	否	否
rank	专业排名	varchar	否	否	否
score	绩点	varchar	否	否	否
authorizedid	状态 ID	int	否	否	是

学生借档预约表如表 13.6 所示。

表 13.6 学生借档预约表

字段名	说明	类型	可否为空	主键	外键
sappoid	学生借档预约 ID	int	否	是	否
studentid	学号	int	否	否	是
date	借阅时间	varchar	否	否	否
reason	借阅理由	text	否	否	否
authorizedid	状态 ID	int	否	否	是

13.4 类库代码实现

13.4.1 实体层设计

实体层实现数据封装,与数据库中的数据表及视图对应,主要类代码设计如下。

(1) 管理员表对应 AdminEntity 类,编辑 AdminEntity.cs 代码如下。

```
public class AdminEntity
{
    public int Adminid{ get; set; }
    public string AdminName{ get; set; }
    public string Password{ get; set; }
    public string Tel{ get; set; }
}
```

(2) 学生表对应 StudentEntity 类,编辑 StudentEntity.cs 代码如下。

```
public class StudentEntity
{
    public string StudentName{ get; set; }
    public string Tel{ get; set; }
    public string Address{ get; set; }
    public string Nation{ get; set; }
    public string Sex{ get; set; }
```

```
    public string Politics{ get; set; }
    public string QQ{ get; set; }
    public string Email{ get; set; }
    public string Height{ get; set; }
    public string Weight{ get; set; }
    public string ID{ get; set; }
    public string Father{ get; set; }
    public string Ftel{ get; set; }
    public string Fjob{ get; set; }
    public string Mother{ get; set; }
    public string Mtel{ get; set; }
    public string Mjob{ get; set; }
    public string Degree{ get; set; }
    public string System{ get; set; }
    public string Photourl{ get; set; }
    public int SchoolId{ get; set; }
    public string SchoolName{ get; set; }
    public int MajorId{ get; set; }
    public string MajorName{ get; set; }
    public int ClassId{ get; set; }
    public string ClassName{ get; set; }
    public int AuthorizedId{ get; set; }
    public string AuthorizedName{ get; set; }
    public int ProvinceId{ get; set; }
    public string  ProvinceName{ get; set; }
    public int CityId{ get; set; }
    public string CityName{ get; set; }
    public int StudentId{ get; set; }
    public string Password{ get; set; }
    public string Lend{ get; set; }
}
```

(3) 教师表对应 TeacherEntity 类,编辑 TeacherEntity.cs 代码如下。

```
public class TeacherEntity
{
    public int TeacherId{ get; set; }
    public string TeacherName{ get; set; }
    public string Tel{ get; set; }
    public int AuthorizedId{ get; set; }
    public string Job{ get; set; }
    public int SchoolId{ get; set; }
    public string Password{ get; set; }
}
```

(4) 借阅表对应 LendEntity 类,编辑 LendEntity.cs 代码如下。

```
public class LendEntity
{
    public int LendId{ get; set; }
    public int StudentId{ get; set; }
    public string StartDate{ get; set; }
    public string EndDate{ get; set; }
```

```
    public string LendMan{ get; set; }
    public string Reason{ get; set; }
    public int AuthorizedId{ get; set; }
    public string ID{ get; set; }
}
```

(5) 奖学金申请表对应 SapplicationEntity 类，编辑 SapplicationEntity.cs 代码如下。

```
public class SapplicationEntity
{
    public int SapplicationId{ get; set; }
    public int ScholarshipId{ get; set; }
    public string Date{ get; set; }
    public string Reason{ get; set; }
    public string Rank{ get; set; }
    public string Score{ get; set; }
    public int AuthorizedId{ get; set; }
    public int StudentId{ get; set; }
}
```

(6) 借档预约表对应 SappoEntity 类，编辑 SappoEntity.cs 代码如下。

```
public class SappoEntity
{
    public int StudentId{ get; set; }
    public string Date{ get; set; }
    public string Reason{ get; set; }
    public int AuthorizedId{ get; set; }
}
```

13.4.2 数据访问层设计

数据访问层与数据库进行交互，实现对数据表的增加、删除、修改、查询操作。数据访问 ArchiveDataAccess 类，编写 ArchiveDataAccess.cs 代码如下。

```
public static class ArchiveDataAccess
{
    static SqlConnection con;
    static SqlCommand cmd;
    static SqlDataAdapter sda;
    static ArchiveDataAccess()
    {
        con=new SqlConnection();
        con.ConnectionString = ConfigurationManager. ConnectionStrings [ "
ArConn"].ConnectionString;
        cmd=new SqlCommand();
        cmd.Connection=con;
        sda=new SqlDataAdapter();
    }
    public static int ExcuteNonSQL(string cmdText, CommandType cmdType, string[]
paramNames, object[] paramValues)
    {
```

```
        cmd.CommandText=cmdText;
        cmd.CommandType=cmdType;
        cmd.Parameters.Clear();
        if (paramNames !=null)
        {
            for (int i=0; i<paramNames.Length; i++)
            {
                cmd.Parameters.AddWithValue(paramNames[i], paramValues[i]);
            }
        }
        if (con.State==ConnectionState.Closed)
        {
            con.Open();
        }
        int n=cmd.ExecuteNonQuery();
        con.Close();
        return n;
    }
    public static DataTable GetDataTable(string cmdText, CommandType cmdType,
string[] paramNames, object[] paramValues)
    {
        sda.SelectCommand=new SqlCommand();
        sda.SelectCommand.CommandText=cmdText;
        sda.SelectCommand.CommandType=cmdType;
        sda.SelectCommand.Connection=con;
        if (paramNames !=null)
        {
            for (int i=0; i<paramNames.Length; i++)
            {
                sda.SelectCommand.Parameters.AddWithValue(paramNames[i],
paramValues[i]);
            }
        }
        DataTable dt=new DataTable();
        sda.Fill(dt);
        return dt;
    }
    public static object ExcuteScalarSQL(string cmdText, CommandType cmdType,
string[] paramNames, object[] paramValues)
    {
        cmd.CommandText=cmdText;
        cmd.CommandType=cmdType;
        cmd.Parameters.Clear();
        if (paramNames !=null)
        {
            for (int i=0; i<paramNames.Length; i++)
            {
                cmd.Parameters.AddWithValue(paramNames[i], paramValues[i]);
            }
        }
        if (con.State==ConnectionState.Closed)
        {
            con.Open();
```

```
            }
            object obj=cmd.ExecuteScalar();
            con.Close();
            return obj;
        }
    }
```

13.4.3 业务逻辑层设计

业务逻辑层实现与档案管理业务相关的功能,对应数据库中数据表的操作,代码设计如下。

(1) 管理管理员表对应业务逻辑 AdminBussiness 类,编辑 AdminBussiness.cs 主要代码如下。

```
using System.Data;
using Entity;

namespace Business
{
    public class AdminBussiness
    {
        public static bool AdminLogin(int aId, string pwd)
        {
            string[] paramNames=new string[] { "Adminid","Password" };
            object[] paramValues=new object[] { aId,pwd };
            object obj = DataAccess. ArchiveDataAccess. ExcuteScalarSQL ( " Up _
AdminLogin", CommandType.StoredProcedure, paramNames, paramValues);
            if (obj ! =null)
                return true;
            else
                return false;
        }
        public static AdminEntity SelectAdminByID(int aId)
        {
            string[] paramNames=new string[] { "Adminid" };
            object[] paramValues=new object[] { aId };
            DataTable dt = DataAccess. ArchiveDataAccess. GetDataTable ( " Up _
SelectAdminByID", CommandType.StoredProcedure, paramNames, paramValues);
            AdminEntity admin=new AdminEntity();
            admin.Adminid=(int)dt.Rows[0]["adminid"];
            admin.AdminName=dt.Rows[0]["adminname"].ToString();
            admin.Password=dt.Rows[0]["password"].ToString();
            admin.Tel=dt.Rows[0]["tel"].ToString();
            return admin;
        }
        public static bool UpdateAdminByID(AdminEntity adminInfo)
        {
            string[] paramNames=new string[] { "Adminid","AdminName","Password","
Tel" };
            object[] paramValues = new object [ ] {adminInfo. Adminid, adminInfo.
AdminName,adminInfo.Password,adminInfo.Tel };
```

```
            int n=DataAccess.ArchiveDataAccess.ExcuteNonSQL("Up_UpdateAdminByID",
CommandType.StoredProcedure, paramNames, paramValues);
            if (n >=1)
                return true;
            else
                return false;
        }
    }
}
```

(2) 管理档案类别表对应业务逻辑 AuthorizedBusiness 类，编辑 AuthorizedBusiness.cs 主要代码如下。

```
using System.Data;

namespace Business
{
    public class AuthorizedBusiness
    {
        public static DataTable SelectStudentAuthorized()
        {
            return DataAccess.ArchiveDataAccess.GetDataTable("Up_
SelectStudentAuthorized", CommandType.StoredProcedure, null, null);
        }
    }
}
```

(3) 管理借阅表对应业务逻辑 LendBusiness 类，编辑 LendBusiness.cs 主要代码如下。

```
using Entity;
using System.Data;
using DataAccess;

namespace Business
{
    public class LendBusiness
    {
        public static DataTable LendInfo()
        {
            return ArchiveDataAccess.GetDataTable("Up_LendInfo", CommandType.
StoredProcedure, null, null); ;
        }
        public static LendEntity SelectLendByID(int lId)
        {
            string[] paramNames=new string[] { "LendId" };
            object[] paramValues=new object[] { lId };
            DataTable dt=ArchiveDataAccess.GetDataTable("Up_SelectLendByID",
CommandType.StoredProcedure, paramNames, paramValues);
            LendEntity lend=new LendEntity();
            lend.StudentId=(int)dt.Rows[0]["StudentId"];
            lend.LendId=(int)dt.Rows[0]["LendId"];
            lend.AuthorizedId=(int)dt.Rows[0]["AuthorizedId"];
            lend.Reason=dt.Rows[0]["Reason"].ToString().Trim();
```

```
            lend.LendMan=dt.Rows[0]["LendMan"].ToString().Trim();
            lend.ID=dt.Rows[0]["ID"].ToString().Trim();
            lend.StartDate=dt.Rows[0]["StartDate"].ToString().Trim();
            lend.EndDate=dt.Rows[0]["EndDate"].ToString().Trim();
            return lend;
        }
        public static bool UpdateLendByID(int lid,LendEntity lendInfo)
        {
            string[] paramNames=new string[] { "LendId", "StudentId", "StartDate",
"EndDate", "LendMan","Reason","AuthorizedId","ID" };
            object[] paramValues = new object [ ] { lid, lendInfo. StudentId,
lendInfo. StartDate, lendInfo. EndDate, lendInfo. LendMan, lendInfo. Reason,
lendInfo.AuthorizedId,lendInfo.ID };
            int n = ArchiveDataAccess. ExcuteNonSQL ( " Up _ UpdateLendByID ",
CommandType.StoredProcedure, paramNames, paramValues);
            if (n >0)
                return true;
            else
                return false;
        }
        public static bool InsertLend(LendEntity lendInfo)
        {
            string[] paramNames = new string [ ] { " StudentId", " StartDate",
"EndDate", "LendMan", "Reason", "AuthorizedId", "ID" };
            object[] paramValues = new object [ ] {lendInfo. StudentId, lendInfo.
StartDate, lendInfo. EndDate, lendInfo. LendMan, lendInfo. Reason, lendInfo.
AuthorizedId, lendInfo.ID };
            int n=ArchiveDataAccess.ExcuteNonSQL("Up_InsertLend", CommandType.
StoredProcedure, paramNames, paramValues);
            if (n >0)
                return true;
            else
                return false;
        }
        public static bool UpdateLendAuthorized(int lId)
        {
            string[] paramNames=new string[] { "LendId" };
            object[] paramValues=new object[] { lId };
            int n=ArchiveDataAccess.ExcuteNonSQL("Up_UpdateLendAuthorized",
CommandType.StoredProcedure, paramNames, paramValues);
            if (n >0)
                return true;
            else
                return false;
        }
        public static DataTable SelectLendByStudentIDLendman (int sId, string
lendMan)
        {
            string[] paramNames=new string[] { "StudentId","LendMan" };
            object[] paramValues=new object[] { sId,lendMan };
            return ArchiveDataAccess.GetDataTable("Up_SelectLendByStudentIDLendMan",
CommandType.StoredProcedure, paramNames, paramValues);
```

```
        }
    }
}
```

(4) 管理奖学金申请表对应业务逻辑 SapplicationBusiness 类，编辑 SapplicationBusiness.cs 主要代码如下。

```
using DataAccess;
using Entity;
using System.Data;

namespace Business
{
    public class SapplicationBusiness
    {
        public static bool UpdateSapplicationAuthorized(int sId)
        {
            string[] paramNames=new string[] { "SapplicationId" };
            object[] paramValues=new object[] { sId };
            int n=ArchiveDataAccess.ExcuteNonSQL("Up_UpdateSapplicationAuthorized",
CommandType.StoredProcedure, paramNames, paramValues);
            if (n >0)
                return true;
            else
                return false;
        }
        public static DataTable SapplicationInfo()
        {
            string[] paramNames=new string[] { };
            object[] paramValues=new object[] { };
            return ArchiveDataAccess. GetDataTable ( " Up _ SapplicationInfo ",
CommandType.StoredProcedure, paramNames, paramValues);
        }
        public static SapplicationInfoEntity SelectSapplicationByID(int sId)
        {
            string[] paramNames=new string[] { "SapplicationId" };
            object[] paramValues=new object[] { sId };
            DataTable dt=ArchiveDataAccess.GetDataTable("Up_SelectSapplicationByID",
CommandType.StoredProcedure, paramNames, paramValues);
            SapplicationInfoEntity sapplication=new SapplicationInfoEntity();
            sapplication.StudentId =int.Parse( dt.Rows[0]["StudentId"].ToString().
Trim());
             sapplication. StudentName= dt. Rows [0] ["studentname"]. ToString ().
Trim();
            sapplication.Tel=dt.Rows[0]["Tel"].ToString().Trim();
            sapplication.Address=dt.Rows[0]["address"].ToString().Trim();
            sapplication.Sex=dt.Rows[0]["sex"].ToString().Trim();
            sapplication.Politics=dt.Rows[0]["politics"].ToString().Trim();
            sapplication.SchoolName=dt.Rows[0]["schoolname"].ToString().Trim();
            sapplication.MajorName=dt.Rows[0]["majorname"].ToString().Trim();
            sapplication.ClassName=dt.Rows[0]["classname"].ToString().Trim();
            sapplication.Rank=dt.Rows[0]["Rank"].ToString().Trim();
```

```
            sapplication.Score=dt.Rows[0]["Score"].ToString().Trim();
            sapplication.Money=dt.Rows[0]["money"].ToString().Trim();
            sapplication.Date=dt.Rows[0]["Date"].ToString().Trim();
            sapplication.ScholarshipName= dt.Rows [0] ["scholarshipname"].ToString ().
Trim ();
            sapplication.Reason=dt.Rows[0]["Reason"].ToString().Trim();
            sapplication.AuthorizedId= dt.Rows [0] ["AuthorizedId"].ToString ().
Trim ();
            sapplication.AuthorizedName= dt.Rows [0] ["AuthorizedName"].ToString ().
Trim ();
            sapplication.MajorId=dt.Rows[0]["MajorId"].ToString().Trim();
            sapplication.MajorName=dt.Rows[0]["MajorName"].ToString().Trim();
            return sapplication;
        }
         public static bool UpdateSapplicationByID (int sId, SapplicationEntity
sapplicationInfo)
        {
            string[] paramNames=new string[] { "SapplicationId","ScholarshipId",
"Date","Reason","Rank","Score","AuthorizedId" };
              object [ ] paramValues = new object [ ] { sId, sapplicationInfo.
ScholarshipId, sapplicationInfo. Date, sapplicationInfo. Reason, sapplicationInfo.
Rank,sapplicationInfo.Score,sapplicationInfo.AuthorizedId };
             int n= ArchiveDataAccess. ExcuteNonSQL ( "Up _UpdateSapplicationByID",
CommandType.StoredProcedure, paramNames, paramValues);
            if (n >0)
                return true;
            else
                return false;
        }
        public static bool InsertSapplication(SapplicationEntity sapplicationInfo)
        {
            string[] paramNames=new string[] { "StudentId","ScholarshipId", "
Date", "Reason", "Rank", "Score", "AuthorizedId" };
            object[] paramValues= new object [] { sapplicationInfo. StudentId,
sapplicationInfo. ScholarshipId, sapplicationInfo. Date, sapplicationInfo.
Reason, sapplicationInfo. Rank, sapplicationInfo. Score, sapplicationInfo.
AuthorizedId };
            int n = ArchiveDataAccess. ExcuteNonSQL ( " Up _ InsertSapplication ",
CommandType.StoredProcedure, paramNames, paramValues);
            if (n==1)
                return true;
            else
                return false;
        }
        public static bool TeacherCheckSapplication(int sId)
        {
            string[] paramNames=new string[] { "SapplicationId" };
            object[] paramValues=new object[] { sId };
            int n=ArchiveDataAccess.ExcuteNonSQL ( "Up _TeacherCheckSapplication ",
CommandType.StoredProcedure, paramNames, paramValues);
            if (n==1)
```

```
                return true;
            else
                return false;
        }
        public static DataTable SelectSapplicationByScholarshipID(int sId)
        {
            string[] paramNames=new string[] { "ScholarshipId" };
            object[] paramValues=new object[] { sId };
            return ArchiveDataAccess. GetDataTable ( " Up _ SelectSapplicationBy-
ScholarshipID", CommandType.StoredProcedure, paramNames, paramValues);
        }
        public static DataTable SelectSapplicationByAuthorizedSchool(int aId,
int sId)
        {
            string[] paramNames=new string[] { "AuthorizedId","schoolid"};
            object[] paramValues=new object[] { aId,sId };
            return ArchiveDataAccess.GetDataTable("Up_SelectSapplicationByAu-
thorizedSchool", CommandType.StoredProcedure, paramNames, paramValues);
        }
        public static DataTable SelectSapplicationNoByStudentID(int sId)
        {
            string[] paramNames=new string[] { "StudentId" };
            object[] paramValues=new object[] { sId };
            return ArchiveDataAccess.GetDataTable("Up_SelectSapplicationNoBy-
StudentID", CommandType.StoredProcedure, paramNames, paramValues);
        }
        public static DataTable SelectSapplicationYesByStudentID(int sId)
        {
            string[] paramNames=new string[] { "StudentId" };
            object[] paramValues=new object[] { sId };
            return ArchiveDataAccess.GetDataTable("Up_SelectSapplicationYesBy-
StudentID", CommandType.StoredProcedure, paramNames, paramValues);
        }
        public static DataTable SelectSapplicationBySchool(int sId)
        {
            string[] paramNames=new string[] { "schoolid" };
            object[] paramValues=new object[] { sId };
            return ArchiveDataAccess. GetDataTable ( " Up _ SelectSapplicationBy-
School", CommandType.StoredProcedure, paramNames, paramValues);
        }
    }
}
```

(5) 管理借档预约表对应业务逻辑 SappoBusiness 类，编辑 SappoBusiness.cs 主要代码如下。

```
using DataAccess;
using Entity;
using System.Data;

namespace Business
{
```

```
    public class SappoBusiness
    {
        public static bool UpdateSappoAuthorized(int sId)
        {
            string[] paramNames=new string[] { "sappoid" };
            object[] paramValues=new object[] { sId };
            int n=ArchiveDataAccess.ExcuteNonSQL("Up_UpdateSappoAuthorized",
CommandType.StoredProcedure, paramNames, paramValues);
            if (n >0)
                return true;
            else
                return false;
        }
        public static bool InsertSappo(SappoEntity sappoInfo)
        {
            string[] paramNames=new string[] { "StudentId", "Date", "Reason",
"AuthorizedId" };
            object[] paramValues=new object[] { sappoInfo.StudentId, sappoInfo.
Date, sappoInfo.Reason, sappoInfo.AuthorizedId };
            int n = ArchiveDataAccess. ExcuteNonSQL ( " Up _ InsertSappo ",
CommandType.StoredProcedure, paramNames, paramValues);
            if (n >0)
                return true;
            else
                return false;
        }
        public static DataTable SappoInfo()
        {
            string[] paramNames=new string[] { };
            object[] paramValues=new object[] {  };
            return ArchiveDataAccess.GetDataTable("Up_SappoInfo", CommandType.
StoredProcedure, paramNames, paramValues);
        }
        public static bool UpdateSappoByID(int sId, SappoEntity sappoInfo)
        {
            string[] paramNames=new string[] { "sappoid", "StudentId", "Date",
"Reason", "AuthorizedId" };
            object[] paramValues = new object [ ] { sId, sappoInfo. StudentId,
sappoInfo.Date, sappoInfo.Reason, sappoInfo.AuthorizedId };
            int n = ArchiveDataAccess. ExcuteNonSQL ( " Up _ UpdateSappoByID ",
CommandType.StoredProcedure, paramNames, paramValues);
            if (n==1)
                return true;
            else
                return false;
        }
        public static SappoInfoEntity SelectSappoByID(int sId)
        {
            string[] paramNames=new string[] { "sappoid" };
            object[] paramValues=new object[] { sId };
            DataTable dt=ArchiveDataAccess.GetDataTable("Up_SelectSappoByID",
CommandType.StoredProcedure, paramNames, paramValues);
            SappoInfoEntity sappoInfo=new SappoInfoEntity();
```

```
            sappoInfo.StudentId=(int)dt.Rows[0]["StudentId"];
            sappoInfo.Date=dt.Rows[0]["Date"].ToString().Trim();
            sappoInfo.Reason=dt.Rows[0]["Reason"].ToString().Trim();
            sappoInfo.StudentName=dt.Rows[0]["studentname"].ToString().Trim();
            sappoInfo.Tel=dt.Rows[0]["Tel"].ToString().Trim();
            sappoInfo.AuthorizedName=dt.Rows[0]["authorizedname"].ToString().Trim();
            return sappoInfo;
        }
        public static DataTable SelectSappoByStudentIDAuthorizedID(int sId,int aId)
        {
            string[] paramNames=new string[] { "StudentId", "authorizedid" };
            object[] paramValues=new object[] { sId, aId };
            return ArchiveDataAccess.GetDataTable("Up_SelectSappoByStudentID-
Authorized", CommandType.StoredProcedure, paramNames, paramValues);
        }
        public static DataTable SelectSappoYesByStudentID(int sId)
        {
            string[] paramNames=new string[] { "StudentId" };
            object[] paramValues=new object[] { sId };
            return ArchiveDataAccess.GetDataTable("Up_SelectSappoYesByStuden-
tID", CommandType.StoredProcedure, paramNames, paramValues);
        }
        public static DataTable SelectSappoNoByStudentID(int sId)
        {
            string[] paramNames=new string[] { "StudentId" };
            object[] paramValues=new object[] { sId };
            return ArchiveDataAccess.GetDataTable("Up_SelectSappoNoByStuden-
tID", CommandType.StoredProcedure, paramNames, paramValues);
        }
    }
}
```

（6）管理学生表对应业务逻辑 StudentBusiness 类，编辑 StudentBusiness.cs 主要代码如下。

```
using DataAccess;
using Entity;
using System.Data;

namespace Business
{
    public class StudentBusiness
    {
        public static bool StudentLogin(int sId, string pwd)
        {
            string[] paramNames=new string[] { "StudentId", "Password" };
            object[] paramValues=new object[] { sId, pwd };
            object obj=ArchiveDataAccess.ExcuteScalarSQL("Up_StudentLogin",
CommandType.StoredProcedure, paramNames, paramValues);
            if (obj !=null)
                return true;
            else
```

```
            return false;
        }
        public static bool IsExistStudentId(int sId)
        {
            string[] paramNames=new string[] { "StudentId" };
            object[] paramValues=new object[] { sId };
            DataTable dt=ArchiveDataAccess.GetDataTable("Up_SelectStudentByID",
CommandType.StoredProcedure, paramNames, paramValues);
            if (dt.Rows.Count >0)
                return true;
            else
                return false;
        }
        public static bool IsLendStudentId(int sId)
        {
            string[] paramNames=new string[] { "StudentId" };
            object[] paramValues=new object[] { sId };
            DataTable dt=ArchiveDataAccess.GetDataTable("Up_SelectStudentByID",
CommandType.StoredProcedure, paramNames, paramValues);
            if (dt.Rows[0]["lend"].ToString().Trim()=="已借出")
                return true;
            else
                return false;
        }
        public static DataTable StudentInfo()
        {
            return ArchiveDataAccess.GetDataTable("Up_StudentInfo", CommandType.
StoredProcedure, null, null);
        }
        public static DataTable SelectStudentByID(int sId)
        {
            string[] paramNames=new string[] {"StudentId" };
            object[] paramValues=new object[] {sId};
            return ArchiveDataAccess. GetDataTable ( " Up _ SelectStudentByID",
CommandType.StoredProcedure, paramNames, paramValues);
        }
        public static bool UpdateStudentByID(int sId,StudentEntity studentInfo)
        {
            string[] paramNames = new string [ ] { " StudentId"," studentname","
Tel","address","nation","sex","politics","QQ","email","height","weight","
ID","father","ftel","fjob","mother","mtel","mjob","degree","schoolid","major-
id"," classid"," system"," AuthorizedId"," photourl"," provinceid"," cityid","
lend" };
            object[] paramValues = new object [ ] { sId, studentInfo. StudentName,
studentInfo.Tel, studentInfo.Address, studentInfo.Nation, studentInfo.Sex, stu-
dentInfo.Politics, studentInfo.QQ, studentInfo.Email, studentInfo.Height, stu-
dentInfo. Weight, studentInfo. ID, studentInfo. Father, studentInfo. Ftel, stu-
dentInfo. Fjob, studentInfo. Mother, studentInfo. Mtel, studentInfo. Mjob, stu-
dentInfo. Degree, studentInfo. SchoolId, studentInfo. MajorId, studentInfo.
ClassId, studentInfo. System, studentInfo. AuthorizedId, studentInfo. Photourl,
studentInfo.ProvinceId,studentInfo.CityId,studentInfo.Lend};
            int n = ArchiveDataAccess. ExcuteNonSQL ( " Up _ UpdateStudentByID",
CommandType.StoredProcedure, paramNames, paramValues);
```

```
            if (n >0)
                return true;
            else
                return false;
        }
        public static object SelectMaxStudentid()
        {
            return ArchiveDataAccess.ExcuteScalarSQL("Up_SelectMaxStudentid",
    CommandType.StoredProcedure, null, null);
        }
        public static bool UpdateStudentAuthorized(int sId)
        {
            string[] paramNames=new string[] { "StudentId" };
            object[] paramValues=new object[] { sId };
            int n= ArchiveDataAccess. ExcuteNonSQL ( " Up _ UpdateStudentAuthorized",
    CommandType.StoredProcedure, paramNames, paramValues);
            if (n >0)
                return true;
            else
                return false;
        }
        public static bool UpdateStudentPasswordByID(int sId,string pwd)
        {
            string[] paramNames=new string[] { "StudentId","Password" };
            object[] paramValues=new object[] { sId ,pwd};
            int n=ArchiveDataAccess.ExcuteNonSQL("Up_UpdateStudentPasswordBy-
    ID", CommandType.StoredProcedure, paramNames, paramValues);
            if (n >0)
                return true;
            else
                return false;
        }
        public static DataTable SelectStudentBySMC(int sId,int mId,int cId)
        {
            string[] paramNames=new string[] { "schoolid","majorid","classid" };
            object[] paramValues=new object[] { sId,mId,cId };
            return ArchiveDataAccess. GetDataTable ( " Up _ SelectStudentBySMC ",
    CommandType.StoredProcedure, paramNames, paramValues);
        }
    }
}
```

（7）管理教师表对应业务逻辑 TeacherBusiness 类，编辑 TeacherBusiness.cs 主要代码如下。

```
using DataAccess;
using Entity;
using System.Data;

namespace Business
{
    public class TeacherBusiness
```

```
    {
        public static bool TeacherLogin(int tId, string pwd)
        {
            string[] paramNames=new string[] { "teacherid", "Password" };
            object[] paramValues=new object[] { tId, pwd };
            object obj = ArchiveDataAccess. ExcuteScalarSQL ( "Up _TeacherLogin",
CommandType.StoredProcedure, paramNames, paramValues);
            if (obj ! =null)
                return true;
            else
                return false;
        }
        public static DataTable TeacherInfo()
        {
            return  ArchiveDataAccess. GetDataTable  (  " Up  _  TeacherInfo ",
CommandType.StoredProcedure, null, null);
        }
        public static bool UpdateTeacherByID(TeacherEntity teacherInfo)
        {
            string[] paramNames = new string [ ]  {  " teacherid"," teachername ","
Password","Tel","job","schoolid"};
            object[] paramValues=new object[] {teacherInfo.TeacherId ,teacherInfo.
TeacherName, teacherInfo. Password, teacherInfo. Tel, teacherInfo. Job, teacherInfo.
SchoolId};
             int  n = ArchiveDataAccess. ExcuteNonSQL ( " Up _ UpdateTeacherByID ",
CommandType.StoredProcedure, paramNames, paramValues);
            if (n==1)
                return true;
            else
                return false;
        }
        public static bool UpdateTeacherAuthorized(int tId)
        {
            string[] paramNames=new string[] { "teacherid"};
            object[] paramValues=new object[] { tId};
            int n = ArchiveDataAccess. ExcuteNonSQL ( " Up _UpdateTeacherAuthorized ",
CommandType .StoredProcedure, paramNames, paramValues);
            if (n==1)
                return true;
            else
                return false;
        }
        public static bool InsertTeacher(TeacherEntity teacherInfo)
        {
            string[] paramNames = new string [ ] { "teacherid", "teachername", "
Password","Tel", "AuthorizedId","job", "schoolid" };
            object[]  paramValues  =  new  object  [  ]  {  teacherInfo. TeacherId,
teacherInfo. TeacherName, teacherInfo. Password, teacherInfo. Tel, teacherInfo.
AuthorizedId, teacherInfo.Job, teacherInfo.SchoolId };
            int  n  =  ArchiveDataAccess. ExcuteNonSQL  (  " Up  _  InsertTeacher ",
CommandType.StoredProcedure, paramNames, paramValues);
            if (n==1)
                return true;
```

```
            else
                return false;
        }
        public static TeacherInfoEntity SelectTeacherByID(int tId)
        {
            string[] paramNames=new string[] { "teacherid" };
            object[] paramValues=new object[] { tId };
            DataTable dt=ArchiveDataAccess.GetDataTable ("Up_SelectTeacherByID",
CommandType.StoredProcedure, paramNames, paramValues);
            TeacherInfoEntity teacherInfo=new TeacherInfoEntity();
            teacherInfo.AuthorizedId=dt.Rows[0]["Authorizedid"].ToString();
            teacherInfo.Expr1=dt.Rows[0]["Expr1"].ToString();
            teacherInfo.Expr2=dt.Rows[0]["Expr2"].ToString();
            teacherInfo.Job=dt.Rows[0]["Job"].ToString();
            teacherInfo.Password=dt.Rows[0]["Password"].ToString();
            teacherInfo.SchoolId =(int) dt.Rows[0]["SchoolId"];
            teacherInfo.SchoolName=dt.Rows[0]["SchoolName"].ToString();
            teacherInfo.TeacherName=dt.Rows[0]["teacherName"].ToString();
            teacherInfo.Tel=dt.Rows[0]["tel"].ToString();
            return teacherInfo;
        }
        public static DataTable SelectTeacherByTeacherIDSchoolID (int tid,int sId)
        {
            string[] paramNames=new string[] { "teacherid","schoolid" };
            object[] paramValues=new object[] {tid, sId };
            return ArchiveDataAccess.GetDataTable("Up_SelectTeacherByTeacheI-
DSchoolID", CommandType.StoredProcedure, paramNames, paramValues);
        }
    }
}
```

13.5 模块实现

13.5.1 登录页

登录时调用表示层中的 Login 类、业务逻辑层中的 AdminBusiness 类，以及数据访问层中的 ArchiveDataAccess 类。登录页运行测试如图 13.14 所示。

登录页视图主要源代码如下。

```
<form id="form1" runat="server" >
        <asp:Image ID="Image1" runat="server" ImageUrl="~/images/logo.png" />
        <asp:Label ID="lbl_name" runat="server"Text="学生档案管理系统"></asp:
Label>
        <table>
            <tr>
                <td>账号:</td>
                <td>
                    <asp:TextBox ID="txtAccount" runat="server" ></asp:TextBox>
```

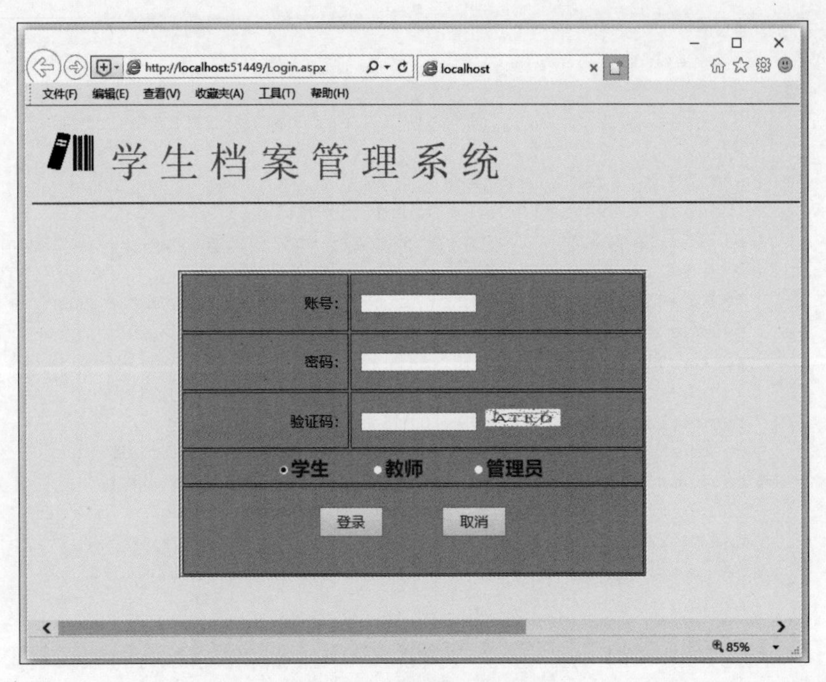

图 13.14　登录页运行测试

```
                <asp:RequiredFieldValidator ID="RequiredFieldValidator1"
runat="server" ControlToValidate="txtAccount" ErrorMessage="*" ForeColor=
"Red"  Display="Dynamic"></asp:RequiredFieldValidator>
            <asp:RegularExpressionValidator ID="RegularExpressionValidator1"
runat="server" ControlToValidate="txtAccount" ErrorMessage="请输入数字账号"
ForeColor="Red" ValidationExpression="^[0-9]*$" Display="Dynamic"></asp:
RegularExpressionValidator>
            </td>
        </tr>
        <tr>
            <td>密码:</td>
            <td>
            <asp:TextBox ID="txtPwd" runat="server" TextMode="Password"></
asp:TextBox>
            <asp:RequiredFieldValidator ID="RequiredFieldValidator2" runat=
"server" ControlToValidate =" txtPwd " ErrorMessage =" * " > </asp:
RequiredFieldValidator>
            </td>
        </tr>
        <tr>
            <td>验证码:</td>
            <td>
                <asp:TextBox ID="txtValidationCode" runat="server"></asp:
TextBox>
                <asp:ImageButton ID="ImageButton1" runat="server" ImageUrl=
"~/picture.aspx" CausesValidation="False" />
            </td>
        </tr>
```

```
            <tr>
                <td colspan="2">
        <asp:RadioButtonList ID="rbtnlLogin" runat="server">
                    <asp:ListItem Selected="True" Value="studentId">学生</asp:
ListItem>
                    <asp:ListItem Value="teacherId">教师</asp:ListItem>
                    <asp:ListItem Value="adminId">管理员</asp:ListItem>
        </asp:RadioButtonList>
                </td>
            </tr>
            <tr>
                <td colspan="2">
        <asp:Button ID="btnLogin" runat="server" Text="登录" OnClick="btnLogin
_Click" />
        <asp:Button ID="btnCancel" runat="server" Text="取消" CausesValidation=
"False" />
                </td>
            </tr>
        </table>
    </div>
</form>
```

登录页对应的主要C#代码如下。

```
protected void btnLogin_Click(object sender, EventArgs e)
{
     if (! String.Equals (txtValidationCode.Text, Session ["validatorCode"].
ToString(), StringComparison.CurrentCultureIgnoreCase))
    {
        AlertInfo.AlertDialog("验证码不正确");
    }
    else
    {
        string loginType=rbtnlLogin.SelectedValue;
        string loginUrl=string.Empty;
        bool isRightLogin=false;
        int loginId=Convert.ToInt32(txtAccount.Text.Trim());
        string pwd=txtPwd.Text.Trim();
        switch (loginType)
        {
            case "studentId":
                isRightLogin=StudentBusiness.StudentLogin(loginId, pwd);
                loginUrl="Student/StudentIndex.aspx";
                break;
            case "teacherId":
                isRightLogin=TeacherBusiness.TeacherLogin(loginId, pwd);
                loginUrl="Teacher/TeacherIndex.aspx";
                break;
           case "adminId":
                isRightLogin=AdminBussiness.AdminLogin(loginId, pwd);
                loginUrl="Admin/AdminIndex.aspx";
                break;
        }
```

```
            if (isRightLogin)
            {
                Session.Add(loginType, loginId);
                AlertInfo.AlertDialog("登录成功", loginUrl);
            }
            else
            {
                AlertInfo.AlertDialog("请检查账号和密码");
            }
        }
    }
```

13.5.2 管理员管理模块

1. 管理员首页

管理员首页显示个人信息管理、教师管理、基本档案管理、奖学金档案管理、借阅记录管理以及借档预约管理等链接。管理员首页测试如图 13.15 所示。

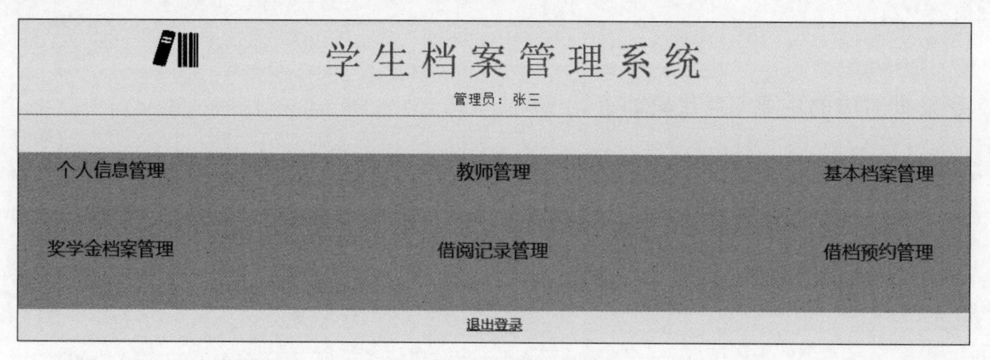

图 13.15 管理员首页测试

管理员主页视图主要源代码如下。

```
<form id="form1" runat="server">
      <asp:Image ID="Image1" runat="server" ImageUrl="~/images/logo.png"/>
      <asp:Label ID="lbl_name" runat="server" Text="学生档案管理系统" ></asp:
Label>
      <asp:Label ID="Label1" runat="server" Text="管理员:" ></asp:Label>
      <asp:Label ID="lblaAdminName" runat="server" ></asp:Label>
      <hr/>
      <table align="center" >
            <tr>
                <td>
        <asp:Menu ID="menuManage" runat="server"  DynamicHorizontalOffset="2"
  Orientation="Horizontal" OnMenuItemClick="menuManage_MenuItemClick" >
            <Items>
                <asp:MenuItem Text="个人信息管理" Value="个人信息管理"></asp:
MenuItem>
                <asp:MenuItem Text="教师管理" Value="教师管理"></asp:MenuItem>
                <asp:MenuItem Text="基本档案管理" Value="基本档案管理"></asp:
MenuItem>
```

```
                <asp:MenuItem Text="奖学金档案管理" Value="奖学金档案管理"></
asp:MenuItem>
                <asp:MenuItem Text="借阅记录管理" Value="借阅记录管理"></asp:
MenuItem>
                <asp:MenuItem Text="借档预约管理" Value="借档预约管理"></asp:
MenuItem>
            </Items>
        </asp:Menu>
                </td>
            </tr>
            <tr>
                <td>
                 <asp: LinkButton ID="lbtnLogout" runat="server" OnClick="
lbtnLogout_Click">退出登录</asp:LinkButton>
                </td>
            </tr>
        </table>
    </div>
</form>
```

管理员首页 C# 代码如下。

```
public partial class AdminIndex : System.Web.UI.Page
{
    protected void Page_Load(object sender, EventArgs e)
    {
        if (Session["adminId"] !=null)
        {
             AdminEntity admin=AdminBussiness.SelectAdminByID(Convert.ToInt16
(Session["Adminid"]));
            lblaAdminName.Text=admin.AdminName.Trim();
        }
         else
        {
            AlertInfo.AlertDialog("请先登录", "../Login.aspx");
        }
    }

    protected void menuManage_MenuItemClick(object sender, MenuEventArgs e)
    {
        string url="";
        switch (menuManage.SelectedValue)
        {
            case "个人信息管理": url="AdminInformation.aspx"; break;
            case "教师管理": url="AdminTeacherManage.aspx"; break;
            case "基本档案管理": url="AdminStudentManage.aspx"; break;
            case "奖学金档案管理": url="AdminScholarshipManage.aspx"; break;
            case "借阅记录管理": url="AdminLendManage.aspx"; break;
            case "借档预约管理": url="AdminAppoManage.aspx"; break;
        }
        Response.Redirect(url);
    }
    protected void lbtnLogout_Click(object sender, EventArgs e)
```

```
        {
            Session["Adminid"]=null;
            Response.Redirect("../Login.aspx");
        }
    }
```

2. 个人信息管理

个人信息管理页可查看个人信息并进行修改。个人信息管理页测试如图 13.16 所示。

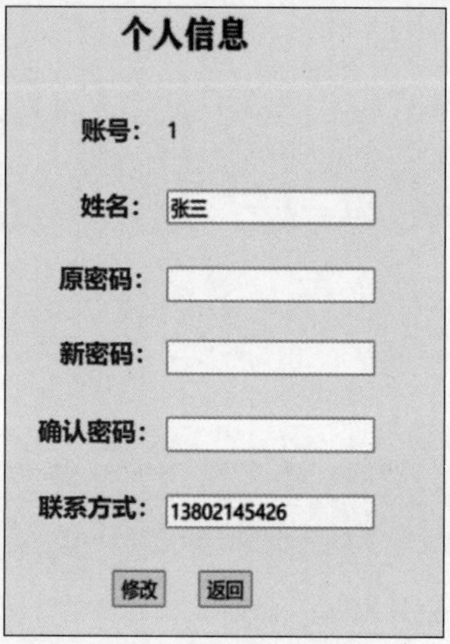

个人信息	
账号：	1
姓名：	张三
原密码：	
新密码：	
确认密码：	
联系方式：	13802145426
修改	返回

图 13.16　个人信息管理页测试

个人信息管理页视图主要源代码如下。

```
<form id="form1" runat="server">
    <asp:Label ID="Label4" runat="server" Text="个人信息"></asp:Label>
    <table>
        <tr>
            <td><asp:Label ID="Label5" runat="server" Text="账号:"></asp:
Label></td>
            <td><asp:Label ID="lblAccount" runat="server"></asp:Label></td>
        </tr>
        <tr>
            <td><asp:Label ID="Label6" runat="server" Text="姓名:"></asp:
Label></td>
            <td>
              <asp:TextBox ID="txtAdminName" runat="server"></asp:TextBox>
              <asp:RequiredFieldValidator ID="RequiredFieldValidator1" runat=
"server" ControlToValidate="txtAdminName" ErrorMessage="*" ForeColor="Red"></asp:
RequiredFieldValidator>
            </td>
        </tr>
        <tr>
        <td><asp:Label ID="Label10" runat="server" Text="原密码:"></asp:Label>
</td>
```

```
            <td>
        <asp:TextBox ID="txtPassword" runat="server"TextMode="Password"></asp:
TextBox>
        <asp:RegularExpressionValidator ID="RegularExpressionValidator1" runat=
"server" ControlToValidate="txtPassword" ErrorMessage="*" ForeColor="Red"
SetFocusOnError="True"></asp:RegularExpressionValidator>
                </td>
            </tr>
            <tr>
              <td><asp:Label ID="Label7" runat="server" Text="新密码:"></asp:
Label></td>
              <td>
                    <asp:TextBox ID="txtNewPassword" runat="server" EnableViewState=
"False" TextMode="Password"></asp:TextBox>
                      <asp:RegularExpressionValidator ID="RegularExpressionValidator2"
runat="server" ControlToValidate="txtNewPassword" ErrorMessage="*" ForeColor=
"Red" SetFocusOnError="True"></asp:RegularExpressionValidator>
                </td>
            </tr>
            <tr>
            <td><asp:Label ID="Label8" runat="server" Text="确认密码:"></asp:
Label></td>
                <td>
                      <asp:TextBox ID="txtRepwd" runat="server" TextMode="Password"></
asp:TextBox>
                      <asp: CompareValidator ID="CompareValidator1" runat="server"
ControlToCompare="txtNewPassword" ControlToValidate="txtRepwd" ErrorMessage="两次
密码不一致" ForeColor="Red"></asp:CompareValidator>
                </td>
            </tr>
            <tr>
            <td><asp:Label ID="Label9" runat="server" Text="联系方式:"></asp:
Label></td>
            <td><asp:TextBox ID="txtTel" runat="server"></asp:TextBox></td>
            </tr>
            <tr>
    <td>
                      <asp:Button ID="btnUpdate" runat="server" Text="修改" OnClick=
"btnUpdate_Click" OnClientClick="return confirm('确认删除?');" />
                      <asp: Button ID="btnReturn" runat="server" Text="返回"
CausesValidation="False" PostBackUrl="~/Admin/AdminIndex.aspx" />
                </td>
            </tr>
        </table>
</form>
```

个人信息管理页 C#代码如下。

```
public partial class AdminInformation : System.Web.UI.Page
{
    int aId;
    protected void Page_Load(object sender, EventArgs e)
    {
```

```
        if (Session["adminId"] !=null)
        {
            if (! IsPostBack)
            {
                lblAccount.Text=Session["Adminid"].ToString().Trim();
                aId=Convert.ToInt16(Session["Adminid"]);
                AdminEntity admin=AdminBussiness.SelectAdminByID(aId);
                txtTel.Text=admin.Tel.Trim();
                txtAdminName.Text=admin.AdminName.Trim();
            }
        }
        else
        {
            AlertInfo.AlertDialog("请先登录", "../Login.aspx");
        }
    }
    protected void btnUpdate_Click(object sender, EventArgs e)
    {
        AdminEntity admin=new AdminEntity();
        admin.AdminName=txtAdminName.Text.Trim();
        admin.Adminid=Convert.ToInt16(lblAccount.Text);
        admin.Tel=txtTel.Text.Trim();
        admin.Password=txtNewPassword.Text.Trim();
        if (AdminBussiness.AdminLogin(admin.Adminid, txtPassword.Text))
        {
            if (AdminBussiness.UpdateAdminByID(admin))
            {
                AlertInfo.AlertDialog("修改成功", "../Login.aspx");
            }
            else
            {
                AlertInfo.AlertDialog("修改失败");
            }
        }
    }
}
```

13.5.3 教师管理模块

1. 教师管理页

在教师管理页可查看所有教师信息，同时可按照教工号或学院进行搜索。教师管理页测试如图 13.17 所示。

教师管理页视图主要源代码如下。

```
<form id="form1" runat="server">
        <table align="center">
            <tr>
            <td><asp:Label ID="Label12" runat="server"Text="教师信息"></asp:
Label></td>
```

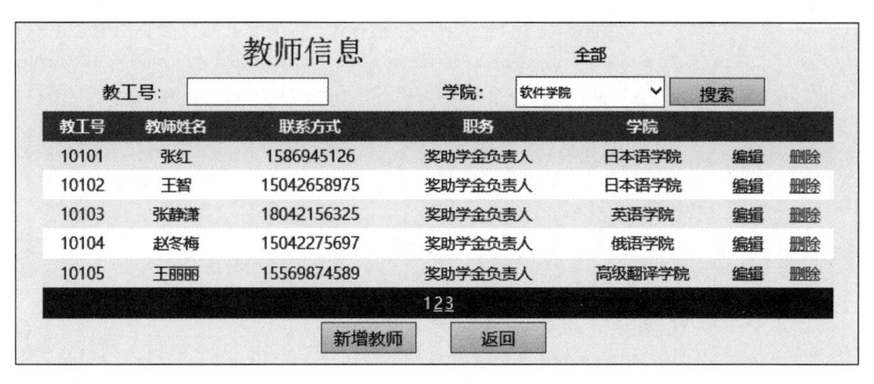

图 13.17　教师管理页测试

```
          <td><asp:LinkButton ID="lbtnAll" runat="server" OnClick="lbtnAll_
Click">全部</asp:LinkButton></td>
          </tr>
          <tr>
           <td><asp:Label ID="Label5" runat="server" Text="教工号:"></asp:
Label></td>
          <td><asp:TextBox ID="txtTeacherId" runat="server"></asp:
TextBox>
          <asp:RegularExpressionValidator ID="RegularExpressionValidator2"
runat="server" ControlToValidate="txtTeacherId" ErrorMessage="*" ForeColor="
Red" ValidationExpression="^[0-9]*$" Display="Dynamic"></asp:
RegularExpressionValidator></td>
           <td><asp:Label ID="Label13" runat="server" Text="学院:"></asp:
Label></td>
          <td><asp:DropDownList ID="ddlSchool" runat="server" AutoPostBack
="True">
           </asp:DropDownList></td>
           <td><asp:Button ID="btnSearch" runat="server" Text="搜索" OnClick
="btnSearch_Click" /></td>
           </tr>
           <tr>
           <td colspan="7">
      <asp:GridView ID="gvTeacher" runat="server" AutoGenerateColumns="
False" CellPadding="4" GridLines="None" OnRowCancelingEdit="gvTeacher_
RowCancelingEdit" OnRowDeleting="gvTeacher_RowDeleting" OnRowEditing="
gvTeacher_RowEditing" OnRowUpdating="gvTeacher_RowUpdating" AllowPaging="
True" OnPageIndexChanging="gvTeacher_PageIndexChanging" PageSize="5" Width="
825px">
          <Columns>
              <asp:BoundField DataField="teacherid" HeaderText="教工号"
ReadOnly="True"></asp:BoundField>
              <asp:TemplateField HeaderText="教师姓名">
                  <EditItemTemplate>
                      <asp:TextBox ID="tbxTeacherName" runat="server" Text='
<%#Bind("teachername")%>'></asp:TextBox>
```

```
                    <asp:RegularExpressionValidator ID="RegularExpression-
Validator3" runat="server" ControlToValidate="tbxTeacherName" ErrorMessage="
*" ForeColor="Red" SetFocusOnError="True"></asp:RegularExpressionValidator>
                </EditItemTemplate>
                <ItemTemplate>
                    <asp:Label ID="Label1" runat="server" Text='<%#Bind("
teachername") %>'></asp:Label>
                </ItemTemplate>
            </asp:TemplateField>
            <asp:TemplateField HeaderText="联系方式">
                <EditItemTemplate>
                    <asp:TextBox ID="tbxTel" runat="server" Text='<%#Bind
("tel") %>'></asp:TextBox>
                    <asp:RegularExpressionValidator ID="RegularExpression-
Validator4" runat="server" ControlToValidate="tbxTel" ErrorMessage="*" ForeC-
olor="Red" SetFocusOnError="True"></asp:RegularExpressionValidator>
                </EditItemTemplate>
                <ItemTemplate>
                    <asp:Label ID="Label2" runat="server" Text='<%#Bind
("tel") %>'></asp:Label>
                </ItemTemplate>
            </asp:TemplateField>
            <asp:TemplateField HeaderText="职务">
                <EditItemTemplate>
                    <asp:DropDownList ID="ddlJob" runat="server" Height=
"21px" Width="175px">
                        <asp:ListItem>奖助学金负责人</asp:ListItem>
                        <asp:ListItem>普通教师</asp:ListItem>
                    </asp:DropDownList>
                </EditItemTemplate>
                <ItemTemplate>
                    <asp:Label ID="lbl_job" runat="server" Text='<%#Eval
("job") %>'></asp:Label>
                </ItemTemplate>
            </asp:TemplateField>
            <asp:TemplateField HeaderText="学院">
                <EditItemTemplate>
                    <asp:DropDownList ID="ddlSchool" runat="server"></asp:
DropDownList>
                </EditItemTemplate>
                <ItemTemplate>
                    <asp:Label ID="lbl_school" runat="server" Text='<%#
Bind("schoolname") %>'></asp:Label>
                </ItemTemplate>
            </asp:TemplateField>
            <asp:CommandField ShowEditButton="True" UpdateText="修改" >
            </asp:CommandField>
            <asp:TemplateField ShowHeader="False">
                <ItemTemplate>
                    <asp:LinkButton ID="LinkButton1" runat="server" Causes-
Validation="False" CommandName="Delete" Text="删除" OnClientClick="return
confirm('确认要删除吗?');"></asp:LinkButton>
                </ItemTemplate>
```

```
            </asp:TemplateField>
        </Columns>
    </asp:GridView>
    <td colspan="3">
    <asp: Button ID =" btnAdd " runat =" server " Text =" 新 增 教 师 "
CausesValidation="False" PostBackUrl="~/Admin/AdminTeacherAdd.aspx" />
    <asp: Button ID =" btnReturn " runat =" server " Text =" 返 回 "
CausesValidation="False" PostBackUrl="~/Admin/AdminIndex.aspx" /></td>
        </table>
    </div>
</form>
```

教师管理页 C#代码如下。

```
using Business;
using Entity;
using System;
using System.Data;
using System.Web.UI.WebControls;

namespace Archive
{
    public partial class AdminTeacherManage : System.Web.UI.Page
    {
        int tId, sId;
        string message;
        protected void Page_Load(object sender, EventArgs e)
        {
            if (Session["adminId"] !=null)
            {
                if (! IsPostBack)
                {
                    DataTable dtsc=SchoolBusiness.GetSchoolInfo();
                    ddlSchool.DataSource=dtsc;
                    ddlSchool.DataValueField="schoolid";
                    ddlSchool.DataTextField="schoolname";
                    ddlSchool.DataBind();
                    LoadTeacher();
                }
            }
            else
            {
                AlertInfo.AlertDialog("请先登录", "../Login.aspx");
            }
        }
        private void LoadTeacher()
        {
            DataTable dt=TeacherBusiness.TeacherInfo();
            gvTeacher.DataSource=dt;
            gvTeacher.DataKeyNames=new string[] { "teacherid" };
            gvTeacher.DataBind();
        }
        protected void gvTeacher_RowEditing(object sender, GridViewEditEventArgs e)
```

```
        {
            gvTeacher.EditIndex=e.NewEditIndex;
            sId=Convert.ToInt16(gvTeacher.Rows[e.NewEditIndex].Cells[7].Text);
            LoadTeacher();
            DropDownList ddlSchool= (DropDownList) gvTeacher.Rows[e.NewEditIndex].
FindControl("ddlSchool");
            DataTable dt=SchoolBusiness.GetSchoolInfo();
            ddlSchool.DataSource=dt;
            ddlSchool.DataValueField="schoolid";
            ddlSchool.DataTextField="schoolname";
            ddlSchool.SelectedValue=sId.ToString(); ;
            ddlSchool.DataBind();
        }
        protected void gvTeacher_RowCancelingEdit(object sender, GridViewCan-
celEditEventArgs e)
        {
            gvTeacher.EditIndex=-1;
            LoadTeacher();
        }
        protected void gvTeacher_RowUpdating(object sender, GridViewUpdateEven-
tArgs e)
        {
            if (e.RowIndex >=0)
            {
                TeacherEntity teacherEntity=new TeacherEntity();
                teacherEntity. TeacherName = ((TextBox) gvTeacher. Rows [e.
RowIndex].FindControl("txtTeachername")).Text.Trim();
                teacherEntity.Tel = ((TextBox) gvTeacher. Rows [e. RowIndex].
FindControl("txtTel")).Text.Trim();
                teacherEntity.Job= ((DropDownList) gvTeacher.Rows[e.RowIndex].
FindControl("ddlJob")).SelectedValue.Trim();
                teacherEntity.SchoolId= Convert.ToInt16(((DropDownList) gvTeacher.
Rows[e.RowIndex].FindControl("ddlSchool")).SelectedValue);
                message="修改失败";
                if (TeacherBusiness.UpdateTeacherByID(teacherEntity))
                {
                    message="修改成功";
                }
                AlertInfo.AlertDialog(message);
                gvTeacher.EditIndex=-1;
                LoadTeacher();
            }
        }
        protected void gvTeacher_RowDeleting(object sender, GridViewDeleteEven-
tArgs e)
        {
            if (e.RowIndex >=0)
            {
                tId=Convert.ToInt16(gvTeacher.DataKeys[e.RowIndex].Value);
                message="修改失败";
                if (TeacherBusiness.UpdateTeacherAuthorized(tId))
                {
                    message="修改成功";
```

```
                }
                AlertInfo.AlertDialog(message);
            }
            gvTeacher.Rows[e.RowIndex].Visible=false;
            LoadTeacher();
        }
        protected void lbtnAll_Click(object sender, EventArgs e)
        {
            LoadTeacher();
        }
        protected void btnSearch_Click(object sender, EventArgs e)
        {
            tId= txtTeacherId.Text != string.Empty? Convert.ToInt32(txtTeacherId.
Text) : 0;
            sId=Convert.ToInt32(ddlSchool.SelectedValue);
            DataTable dt = TeacherBusiness.SelectTeacherByTeacherIDSchoolID (tId,
sId);
            gvTeacher.DataSource=dt;
            gvTeacher.DataKeyNames=new string[] { "teacherid" };
            gvTeacher.DataBind();
        }
            protected void gvTeacher _ PageIndexChanging ( object sender,
GridViewPageEventArgs e)
        {
            gvTeacher.PageIndex=e.NewPageIndex;
            LoadTeacher();
        }
    }
}
```

2. 新增教师页

在新增教师页可输入教师信息。新增教师页测试如图 13.18 所示。

教师信息

姓名：

联系方式：

职务：奖助学金负责人

学院：软件学院

保存 返回

图 13.18 新增教师页测试

新增教师页视图主要源代码如下。

```
<form id="form1" runat="server">
      <asp:Label ID="Label12" runat="server"  Text="教师信息"></asp:Label>
       <table>
           <tr>
            <td><asp:Label ID="Label13" runat="server" Text="姓名:"></asp:
Label></td>
              <td><asp:TextBox ID="txtTeacherName" runat="server"></asp:
TextBox>
            <asp:RequiredFieldValidator ID="RequiredFieldValidator1" runat=
"server" ControlToValidate="txtTeacherName" ErrorMessage="*" ForeColor="Red">
</asp:RequiredFieldValidator></td>
           </tr>
           <tr>
           <td><asp:Label ID="Label14" runat="server" Text="联系方式:"></asp:
Label></td>
           <td><asp:TextBox ID="txtTel" runat="server" ></asp:TextBox></td>
           </tr>
           <tr>
            <td><asp:Label ID="Label15" runat="server" Text="职务:"></asp:
Label></td>
           <td>
               <asp:DropDownList ID="ddlJob" runat="server">
                   <asp:ListItem>奖助学金负责人</asp:ListItem>
                   <asp:ListItem>普通教师</asp:ListItem>
               </asp:DropDownList>
           </td>
           </tr>
           <tr>
            <td><asp:Label ID="Label16" runat="server" Text="学院:"></asp:
Label></td>
            <td><asp:DropDownList ID="ddlSchool" runat="server">
                  </asp:DropDownList></td>
           </tr>
           <tr>
               <td colspan="2">
                   <asp:Button ID="btnSave" runat="server" OnClick="btnSave_
Click" Text="保存"  OnClientClick="return confirm('确认保存?');" />
                   <asp:Button ID="btnReturn" runat="server" Text="返回"
CausesValidation="False" PostBackUrl="~/Admin/AdminTeacherManage.aspx" />
               </td>
           </tr>
       </table>
</form>
```

新增教师页 C#代码如下。

```
using Business;
using Entity;
using System;
using System.Data;

namespace Archive
```

```
{
    public partial class AdminTeacherAdd : System.Web.UI.Page
    {
        protected void Page_Load(object sender, EventArgs e)
        {
            if (Session["adminId"] ! =null)
            {
                DataTable dt=SchoolBusiness.GetSchoolInfo();
                ddlSchool.DataSource=dt;
                ddlSchool.DataValueField="schoolid";
                ddlSchool.DataTextField="schoolname";
                ddlSchool.DataBind();
            }
            else
            {
                AlertInfo.AlertDialog("请先登录", "../Login.aspx");
            }
        }
        protected void btnSave_Click(object sender, EventArgs e)
        {
            TeacherEntity teacherEntity=new TeacherEntity();
            teacherEntity.TeacherName=txtTeacherName.Text.Trim();
            teacherEntity.Tel=txtTel.Text.Trim();
            teacherEntity.AuthorizedId=7;
            teacherEntity.Job=ddlJob.Text.Trim();
            teacherEntity.SchoolId=Convert.ToInt16(ddlSchool.SelectedValue);
            if (TeacherBusiness.InsertTeacher(teacherEntity))
            {
                AlertInfo.AlertDialog("新增成功","AdminTeacherManage.aspx");
            }
            else
            {
                AlertInfo.AlertDialog("新增失败");
            }
        }
    }
}
```

13.5.4 基本档案管理

1. 基本档案管理页

在基本档案管理页可查看学生基本档案，可按照学号或班级对学生基本档案进行搜索。基本档案管理页测试如图 13.19 所示。

基本档案管理页视图主要源代码如下。

```
<form id="form1" runat="server">
      <asp:Label ID="Label12" runat="server" Text="学生基本档案"></asp:Label>
      <asp:LinkButton ID="lbtnAll" runat="server"  OnClick="lbtnAll_Click">全
部</asp:LinkButton>
          <asp:Label ID="Label4" runat="server" Text="学号:"></asp:Label>
          <asp:TextBox ID="txtStudentId" runat="server" ></asp:TextBox>
```

图 13.19 基本档案管理页测试

```
            <asp:Button ID="btnSearch" runat="server" Text="搜索" OnClick="
btnSearch_Click" />
            <asp:DropDownList ID="ddlSchool" runat="server"  OnTextChanged="
ddlSchool_TextChanged" ></asp:DropDownList>
            <asp:DropDownList ID="ddlMajor" runat="server" AutoPostBack="True"
OnTextChanged="ddlMajor_TextChanged"></asp:DropDownList>
                < asp: DropDownList  ID =" ddlClass "  runat =" server " > </asp:
DropDownList>
            <asp:Button ID="btnSearch1" runat="server" Text="搜索" OnClick="
btnSearch1_Click"  />
             <asp:GridView ID="gvStudent" runat="server" AutoGenerateColumns
=" False "  CellPadding =" 4 "    GridLines =" None "  OnRowEditing =" gvStudent _
RowEditing"  OnRowDeleting =" gvStudent _ RowDeleting "  AllowPaging =" True "
OnPageIndexChanging="gvStudent_PageIndexChanging" PageSize="5">
            <Columns>
                <asp: BoundField  DataField =" studentid "  HeaderText =" 学 号 "
ReadOnly="True" >
                </asp:BoundField>
                <asp:BoundField DataField="studentname" HeaderText="姓名" >
                </asp:BoundField>
                <asp:BoundField DataField="schoolname" HeaderText="学院" >
                </asp:BoundField>
                <asp:BoundField DataField="majorname" HeaderText="专业" >
                </asp:BoundField>
                <asp:BoundField DataField="classname" HeaderText="班级" >
                </asp:BoundField>
                <asp:BoundField DataField="lend" HeaderText="档案状态" >
                </asp:BoundField>
                <asp:CommandField ShowEditButton="True" >
                </asp:CommandField>
                <asp:TemplateField ShowHeader="False">
                    <ItemTemplate>
                        <asp: LinkButton  ID =" LinkButton1 "  runat =" server "
CausesValidation="False" CommandName="Delete" Text="删除" OnClientClick="
return confirm('确认要删除吗?');"></asp:LinkButton>
                    </ItemTemplate>
                </asp:TemplateField>
            </Columns>
        </asp:GridView>
```

```
        <asp:Button ID="btnAdd" runat="server" Text="新增学生基本档案"
PostBackUrl="AdminStudentAdd.aspx" />
        <asp:Button ID="btnReturn" runat="server" Text="返回"
CausesValidation="False" PostBackUrl="~/Admin/AdminIndex.aspx" />
    </div>
</form>
```

基本档案管理页 C#代码如下。

```
using Business;
using System;
using System.Data;
using System.Web.UI.WebControls;

namespace Archive
{
    public partial class AdminStudentManage : System.Web.UI.Page
    {
        int sId,mId,cId;
        protected void Page_Load(object sender, EventArgs e)
        {
            Session["adminId"]=20220001;
            if (Session["adminId"] !=null)
            {
                if (!IsPostBack)
                {
                    DataTable dtsc=SchoolBusiness.GetSchoolInfo();
                    ddlSchool.DataSource=dtsc;
                    ddlSchool.DataValueField="schoolid";
                    ddlSchool.DataTextField="schoolname";
                    ddlSchool.DataBind();
                    if (ddlSchool.Text !=string.Empty)
                    {
                        int sid=Convert.ToInt16(ddlSchool.SelectedValue);
                        DataTable dtm=MajorBusiness.SelectMajorBySchoolID(sid);
                        ddlMajor.DataSource=dtm;
                        ddlMajor.DataValueField="majorid";
                        ddlMajor.DataTextField="majorname";
                        ddlMajor.DataBind();
                    }
                    if (ddlSchool.Text !=string.Empty && ddlMajor.Text !=string.Empty)
                    {
                        int mid=Convert.ToInt16(ddlMajor.SelectedValue);
                        int sid=Convert.ToInt16(ddlSchool.SelectedValue);
                        DataTable dtc = ClassBusiness.SelectClassByMajorSchoolID
(mid, sid);
                        ddlClass.DataSource=dtc;
                        ddlClass.DataValueField="classid";
                        ddlClass.DataTextField="classname";
                        ddlClass.DataBind();
                    }
                    LoadStudent();
                }
```

```
            }
            else
            {
                AlertInfo.AlertDialog("请先登录", "../Login.aspx");
            }
        }
        private void LoadStudent()
        {
            DataTable dt=StudentBusiness.StudentInfo();
            gvStudent.DataSource=dt;
            gvStudent.DataKeyNames=new string[] { "StudentId" };
            gvStudent.DataBind();
        }
        protected void gvStudent_RowEditing(object sender, GridViewEditEventArgs e)
        {
            Session["StudentId"] = gvStudent. DataKeys [e. NewEditIndex]. Value.
ToString().
Trim();
            Response.Redirect("AdminStudentUpdate.aspx");
        }
        protected void gvStudent_RowDeleting(object sender, GridViewDeleteEventArgs e)
        {
            if (e.RowIndex >=0)
            {
                sId=Convert.ToInt32(gvStudent.DataKeys[e.RowIndex].Value);
                string message="删除失败";
                if (StudentBusiness.UpdateStudentAuthorized(sId))
                {
                    message="删除成功";
                }
                AlertInfo.AlertDialog(message);
            }
            gvStudent.Rows[e.RowIndex].Visible=false;
            LoadStudent();
        }
        protected void btnSearch_Click(object sender, EventArgs e)
        {
            if (txtStudentId.Text.Trim()==string.Empty)
            {
                AlertInfo.AlertDialog("请输入学号");
            }
            else
            {
                sId=Convert.ToInt32(txtStudentId.Text);
                if (! StudentBusiness.IsExistStudentId(sId))
                {
                    AlertInfo.AlertDialog("该学号不存在,请重新输入");
                }
                else
                {
                    gvStudent.DataSource=StudentBusiness.SelectStudentByID(sId); ;
                    gvStudent.DataKeyNames=new string[] { "StudentId" };
                    gvStudent.DataBind();
```

```
            }
        }
    }
    protected void ddlSchool_TextChanged(object sender, EventArgs e)
    {
        if (ddlSchool.Text ! =string.Empty)
        {
            sId=Convert.ToInt16(ddlSchool.SelectedValue);
            DataTable dtm=MajorBusiness.SelectMajorBySchoolID(sId);
            ddlMajor.DataSource=dtm;
            ddlMajor.DataValueField="majorid";
            ddlMajor.DataTextField="majorname";
            ddlMajor.DataBind();
        }
    }
    protected void ddlMajor_TextChanged(object sender, EventArgs e)
    {
        if (ddlSchool.Text ! =string.Empty && ddlMajor.Text ! =string.Empty)
        {
            mId=Convert.ToInt16(ddlMajor.SelectedValue);
            sId=Convert.ToInt16(ddlSchool.SelectedValue);
            DataTable dtc =ClassBusiness.SelectClassByMajorSchoolID(mId, sId);
            ddlClass.DataSource=dtc;
            ddlClass.DataValueField="classid";
            ddlClass.DataTextField="classname";
            ddlClass.DataBind();
        }
    }
    protected void btnSearch1_Click(object sender, EventArgs e)
    {
        sId=Convert.ToInt16(ddlSchool.SelectedValue);
        mId=Convert.ToInt16(ddlMajor.SelectedValue);
        cId=Convert.ToInt16(ddlClass.SelectedValue);
        DataTable dt=StudentBusiness.SelectStudentBySMC(sId,mId,cId);
        gvStudent.DataSource=dt;
        gvStudent.DataKeyNames=new string[] { "StudentId" };
        gvStudent.DataBind();
    }
    protected void lbtnAll_Click(object sender, EventArgs e)
    {
        LoadStudent();
    }
       protected  void  gvStudent _ PageIndexChanging ( object  sender,
GridViewPageEventArgs e)
    {
        gvStudent.PageIndex=e.NewPageIndex;
        LoadStudent();
    }
  }
}
```

2. 修改基本档案

在修改基本档案页可对档案进行修改。修改基本档案页测试如图 13.20 所示。

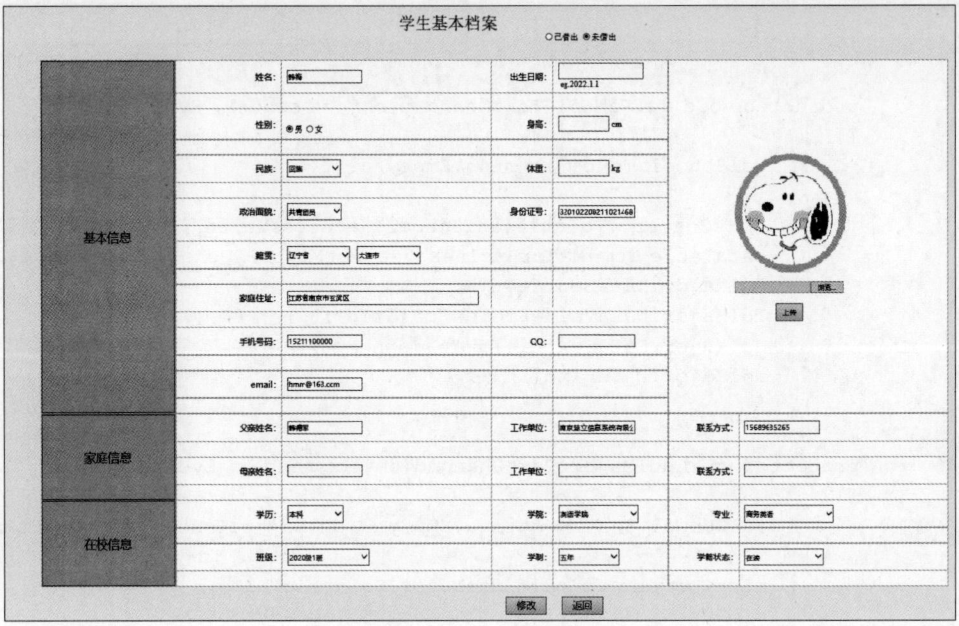

图 13.20 修改基本档案页测试

修改基本档案页视图主要源代码如下。

```
<form id="form1" runat="server">
        <asp:Label ID="Label12" runat="server" Text="学生基本档案"></asp:
Label>
        <asp:RadioButtonList ID="rbllend" runat="server">
            <asp:ListItem>已借出</asp:ListItem>
            <asp:ListItem Selected="True">未借出</asp:ListItem>
        </asp:RadioButtonList>
        <asp:Label ID="Label40" runat="server" Text="基本信息"></asp:
Label>
        <asp:Label ID="Label13" runat="server" Text="姓名:"></asp:Label>
        <asp:TextBox ID="txtStudentName" runat="server" ></asp:TextBox>
        <asp:RequiredFieldValidator ID="RequiredFieldValidator1" runat=
"server" ControlToValidate =" txtStudentName " ErrorMessage =" * " > </asp:
RequiredFieldValidator>
        <asp:Label ID="Label14" runat="server" Text="出生日期:"></asp:
Label>
        <asp:TextBox ID="txtBirthday" runat="server"></asp:TextBox>
        <asp:RequiredFieldValidator ID="RequiredFieldValidator2" runat=
"server" ControlToValidate =" txtBirthday " ErrorMessage =" * " > </asp:
RequiredFieldValidator>
        <asp:Label ID="Label41" runat="server" Text="eg.2022.1.1"></asp:
Label>
        <asp:Image ID="imgStudent" runat="server"/>
        <asp:FileUpload ID="FileUpload1" runat="server" />
        <asp:Button ID="btnUpload" runat="server" OnClick="btnUpload_
Click" Text="上传"/>
```

```
        <asp:Label ID="Label16" runat="server" Text="性别:"></asp:Label>
        <asp:RadioButtonList ID="rblSex" runat="server" RepeatDirection=
"Horizontal">
                <asp:ListItem Selected="True">男</asp:ListItem>
                <asp:ListItem>女</asp:ListItem>
                </asp:RadioButtonList></td>
            <asp:Label ID="Label15" runat="server"Text="身高:"></asp:Label>
            <asp:TextBox ID="txtHeight" runat="server"></asp:TextBox>cm
            <asp:Label ID="Label17" runat="server" Text="民族:"></asp:
Label>
            <asp:DropDownList ID="ddlNation" runat="server" >
                    <asp:ListItem>汉族</asp:ListItem>
                    <asp:ListItem>壮族</asp:ListItem>
                    <asp:ListItem>满族</asp:ListItem>
                    <asp:ListItem>回族</asp:ListItem>
                </asp:DropDownList>
            <asp:Label ID="Label18" runat="server"Text="体重:"></asp:
Label>
            <asp:TextBox ID="txtWeight" runat="server"></asp:TextBox>kg
            <asp:Label ID="Label20" runat="server" Text="政治面貌:"></asp:
Label>
            <asp:DropDownList ID="ddlPolitics" runat="server" >
                    <asp:ListItem>中共党员</asp:ListItem>
                    <asp:ListItem>中共预备党员</asp:ListItem>
                    <asp:ListItem>共青团员</asp:ListItem>
                    <asp:ListItem>群众</asp:ListItem>
                    <asp:ListItem>其他</asp:ListItem>
            </asp:DropDownList>
            <asp:Label ID="Label19" runat="server" Text="身份证号:"></asp:
Label>
            <asp:TextBox ID="txtID" runat="server""></asp:TextBox>
            <asp:RequiredFieldValidator ID="RequiredFieldValidator3" runat
="server" ControlToValidate="txtID" ErrorMessage="*" ForeColor="Red"></asp:
RequiredFieldValidator>
            <asp:Label ID="Label21" runat="server" Text="籍贯:" ></asp:
Label>
                <asp:DropDownList ID="ddlProvince" runat="server"
OnTextChanged="ddlProvince_TextChanged" AutoPostBack="True">
                </asp:DropDownList>
                <asp:DropDownList ID="ddlCity" runat="server"AutoPostBack
="True">
                </asp:DropDownList>
            <asp:Label ID="Label22" runat="server" Text="家庭住址:" ></asp:
Label>
            <asp:TextBox ID="txtAddress" runat="server"></asp:TextBox>
            <asp:RequiredFieldValidator ID="RequiredFieldValidator4" runat
="server" ControlToValidate="txtAddress" ErrorMessage="*" ></asp:
RequiredFieldValidator>
            <asp:Label ID="Label23" runat="server" Text="手机号码:" ></asp:
Label>
            <asp:TextBox ID="txtTel" runat="server"></asp:TextBox>
            <asp:RequiredFieldValidator ID="RequiredFieldValidator5" runat
="server" ControlToValidate="txtTel" ErrorMessage="*" ></asp:
RequiredFieldValidator>
```

```
            <asp:Label ID="Label25" runat="server" Text="QQ:" ></asp:Label>
            <asp:TextBox ID="txtQQ" runat="server"></asp:TextBox>
             <asp:Label ID="Label24" runat="server" Text="email:" ></asp:
Label>
            <asp:TextBox ID="txtEmail" runat="server"></asp:TextBox>
                        < asp: RegularExpressionValidator ID ="
RegularExpressionValidator1" runat =" server " ControlToValidate =" txtEmail "
ErrorMessage="*" ValidationExpression=
"\w+([-+.']\w+)*@\w+([-.]\w+)*\.\w+([-.]\w+)*"></asp:
RegularExpressionValidator>
             <asp:Label ID="Label39" runat="server" Text="家庭信息"></asp:
Label>
            <asp:Label ID="Label26" runat="server" Text="父亲姓名:" ></asp:
Label>
            <asp:TextBox ID="txtFatherName" runat="server"></asp:TextBox>
            <asp:Label ID="Label28" runat="server" Text="工作单位:" ></asp:
Label>
            <asp:TextBox ID="txtFatherJob" runat="server"></asp:TextBox>
            <asp:Label ID="Label30" runat="server" Text="联系方式:" ></asp:
Label>
            <asp:TextBox ID="txtFatherTel" runat="server"></asp:TextBox>
            <asp:Label ID="Label27" runat="server" Text="母亲姓名:" ></asp:
Label>
            <asp:TextBox ID="txtMatherName" runat="server"></asp:TextBox>
            <asp:Label ID="Label29" runat="server" Text="工作单位:" ></asp:
Label>
            <asp:TextBox ID="txtMatherJob" runat="server" ></asp:TextBox>
            <asp:Label ID="Label31" runat="server" Text="联系方式:" ></asp:
Label>
            <asp:TextBox ID="txtMatherTel" runat="server"></asp:TextBox>
             <asp:Label ID="Label38" runat="server" Text="在校信息"></asp:
Label>
             <asp:Label ID="Label34" runat="server" Text="学历:"></asp:
Label>
            <asp:DropDownList ID="ddlDegree" runat="server">
                <asp:ListItem>本科</asp:ListItem>
                <asp:ListItem>研究生</asp:ListItem>
              </asp:DropDownList>
              <asp:Label ID="Label33" runat="server" Text="学院:" ></asp:
Label>
             <asp:DropDownList ID="ddlSchool" runat="server" OnTextChanged
="ddlSchool_TextChanged" AutoPostBack="True"></asp:DropDownList>
              <asp:Label ID="Label32" runat="server" Text="专业:" ></asp:
Label>
             <asp:DropDownList ID="ddlMajor" runat="server" OnTextChanged="
ddlMajor_TextChanged" AutoPostBack="True"></asp:DropDownList>
              <asp:Label ID="Label35" runat="server" Text="班级:" ></asp:
Label>
             <asp:DropDownList ID="ddlClass" runat="server" AutoPostBack="
True"></asp:DropDownList>
             <asp:Label ID="Label36" runat="server" Text="学制:" ></asp:
Label>
                <asp:DropDownList ID="ddlSystem" runat="server">
```

```
                    <asp:ListItem>五年</asp:ListItem>
                    <asp:ListItem>四年</asp:ListItem>
                    <asp:ListItem>三年</asp:ListItem>
                    <asp:ListItem>两年半</asp:ListItem>
                    <asp:ListItem>两年</asp:ListItem>
                </asp:DropDownList>
        <asp:Label ID="Label37" runat="server" Text="学籍状态:" ></asp:
Label>
                <asp:DropDownList ID="ddlAauthorized" runat="server" >
                </asp:DropDownList>
        <asp:Button ID="btnUpdate" runat="server" Text="修改" OnClick="
btnUpdate_Click" OnClientClick="return confirm('确认修改?');" />
            <asp:Button ID="btnReturn" runat="server" Text="返回"
CausesValidation="False" PostBackUrl="~/Admin/AdminStudentManage.aspx" />
</form>
```

修改基本档案页 C#代码如下。

```
using Business;
using Entity;
using System;
using System.Data;
using System.IO;

namespace Archive
{
    public partial class AdminUpdateStudent : System.Web.UI.Page
    {
        int sId, pId;
        protected void Page_Load(object sender, EventArgs e)
        {
            if (Session["adminId"] !=null)
            {
                if (! IsPostBack && Session["StudentId"] !=null)
                {
                    LoadStudent();
                }
            }
            else
            {
                AlertInfo.AlertDialog("请先登录", "../Login.aspx");
            }
        }
        private void LoadStudent()
        {
            StudentEntity student=StudentBusiness.SelectStudentByID2(Convert.
ToInt32(Session["StudentId"]));
            txtStudentName.Text=student.StudentName;
            rblSex.SelectedValue=student.Sex;
            txtHeight.Text=student.Height.ToString();
            txtWeight.Text=student.Weight;
            ddlNation.Text=student.Nation;
```

```
ddlPolitics.Text=student.Politics;
txtID.Text=student.ID;
DataTable dtp=ProvinceBusiness.ProvinceInfo();
ddlProvince.DataSource=dtp;
ddlProvince.DataValueField="provinceid";
ddlProvince.DataTextField="provincename";
ddlProvince.DataBind();
ddlProvince.SelectedValue=student.ProvinceId.ToString();
if (ddlProvince.Text ! =string.Empty)
{
    pId=Convert.ToInt32(ddlProvince.SelectedValue);
    DataTable dtc=CityBusiness.SelectCityByProvinceID(pId);
    ddlCity.DataSource=dtc;
    ddlCity.DataValueField="cityid";
    ddlCity.DataTextField="cityname";
    ddlCity.DataBind();
}
ddlCity.SelectedValue=student.CityId.ToString();
txtAddress.Text=student.Address;
txtTel.Text=student.Tel;
txtQQ.Text=student.QQ;
txtEmail.Text=student.Email;
txtFatherName.Text=student.Father;
txtFatherJob.Text=student.Fjob;
txtFatherTel.Text=student.Ftel;
txtMatherName.Text=student.Mother;
txtMatherJob.Text=student.Mjob;
txtMatherTel.Text=student.Mtel;
ddlDegree.Text=student.Degree;
DataTable dtsc=SchoolBusiness.GetSchoolInfo();
ddlSchool.DataSource=dtsc;
ddlSchool.DataValueField="schoolid";
ddlSchool.DataTextField="schoolname";
ddlSchool.DataBind();
ddlSchool.SelectedValue=student.SchoolId.ToString();
if (ddlSchool.Text ! =string.Empty)
{
    int sid=Convert.ToInt16(ddlSchool.SelectedValue);
    DataTable dtm=MajorBusiness.SelectMajorBySchoolID(sid);
    ddlMajor.DataSource=dtm;
    ddlMajor.DataValueField="majorid";
    ddlMajor.DataTextField="majorname";
    ddlMajor.DataBind();
}
ddlMajor.SelectedValue=student.MajorId.ToString();
if (ddlSchool.Text ! =string.Empty && ddlMajor.Text ! =string.Empty)
{
    int mid=Convert.ToInt16(ddlMajor.SelectedValue);
    int sid=Convert.ToInt16(ddlSchool.SelectedValue);
    DataTable dtc=ClassBusiness.SelectClassByMajorSchoolID(mid, sid);
    ddlClass.DataSource=dtc;
    ddlClass.DataValueField="classid";
    ddlClass.DataTextField="classname";
```

```
            ddlClass.DataBind();
        }
        ddlClass.SelectedValue=student.ClassId.ToString();
        ddlSystem.Text=student.System;
        DataTable dta=AuthorizedBusiness.SelectStudentAuthorized();
        ddlAauthorized.DataSource=dta;
        ddlAauthorized.DataValueField="AuthorizedId";
        ddlAauthorized.DataTextField="authorizedname";
        ddlAauthorized.DataBind();
        ddlAauthorized.SelectedValue=student.AuthorizedId.ToString();
        rbllend.SelectedValue=student.Lend;
        imgSudent.ImageUrl=student.Photourl;
    }
    protected void btnUpdate_Click(object sender, EventArgs e)
    {
        sId=Convert.ToInt32(Session["StudentId"]);
        StudentEntity studentEntity=new StudentEntity();
        studentEntity.StudentName=txtStudentName.Text.Trim();
        studentEntity.Tel=txtTel.Text.Trim();
        studentEntity.Address=txtAddress.Text.Trim();
        studentEntity.Nation=ddlNation.Text.Trim();
        studentEntity.Sex=rblSex.SelectedValue;
        studentEntity.Politics=ddlPolitics.Text.Trim();
        studentEntity.QQ=txtQQ.Text.Trim();
        studentEntity.Email=txtEmail.Text.Trim();
        studentEntity.Height=txtHeight.Text.Trim();
        studentEntity.Weight=txtWeight.Text.Trim();
        studentEntity.ID=txtID.Text.Trim();
        studentEntity.Father=txtFatherName.Text.Trim();
        studentEntity.Ftel=txtFatherTel.Text.Trim();
        studentEntity.Fjob=txtFatherJob.Text.Trim();
        studentEntity.Mother=txtMatherName.Text.Trim();
        studentEntity.Mtel=txtMatherTel.Text.Trim();
        studentEntity.Mjob=txtMatherJob.Text.Trim();
        studentEntity.Degree=ddlDegree.Text.Trim();
        studentEntity.SchoolId=Convert.ToInt16(ddlSchool.SelectedValue);
        studentEntity.MajorId=Convert.ToInt16(ddlMajor.SelectedValue);
        studentEntity.ClassId=Convert.ToInt16(ddlClass.SelectedValue);
        studentEntity.System=ddlSystem.Text.Trim();
        studentEntity. AuthorizedId = Convert. ToInt16 ( ddlAauthorized.
SelectedValue);
        studentEntity.Photourl=imgStudent.ImageUrl;
        studentEntity.ProvinceId=Convert.ToInt16(ddlProvince.SelectedValue);
        studentEntity.CityId=Convert.ToInt16(ddlCity.SelectedValue);
        studentEntity.Lend=rbllend.SelectedValue;
        string message="修改失败";
        if (StudentBusiness.UpdateStudentByID(sId, studentEntity))
        {
            message="修改成功";
        }
        AlertInfo.AlertDialog(message);
    }
```

```
protected void btnUpload_Click(object sender, EventArgs e)
{
    string uploadfile=FileUpload1.FileName;
    string fileextn=Path.GetExtension(uploadfile);
    string filename=txtStudentName.Text+fileextn;
    if (fileextn.ToLower()==".jpg" || fileextn.ToLower()==".jpeg")
    {
        string dcty="~/images/student";
        if (! Directory.Exists(Server.MapPath(dcty)))
        {
            Directory.CreateDirectory(Server.MapPath(dcty));
        }
        FileUpload1.SaveAs(Server.MapPath(dcty+"/")+filename);
        imgStudent.ImageUrl=dcty+filename;
    }
}
protected void ddlProvince_TextChanged(object sender, EventArgs e)
{
    if (ddlProvince.Text !=string.Empty)
    {
        pId=Convert.ToInt16(ddlProvince.SelectedValue);
        DataTable dtc=CityBusiness.SelectCityByProvinceID(pId);
        ddlCity.DataSource=dtc;
        ddlCity.DataValueField="cityid";
        ddlCity.DataTextField="cityname";
        ddlCity.DataBind();
    }
}
protected void ddlSchool_TextChanged(object sender, EventArgs e)
{
    if (ddlSchool.Text !=string.Empty)
    {
        sId=Convert.ToInt16(ddlSchool.SelectedValue);
        DataTable dtm=MajorBusiness.SelectMajorBySchoolID(sId);
        ddlMajor.DataSource=dtm;
        ddlMajor.DataValueField="majorid";
        ddlMajor.DataTextField="majorname";
        ddlMajor.DataBind();
    }
}
protected void ddlMajor_TextChanged(object sender, EventArgs e)
{
    if (ddlSchool.Text !=string.Empty && ddlMajor.Text !=string.Empty)
    {
        int mid=Convert.ToInt16(ddlMajor.SelectedValue);
        int sid=Convert.ToInt16(ddlSchool.SelectedValue);
        DataTable dtc=ClassBusiness.SelectClassByMajorSchoolID(mid, sid);
        ddlClass.DataSource=dtc;
        ddlClass.DataValueField="classid";
        ddlClass.DataTextField="classname";
        ddlClass.DataBind();
    }
}
```

```
    }
}
```

13.5.5 奖学金档案管理

1. 奖学金档案管理页

在奖学金档案管理页可查看所有奖学金档案，同时可按照奖学金名称进行搜索。奖学金档案管理页测试如图 13.21 所示。

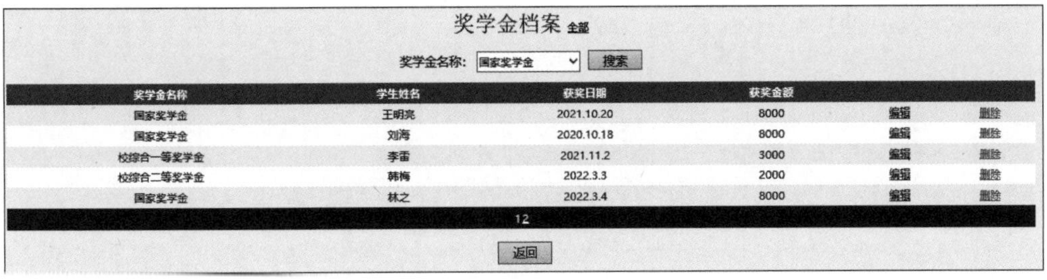

奖学金档案 全部

奖学金名称: 国家奖学金 搜索

奖学金名称	学生姓名	获奖日期	获奖金额		
国家奖学金	王明亮	2021.10.20	8000	编辑	删除
国家奖学金	刘海	2020.10.18	8000	编辑	删除
校综合一等奖学金	李雷	2021.11.2	3000	编辑	删除
校综合二等奖学金	韩梅	2022.3.3	2000	编辑	删除
国家奖学金	林之	2022.3.4	8000	编辑	删除
12					

返回

图 13.21 奖学金档案管理页测试

奖学金档案管理页视图主要源代码如下。

```
<form id="form1" runat="server">
      <asp:Label ID="Label12" runat="server"  Text="奖学金档案"></asp:Label>
      <asp:LinkButton ID="lbtnAll" runat="server"  OnClick="lbtnAll_Click">全部</asp:LinkButton>
       <asp:Label ID="Label13" runat="server" Text="奖学金名称:"></asp:Label>
       <asp:DropDownList ID="ddlScholarship" runat="server" Height="30px" >
       </asp:DropDownList>
        <asp:Button ID="btnSearch" runat="server" Text="搜索" OnClick="btnSearch_Click"/>
       <asp:GridView ID="gvScholarship" runat="server" AutoGenerateColumns="False" OnRowDeleting="gvScholarship_RowDeleting" OnRowEditing="gvScholarship_RowEditing" AllowPaging ="True" OnPageIndexChanging ="gvScholarship_PageIndexChanging" PageSize="5">
                <AlternatingRowStyle BackColor="White" ForeColor="#284775" />
                <Columns>
<asp:BoundField DataField="sapplicationid" HeaderText="sapplicationid" ></asp:BoundField>
<asp:BoundField DataField="scholarshipname" HeaderText="奖学金名称"></asp:BoundField>
<asp:BoundField DataField="studentname" HeaderText="学生姓名" ></asp:BoundField>
<asp:BoundField DataField="date" HeaderText="获奖日期" ></asp:BoundField>
<asp:BoundField DataField="money" HeaderText="获奖金额" ></asp:BoundField>
<asp:CommandField ShowEditButton="True" ></asp:CommandField>
<asp:TemplateField ShowHeader="False"><ItemTemplate>
<asp:LinkButton ID="LinkButton1" runat="server" CausesValidation="False" CommandName="Delete" Text="删除" OnClientClick="return confirm('确认要删除吗?');"></asp:LinkButton></asp:TemplateField>
```

```
<asp:Button ID="btnReturn" runat="server" Text="返回" PostBackUrl="~/Admin/
AdminIndex.aspx" />
</form>
```

奖学金档案管理页C#代码如下：

```
using Business;
using System;
using System.Data;
using System.Web.UI.WebControls;

namespace Archive
{
    public partial class AdminScholarshipManage : System.Web.UI.Page
    {
        int sId;
        protected void Page_Load(object sender, EventArgs e)
        {
            if (Session["adminId"] !=null)
            {
                if (! IsPostBack)
                {
                    DataTable dt=ScholarshipBusiness.ScholarshipInfo();
                    ddlScholarship.DataSource=dt;
                    ddlScholarship.DataTextField="scholarshipname";
                    ddlScholarship.DataValueField="ScholarshipId";
                    ddlScholarship.DataBind();
                    LoadScholarship();
                }
            }
            else
            {
                AlertInfo.AlertDialog("请先登录", "../Login.aspx");
            }
        }
        private void LoadScholarship()
        {
            DataTable dt=SapplicationBusiness.SapplicationInfo();
            gvScholarship.DataSource=dt;
            gvScholarship.DataKeyNames=new string[] { "SapplicationId" };
            gvScholarship.DataBind();
        }
        protected void gvScholarship_RowEditing(object sender, GridViewEd-
itEventArgs e)
        {
            Session["SapplicationId"]=gvScholarship.DataKeys[e.NewEditIndex].
Value.ToString().Trim();
            Response.Redirect("AdminScholarshipUpdate.aspx");
        }
        protected void gvScholarship_RowDeleting(object sender, GridViewDele-
teEventArgs e)
        {
```

```
            if (e.RowIndex >=0)
            {
                sId=Convert.ToInt16(gvScholarship.DataKeys[e.RowIndex].Value);
                string message="删除失败";
                if ( SapplicationBusiness.UpdateSapplicationAuthorized(sId))
                {
                   message="删除成功";
                }
                AlertInfo.AlertDialog(message);
            }
            gvScholarship.Rows[e.RowIndex].Visible=false;
            LoadScholarship();
        }
        protected void btnSearch_Click(object sender, EventArgs e)
        {
            sId=Convert.ToInt16(ddlScholarship.SelectedValue);
            DataTable dt=SapplicationBusiness.SelectSapplicationByScholarshipID(sId);
            gvScholarship.DataSource=dt;
            gvScholarship.DataKeyNames=new string[] { "SapplicationId" };
            gvScholarship.DataBind();
        }
        protected void lbtnAll_Click(object sender, EventArgs e)
        {
           LoadScholarship();
        }
        protected  void  gvScholarship _ PageIndexChanging ( object  sender,
    GridViewPageEventArgs e)
        {
            gvScholarship.PageIndex=e.NewPageIndex;
            LoadScholarship();
        }
    }
}
```

2. 修改奖学金申请

在修改奖学金申请页可对奖学金档案进行修改。修改奖学金申请页测试如图 13.22 所示。

修改奖学金申请页视图主要源代码如下。

```
<form id="form1" runat="server">
     <asp:Label ID="Label12" runat="server" Text="奖学金申请"></asp:Label>
     <asp:Label ID="lblAuthorized" runat="server" ></asp:Label>
     <asp:Label ID="Label13" runat="server" Text="学号:"></asp:Label>
     <asp:Label ID="lblStudentid" runat="server" ></asp:Label>
     <asp:Label ID="Label14" runat="server" Text="性别:"></asp:Label>
      <asp:Label ID="lblSex" runat="server" ></asp:Label>
      <asp:Label ID="Label15" runat="server" Text="学生姓名:"></asp:Label>
      <asp:Label ID="lblStudentName" runat="server"></asp:Label>
      <asp:Label ID="Label16" runat="server"Text="政治面貌:"></asp:Label>
      <asp:Label ID="lblPolitics" runat="server"></asp:Label>
      <asp:Label ID="Label17" runat="server"Text="联系方式:"></asp:Label>
```

奖学金申请

学号： 20210002 性别： 女

学生姓名： 林之 政治面貌： 中共党员

联系方式： 15600000000

家庭住址： 四川省成都市锦江区

学院、专业、班级： 0 2 2018级1班

专业排名： 2/300 eg.1/20 绩点： 3.88

奖学金名称： 国家奖学金 金额： 8000

获奖日期： 2022.3.4 eg.2022.1.1

申请理由：

修改 返回

图 13.22 修改奖学金申请页测试

```
        <asp:Label ID="Label18" runat="server"Text="家庭住址:"></asp:Label>
        <asp:Label ID="lblAddress" runat="server"></asp:Label>
        <asp:Label ID="Label19" runat="server"Text="学院、专业、班级:"></asp:
Label>
        <asp:Label ID="lblSchool" runat="server" ></asp:Label>
        <asp:Label ID="lblMajor" runat="server"></asp:Label>
        <asp:Label ID="lblClass" runat="server"></asp:Label>
        <asp:Label ID="Label20" runat="server" Text="专业排名:"></asp:Label>
        <asp:TextBox ID="txtRank" runat="server"></asp:TextBox>
        <asp:RequiredFieldValidator ID="RequiredFieldValidator1" runat="server"
ControlToValidate="txtRank" ErrorMessage="*" ></asp:RequiredFieldValidator>
        <asp:Label ID="Label26" runat="server" Text="eg.1/20"></asp:Label>
        <asp:Label ID="Label23" runat="server" Text="绩点:"></asp:Label>
        <asp:TextBox ID="txtScore" runat="server"></asp:TextBox>
         <asp:RequiredFieldValidator ID="RequiredFieldValidator2" runat=
"server" ControlToValidate="txtScore" ErrorMessage="*" ForeColor="Red"></asp:
RequiredFieldValidator>
         <asp:Label ID="Label21" runat="server"Text="奖学金名称:"></asp:
Label>
        <asp:DropDownList ID="ddlScholarship" runat="server" AutoPostBack=
"True" Height="32px" OnTextChanged="ddlScholarship_TextChanged" ></asp:
DropDownList>
        <asp:Label ID="Label24" runat="server"Text="金额:"></asp:Label>
        <asp:Label ID="lblMoney" runat="server"></asp:Label>
        <asp:Label ID="Label22" runat="server"Text="获奖日期:"></asp:Label>
        <asp:TextBox ID="txtDate" runat="server" TextMode="Date" ></asp:
TextBox>
```

```
        <asp:RequiredFieldValidator ID="RequiredFieldValidator3" runat=
"server" ControlToValidate="txtDate" ErrorMessage="*" ForeColor="Red"></asp:
RequiredFieldValidator>
        <asp:Label ID="Label11" runat="server" Text="eg.2022.1.1"></asp:
Label>
        <asp:Label ID="Label25" runat="server" Text="申请理由:"></asp:
Label>
        <asp:TextBox ID="txtReason" runat="server" TextMode="MultiLine" >
</asp:TextBox>
        <asp:RequiredFieldValidator ID="RequiredFieldValidator4" runat=
"server" ControlToValidate =" txtReason " ErrorMessage =" * " > </asp:
RequiredFieldValidator>
        <asp:Button ID="btnUpdate" runat="server" OnClick="btnUpdate_Click"
Text="修改" OnClientClick="return confirm('确认修改?');" />
        <asp:Button ID="btnReturn" runat="server" Text="返回" CausesValidation
="False" PostBackUrl="~/Admin/AdminScholarshipManage.aspx" />
</form>
```

修改奖学金申请页 C#代码如下:

```
using Business;
using Entity;
using System;
using System.Data;

namespace Archive
{
    public partial class AdminScholarshipUpdate : System.Web.UI.Page
    {
        int sId;
        protected void Page_Load(object sender, EventArgs e)
        {
            if (Session["adminId"] !=null)
            {
                if (! IsPostBack && Session["SapplicationId"] !=null)
                {
                    LoadSapplication();
                }
            }
            else
            {
                AlertInfo.AlertDialog("请先登录", "../Login.aspx");
            }
        }
        private void LoadSapplication()
        {
            SapplicationInfoEntity sapplication = SapplicationBusiness.
SelectSapplicationByID(Convert.ToInt16(Session["SapplicationId"]));
            lblAuthorized.Text=sapplication.AuthorizedId.Trim();
            lblStudentid.Text=sapplication.StudentId.ToString().Trim();
            lblStudentName.Text=sapplication.StudentName.Trim();
            lblTel.Text=sapplication.Tel.Trim();
            lblAddress.Text=sapplication.Address.Trim();
```

```
            lblSex.Text=sapplication.Sex.Trim();
            lblPolitics.Text=sapplication.Politics.Trim();
            lblSchool.Text=sapplication.SchoolId.ToString().Trim();
            lblMajor.Text=sapplication.MajorId.Trim();
            lblClass.Text=sapplication.ClassName.Trim();
            txtRank.Text=sapplication.Rank.Trim();
            txtScore.Text=sapplication.Score.Trim();
            lblMoney.Text=sapplication.Money.Trim();
            txtDate.Text=sapplication.Date.Trim();
            DataTable dts=ScholarshipBusiness.ScholarshipInfo();
            ddlScholarship.DataSource=dts;
            ddlScholarship.DataValueField="ScholarshipId";
            ddlScholarship.DataTextField="scholarshipname";
            ddlScholarship.DataBind();
            ddlScholarship.SelectedValue=sapplication.ScholarshipId.ToString().
Trim();
            txtReason.Text=sapplication.Reason.Trim();
        }
        protected void btnUpdate_Click(object sender, EventArgs e)
        {
            sId=Convert.ToInt16(Session["SapplicationId"]);
            SapplicationEntity sapplicationEntity=new SapplicationEntity();
            sapplicationEntity.ScholarshipId=Convert.ToInt16(ddlScholarship.
SelectedValue);
            sapplicationEntity.Score=txtScore.Text.Trim();
            sapplicationEntity.Rank=txtRank.Text.Trim();
            sapplicationEntity.Reason=txtReason.Text.Trim();
            sapplicationEntity.Date=txtDate.Text.Trim();
            sapplicationEntity.AuthorizedId=9;
            string message="修改成功";
            if (SapplicationBusiness.UpdateSapplicationByID(sId, sapplica-
tionEntity))
            {
                message="修改失败";
            }
            AlertInfo.AlertDialog(message);
            LoadSapplication();
        }
        protected void ddlScholarship_TextChanged(object sender, EventArgs e)
        {
            string money;
            switch (ddlScholarship.SelectedValue)
            {
                case "1": money="8000"; break;
                case "2": money="5000"; break;
                case "3": money="3000"; break;
                case "4": money="2000"; break;
                default: money="1000"; break;
            }
            lblMoney.Text=money;
        }
    }
}
```

13.5.6 借阅记录管理

1. 借阅记录管理页

在借阅记录管理页可查看所有借阅记录，同时可按照所借档案号进行搜索，也可按照借阅人进行模糊搜索。借阅记录管理页测试如图 13.23 所示。

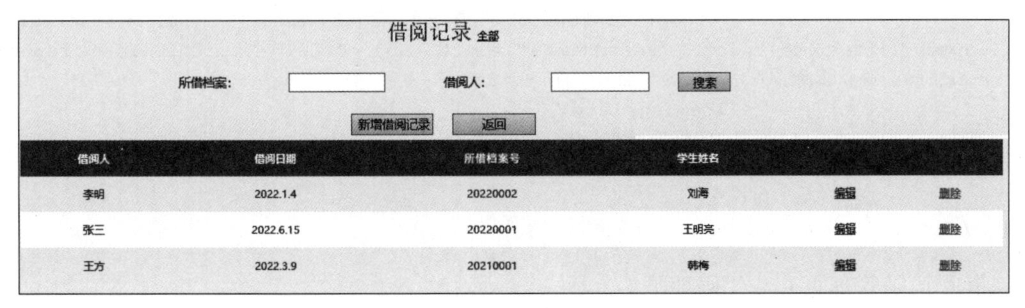

图 13.23 借阅记录管理页测试

借阅记录管理页视图主要源代码如下。

```
<form id="form1" runat="server">
    <asp:Label ID="Label4" runat="server"Text="借阅记录"></asp:Label>
    <asp:LinkButton ID="lbtnAll" runat="server" OnClick="lbtnAll_Click">全部
</asp:LinkButton>
    <table align="center">
        <tr>
        <td><asp:Label ID="Label5" runat="server"  Text="所借档案:"></asp:
Label></td>
        <td><asp:TextBox ID="txtStudentId" runat="server"></asp:TextBox></td>
        <td><asp:Label ID="Label6" runat="server"Text="借阅人:"></asp:Label></
td>
        <td><asp:TextBox ID="txtLendman" runat="server"></asp:TextBox></td>
        <td><asp:Button ID="btnSearch" runat="server"Text="搜索" OnClick="
btnSearch_Click" /></td>
        </tr>
      </table>
      < asp: Button  ID =" btnAdd"  runat =" server"  Text =" 新 增 借 阅 记 录"
CausesValidation="False" PostBackUrl="~/Admin/AdminLendAdd.aspx" />
      <asp:Button ID="btnReturn" runat="server" Text="返回" CausesValidation="
False" PostBackUrl="~/Admin/AdminIndex.aspx" />
      <asp:GridView ID="gvLend" runat="server" AutoGenerateColumns="False"
OnRowEditing="gvLend_RowEditing" OnRowDeleting="gvLend_RowDeleting" AllowPaging="
True" OnPageIndexChanging="gvLend_PageIndexChanging" PageSize="5">
          <Columns>
              <asp:BoundField DataField="lendman" HeaderText="借阅人" >
              </asp:BoundField>
              <asp:BoundField DataField="startdate" HeaderText="借阅日期" >
              </asp:BoundField>
              <asp:BoundField DataField="studentid" HeaderText="所借档案">
              </asp:BoundField>
              <asp:BoundField DataField="studentname" HeaderText="学生姓名" >
```

```
                </asp:BoundField>
                <asp:BoundField DataField="lendid" HeaderText="lendid" >
                </asp:BoundField>
                <asp:CommandField ShowEditButton="True" >
                </asp:CommandField>
                <asp:TemplateField ShowHeader="False">
                    <ItemTemplate>
                        <asp: LinkButton  ID =" LinkButton1 "  runat =" server "
CausesValidation="False" CommandName="Delete" Text="删除" OnClientClick="return
confirm('确认要删除吗?');"></asp:LinkButton>
                    </ItemTemplate>
                </asp:TemplateField>
            </Columns>
        </asp:GridView>
</form>
```

借阅记录管理页C#代码如下。

```
using Business;
using System;
using System.Data;
using System.Web.UI.WebControls;

namespace Archive
{
    public partial class AdminLendManage : System.Web.UI.Page
    {
        int sId, lId;
        protected void Page_Load(object sender, EventArgs e)
        {
            if (Session["adminId"] !=null)
            {
                if (! IsPostBack)
                {
                    LoadLend();
                }
            }
            else
            {
                AlertInfo.AlertDialog("请先登录", "../Login.aspx");
            }
        }
        private void LoadLend()
        {
            DataTable dt=LendBusiness.LendInfo();
            gvLend.DataSource=dt;
            gvLend.DataKeyNames=new string[] { "LendId" };
            gvLend.DataBind();
        }
        protected void gvLend_RowEditing(object sender, GridViewEditEventArgs e)
        {
            Session["LendId"]=gvLend.DataKeys[e.NewEditIndex].Value.ToString().
Trim();
```

```
            Response.Redirect("AdminLendUpdate.aspx");
        }
        protected void gvLend_RowDeleting(object sender, GridViewDeleteEventArgs e)
        {
            if (e.RowIndex >=0)
            {
                lId=Convert.ToInt16(gvLend.DataKeys[e.RowIndex].Value);
                string message="删除失败";
                if (LendBusiness.UpdateLendAuthorized(lId))
                {
                    message="删除成功";
                }
                AlertInfo.AlertDialog(message);
            }
            gvLend.Rows[e.RowIndex].Visible=false;
            LoadLend();
        }

        protected void btnSearch_Click(object sender, EventArgs e)
        {
            sId=Convert.ToInt32(txtStudentId.Text);
            string lendMan=txtLendman.Text.Trim();
            if (! StudentBusiness.IsExistStudentId(sId))
            {
                AlertInfo.AlertDialog("该学号不存在,请重新输入");
            }
            else
            {
                DataTable dt = LendBusiness. SelectLendByStudentIDLendman ( sId,
    lendMan);
                gvLend.DataSource=dt;
                gvLend.DataKeyNames=new string[] { "LendId" };
                gvLend.DataBind();
            }
        }
        protected void lbtnAll_Click(object sender, EventArgs e)
        {
            LoadLend();
        }
        protected void gvLend_PageIndexChanging(object sender, GridViewPageEventArgs
    e)
        {
            gvLend.PageIndex=e.NewPageIndex;
            LoadLend();
        }
    }
}
```

2. 修改借阅记录

在修改借阅记录页可修改借阅记录。修改借阅记录页测试如图 13.24 所示。

修改借阅记录页视图主要源代码如下。

图 13.24　修改借阅记录页测试

```
<form id="form1" runat="server">
        <asp:Label ID="Label4" runat="server" Text="借阅信息"></asp:Label>
        <table align="center">
            <tr>
            <td><asp:Label ID="Label5" runat="server" Text="所借档案:"></asp:
Label></td>
            <td><asp:TextBox ID="txtStudentId" runat="server" AutoPostBack="
True" CausesValidation="True" OnTextChanged="txtStudentId_TextChanged" ></
asp:TextBox>
                <asp:RequiredFieldValidator ID="RequiredFieldValidator1" runat
="server" ControlToValidate ="txtStudentId" ErrorMessage =" *" > </asp:
RequiredFieldValidator>
                <asp:RegularExpressionValidator ID="RegularExpressionValidator2"
runat =" server " ControlToValidate =" txtStudentId " ErrorMessage =
"RegularExpressionValidator" ValidationExpression="^[0-9]*$">请输入学号
</asp:RegularExpressionValidator>
                    (<asp:Label ID="lblStudentName" runat="server"></asp:Label>)
</td>
            <td><asp:Label ID="Label10" runat="server" Font-Names="微软雅黑"
Font-Size="Large" Text="借阅理由:"></asp:Label></td>
            <td><asp:TextBox ID="txtReason" runat="server" ></asp:TextBox>
                <asp: RequiredFieldValidator ID =" RequiredFieldValidator6 "
runat="server" ControlToValidate="txtReason" ErrorMessage="*" Font-Names="宋
体" ForeColor="Red"></asp:RequiredFieldValidator></td>
            </tr>
            <tr>
            <td><asp:Label ID="Label6" runat="server" Text="借阅人:"></asp:
Label></td>
            <td><asp:TextBox ID="txtLendman" runat="server"></asp:TextBox>
                <asp:RequiredFieldValidator ID="RequiredFieldValidator2" runat
="server" ControlToValidate="txtLendman" ErrorMessage="*" Font-Names="宋体"
ForeColor="Red"></asp:RequiredFieldValidator></td>
            <td><asp:Label ID="Label9" runat="server"Text="借阅人身份证号:">
</asp:Label></td>
            <td><asp:TextBox ID="txtID" runat="server"></asp:TextBox>
```

```
            <asp: RequiredFieldValidator  ID =" RequiredFieldValidator5 " 
runat =" server " ControlToValidate =" txtID "  ErrorMessage =" * "  > </asp:
RequiredFieldValidator></td>
        </tr>
        <tr>
         <td><asp:Label ID="Label7" runat="server"Text="借阅日期:"></asp:
Label></td>
         <td><asp:TextBox ID="txtStartdate" runat="server" TextMode="Date">
</asp:TextBox>
           asp: RequiredFieldValidator ID =" RequiredFieldValidator3" runat =
"server"  ControlToValidate =" txtStartdate "  ErrorMessage =" *" > </asp:
RequiredFieldValidator>
         <asp:Label ID="Label11" runat="server" Text="eg.2022.1.1"></asp:
Label></td>
         <td><asp:Label ID="Labe8" runat="server" Text="归还日期:"></asp:
Label></td>
         <td><asp:TextBox ID="txtEnddate" runat="server" TextMode="Date"></
asp:TextBox>
           <asp: RequiredFieldValidator ID =" RequiredFieldValidator4" runat =
"server"  ControlToValidate =" txtEnddate "  ErrorMessage =" * " > </asp:
RequiredFieldValidator>
           <asp:Label ID="Lbl1" runat="server" Text="eg.2022.12.31"></asp:
Label></td>
        </tr>
        <tr>
            <td colspan="4">
                <asp:Button ID="btnUpdate" runat="server" OnClick="btnUpdate_
Click" Text="修改"  OnClientClick="return confirm('确认修改?');" />
                <asp:Button ID =" btnReturn " runat =" server " Text =" 返 回 "
CausesValidation="False" PostBackUrl="~/Admin/AdminLendManage.aspx" />
            </td>
        </tr>
    </table>
</form>
```

修改借阅记录页C#代码如下。

```
using Business;
using Entity;
using System;

namespace Archive
{
    public partial class AdminLendUpdate : System.Web.UI.Page
    {
        int lId,  sId;
        protected void Page_Load(object sender, EventArgs e)
        {
            if (Session["adminId"] !=null)
            {
                if (! IsPostBack && Session["LendId"] !=null)
                {
                    LoadLend();
```

```
                }
            }
            else
            {
                AlertInfo.AlertDialog("请先登录", "../Login.aspx");
            }
        }
        private void LoadLend()
        {
             LendEntity lend= LendBusiness.SelectLendByID(Convert.ToInt16(Session
["LendId"]));
            txtStudentId.Text =lend.StudentId.ToString();
            StudentEntity student= StudentBusiness.SelectStudentByID2(Convert.
ToInt32(txtStudentId.Text));
            lblStudentName.Text=student.StudentName;
            txtReason.Text=lend.Reason.Trim();
            txtLendman.Text=lend.LendMan.Trim();
            txtID.Text=lend.ID.Trim();
            txtStartdate.Text=lend.StartDate.Trim();
            txtEnddate.Text=lend.EndDate.Trim();
        }
        protected void btnUpdate_Click(object sender, EventArgs e)
        {
            lId=Convert.ToInt16(Session["LendId"]);
            LendEntity lendInfo=new LendEntity();
            lendInfo.StudentId=Convert.ToInt32(txtStudentId.Text);
            lendInfo.Reason =txtReason.Text.Trim();
            lendInfo.LendMan=txtLendman.Text.Trim();
            lendInfo.ID=txtID.Text.Trim();
            lendInfo.StartDate =txtStartdate.Text.Trim();
            lendInfo.EndDate=txtEnddate.Text.Trim();
            lendInfo.AuthorizedId= 7;
            string message="修改失败";
            if (LendBusiness.UpdateLendByID(lId, lendInfo))
            {
                message="修改成功";
            }
            AlertInfo.AlertDialog(message);
            LoadLend();
        }
        protected void txtStudentId_TextChanged(object sender, EventArgs e)
        {
            sId=Convert.ToInt32(txtStudentId.Text.Trim());
            if (StudentBusiness.IsExistStudentId(sId))
            {
                StudentEntity student=StudentBusiness.SelectStudentByID2(sId);
                lblStudentName.Text=student.StudentName;
            }
            else
            {
                AlertInfo.AlertDialog("该学号不存在,请重新输入");
```

```
            }
        }
    }
}
```

13.5.7 借档预约管理

在借档预约管理页可查看所有借档预约申请,同时可按照学生学号搜索学生借档预约申请,按照教工号搜索教师借档预约申请,按照状态搜索全部借档预约申请。借档预约管理页测试如图 13.25 所示。

图 13.25 借档预约管理页测试

借档预约管理页主要源代码如下。

```
<form id="form1" runat="server">
    <div>
        <asp:Label ID="Label1" runat="server" Text="借档预约记录"></asp:Label>
        <asp:LinkButton ID="lbtnAll" runat="server" OnClick="lbtnAll_Click">全部
</asp:LinkButton>
        <asp:Label ID="Label4" runat="server" Text="学生学号:"></asp:Label>
        <asp:TextBox ID="txtStudentId" runat="server" ></asp:TextBox>
        <asp:RegularExpressionValidator ID="RegularExpressionValidator1" runat=
"server" ControlToValidate="txtStudentId" ErrorMessage="请输入学号" ForeColor=
"Red" ValidationExpression =" ^ [0 - 9] * $ " Display =" Dynamic " > </asp:
RegularExpressionValidator>
            <asp:Label ID="Label6" runat="server" Text="状态:"></asp:Label>
             <asp:DropDownList ID="ddlAuthorized" runat="server" AutoPostBack=
"True">
                    <asp:ListItem Value="11">未同意</asp:ListItem>
                    <asp:ListItem Value="10">已同意</asp:ListItem>
                </asp:DropDownList>
```

```
            <asp:Button ID="btnSearchs" runat="server" Text="搜索" Width="80px"
OnClick="btnSearchs_Click" />
            <asp:GridView ID="gvSappo" runat="server" AutoGenerateColumns="False"
OnRowEditing="gvSappo_RowEditing" OnRowDeleting="gvSappo_RowDeleting" AllowPaging=
"True" OnPageIndexChanging="gvSappo_PageIndexChanging" PageSize="4">
                <Columns>
                    <asp:BoundField DataField="studentid" HeaderText="学号" >
                    </asp:BoundField>
                    <asp:BoundField DataField="studentname" HeaderText="姓名" >
                    </asp:BoundField>
                    <asp:BoundField DataField="date" HeaderText="借阅时间" >
                    </asp:BoundField>
                    <asp:BoundField DataField="reason" HeaderText="理由" >
                    </asp:BoundField>
                    <asp:BoundField DataField="authorizedname" HeaderText="状态" >
                    </asp:BoundField>
                    <asp:BoundField DataField="sappoid" HeaderText="sappoid">
                    </asp:BoundField>
                    <asp:CommandField ShowEditButton="True">
                    </asp:CommandField>
                        <ItemTemplate>
                                    <asp:LinkButton ID="LinkButton1" runat="server"
CausesValidation="False" CommandName="Delete" Text="删除" OnClientClick=
"return confirm('确认要删除吗?');"></asp:LinkButton>
                        </ItemTemplate>
                    </asp:TemplateField>
                </Columns>
            </asp:GridView>
                <asp:Label ID="Label5" runat="server" Text="教工号:"></asp:Label>
                <asp:TextBox ID="txtTeacherId" runat="server"></asp:TextBox>
                <asp:RegularExpressionValidator ID="RegularExpressionValidator2"
runat="server" ControlToValidate="txtTeacherId" ErrorMessage="请输入教工号"
ForeColor="Red" ValidationExpression="^[0-9]*$" Display="Dynamic"></asp:
RegularExpressionValidator>
                <asp:Label ID="Label7" runat="server"  Text="状态:"></asp:Label>
                <asp:DropDownList ID="ddlAuthorized0" runat="server" AutoPostBack=
"True" >
                        <asp:ListItem Value="11">未同意</asp:ListItem>
                        <asp:ListItem Value="10">已同意</asp:ListItem>
                </asp:DropDownList>
        <asp:Button ID="btnSearcht" runat="server" OnClick="btnSearcht_Click" />
                <asp:GridView ID="gvTappo" runat="server" AutoGenerateColumns=
"False" CellPadding="4" OnRowDeleting="gvTappo_RowDeleting" OnRowEditing=
"gvTappo_RowEditing" AllowPaging="True" AllowSorting="True"  PageSize="4"
OnPageIndexChanging="gvTappo_PageIndexChanging" >
                    <Columns>
                        <asp:BoundField DataField="teacherid" HeaderText="教工号" >
                        </asp:BoundField>
                        <asp:BoundField DataField="teachername" HeaderText="姓名" >
                        </asp:BoundField>
                        <asp:BoundField DataField="date" HeaderText="借阅时间" >
                        </asp:BoundField>
                        <asp:BoundField DataField="reason" HeaderText="理由" >
```

```
                    </asp:BoundField>
                    <asp:BoundField DataField="authorizedname" HeaderText="状态" >
                    </asp:BoundField>
                    <asp:BoundField DataField="tappoid" HeaderText="tappoid">
                    </asp:BoundField>
                    </asp:CommandField>
                    <asp:TemplateField ShowHeader="False">
                        <ItemTemplate>
                    <asp:CommandField ShowEditButton= "True">
                    </asp:CommandField>
                    <asp: LinkButton  ID =" LinkButton1 "  runat =" server "
CausesValidation="False" CommandName="Delete" Text="删除" OnClientClick="return
confirm('确认要删除吗?');"></asp:LinkButton>
                        </ItemTemplate>
                    </asp:TemplateField>
                </Columns>
            </asp:GridView>
            <asp: Button  ID =" btnReturn "  runat =" server "  Text =" 返 回 "
CausesValidation="False" PostBackUrl="~/Admin/AdminIndex.aspx" />
        </div>
</form>
```

借档预约管理页 C#代码如下。

```
using Business;
using System;
using System.Data;
using System.Web.UI.WebControls;

namespace Archive
{
    public partial class AdminAppoManage : System.Web.UI.Page
    {
        int tId, aId,sId;
        protected void Page_Load(object sender, EventArgs e)
        {
            if (Session["adminId"] !=null)
            {
                if (! IsPostBack)
                {
                    LoadSappo();
                    LoadTappo();
                }
            }
            else
            {
                AlertInfo.AlertDialog("请先登录", "../Login.aspx");
            }
        }
        private void LoadSappo()
        {
            DataTable dt=SappoBusiness.SappoInfo();
```

```
            gvSappo.DataSource=dt;
            gvSappo.DataKeyNames=new string[] { "sappoid" };
            gvSappo.DataBind();
        }
        private void LoadTappo()
        {
            DataTable dt=TappoBusiness.TappoInfo();
            gvTappo.DataSource=dt;
            gvTappo.DataKeyNames=new string[] { "tappoid" };
            gvTappo.DataBind();
        }
        protected void gvSappo_RowEditing(object sender, GridViewEditEventArgs e)
        {
            Session["sappoid"] = gvSappo. DataKeys [ e. NewEditIndex ]. Value.
ToString().Trim();
            Response.Redirect("AdminSappoUpdate.aspx");
        }
        protected void gvTappo_RowEditing(object sender, GridViewEditEventArgs e)
        {
            Session["tappoid"] = gvTappo. DataKeys [ e. NewEditIndex ]. Value.
ToString().Trim();
            Response.Redirect("AdminTappoUpdate.aspx");
        }
        protected void gvSappo_RowDeleting(object sender, GridViewDeleteEventArgs e)
        {
            if (e.RowIndex >=0)
            {
                sId=Convert.ToInt16(this.gvSappo.DataKeys[e.RowIndex].Value);
                string message="删除失败";
                if (SappoBusiness.UpdateSappoAuthorized(sId))
                {
                    message="删除成功";
                }
                AlertInfo.AlertDialog(message);
            }
            gvSappo.Rows[e.RowIndex].Visible=false;
            LoadSappo();
        }
        protected void gvTappo_RowDeleting(object sender, GridViewDeleteEventArgs e)
        {
            if (e.RowIndex >=0)
            {
                tId=Convert.ToInt16(gvTappo.DataKeys[e.RowIndex].Value);
                string message="删除失败";
                if (TappoBusiness.UpdateTappoAuthorized(tId))
                {
                    message="删除成功";
                }
                AlertInfo.AlertDialog(message);
            }
            gvTappo.Rows[e.RowIndex].Visible=false;
            LoadTappo();
        }
```

```
        protected void btnSearchs_Click(object sender, EventArgs e)
        {
            sId= txtStudentId. Text ! = string. Empty Convert. ToInt 32 ( txtStudentId.
    Text) : 0;
            aId=Convert.ToInt32(ddlAuthorized0.SelectedValue);
            DataTable dts = SappoBusiness. SelectSappoByStudentIDAuthorizedID
    (sId, aId);
            gvSappo.DataSource=dts;
            gvSappo.DataKeyNames=new string[] { "sappoid" };
            gvSappo.DataBind();
        }
        protected void btnSearcht_Click(object sender, EventArgs e)
        {
            tId=txtTeacherId.Text ! =string.Empty? Convert. ToInt 32 ( txtTeacherId.
    Text) : 0;
            aId=Convert.ToInt32(ddlAuthorized0.SelectedValue);
            DataTable dt = TappoBusiness. SelectTappoByTeacherIDAuthorizedID
    (tId, aId);
            gvTappo.DataSource=dt;
            gvTappo.DataKeyNames=new string[] { "tappoid" };
            gvTappo.DataBind();
        }
        protected void lbtnAll_Click(object sender, EventArgs e)
        {
            LoadSappo();
            LoadTappo();
        }
        protected void gvSappo_PageIndexChanging(object sender, GridViewPageEventArgs e)
        {
            gvSappo.PageIndex=e.NewPageIndex;
            LoadSappo();
        }
        protected void gvTappo_PageIndexChanging(object sender, GridViewPageEventArgs e)
        {
            gvTappo.PageIndex=e.NewPageIndex;
            LoadTappo();
        }
    }
}
```

参考文献

[1] 朱晔. ASP. NET 第一步：基于 C# 和 ASP. NET 2. 0[M]. 北京：清华大学出版社，2007.
[2] 张跃廷，顾彦玲. ASP. NET 从入门到精通[M]. 北京：清华大学出版社，2008.
[3] 陈伟，卫琳. ASP. NET 3. 5 网站开发实例教程[M]. 北京：清华大学出版社，2009.
[4] 周文琼，王乐球. 数据库应用与开发教程[M]. 北京：中国铁道出版社，2014.
[5] 喻均，田喜群，唐俊勇. AS 程序设计循序渐进教程[M]. 北京：清华大学出版社，2009.
[6] 李春葆. ASP. NET 4. 5 动态网站设计教程[M]. 北京：清华大学出版社，2016.
[7] 郭鹏，门璐瑶. VS 2012. NET Web 高级编程开发[M]. 大连：大连理工大学出版社，2014.
[8] PENBERTHY W. ASP. NET 入门经典：基于 Visual Studio 2015[M]. 9 版. 北京：清华大学出版社，2016.
[9] 康拉德・科克萨. NET 内存管理宝典：提高代码质量、性能和可扩展性[M]. 叶伟民，涂曙光，译. 北京：清华大学出版社，2021.

图书资源支持

感谢您一直以来对清华版图书的支持和爱护。为了配合本书的使用，本书提供配套的资源，有需求的读者请扫描下方的“书圈”微信公众号二维码，在图书专区下载，也可以拨打电话或发送电子邮件咨询。

如果您在使用本书的过程中遇到了什么问题，或者有相关图书出版计划，也请您发邮件告诉我们，以便我们更好地为您服务。

我们的联系方式：

地　　址：北京市海淀区双清路学研大厦 A 座 714

邮　　编：100084

电　　话：010-83470236　010-83470237

客服邮箱：2301891038@qq.com

QQ：2301891038（请写明您的单位和姓名）

资源下载：关注公众号“书圈”下载配套资源。

资源下载、样书申请

书圈

图书案例

清华计算机学堂

观看课程直播